ESSENTIALS OF FIRE FIGHTING

SECOND EDITION

VALIDATED BY
**THE INTERNATIONAL FIRE SERVICE
TRAINING ASSOCIATION**

PUBLISHED BY
**FIRE PROTECTION PUBLICATIONS
OKLAHOMA STATE UNIVERSITY**

Dedication

This manual is dedicated to the members of that unselfish organization of men and women who hold devotion to duty above personal risk, who count sincerity of service above personal comfort and convenience, who strive unceasingly to find better ways of protecting the lives, homes and property of their fellow citizens from the ravages of fire and other disasters . . .

The Firefighters of All Nations.

Dear Firefighter:

The International Fire Service Training Association (IFSTA) is an organization that exists for the purpose of serving firefighters' training needs. IFSTA is a member of the Joint Council of National Fire Organizations. Fire Protection Publications is the publisher of IFSTA materials. Fire Protection Publications staff members participate in the National Fire Protection Association and the International Society of Fire Service Instructors.

If you need additional information concerning these organizations or assistance with manual orders, contact:

For assistance with training materials, recommended material for inclusion in a manual, or questions on manual content, contact:

Customer Services
Fire Protection Publications
Oklahoma State University
Stillwater, OK 74078
(800) 654-4055 in Continental United States

Technical Services
Fire Protection Publications
Oklahoma State University
Stillwater, OK 74078
(405) 624-5723

First Printing - June 1983
Second Printing - September 1983

Table of Contents

List of Tables

THE INTERNATIONAL FIRE SERVICE TRAINING ASSOCIATION

The International Fire Service Training Association is an educational alliance organized to develop training material for the fire service. The annual meeting of its membership consists of a workshop conference which has several objectives —

> ... to develop training material for publication
> ... to validate training material for publication
> ... to check proposed rough drafts for errors
> ... to add new techniques and developments
> ... to delete obsolete and outmoded methods
> ... to upgrade the fire service through training

This training association was formed in November 1934, when the Western Actuarial Bureau sponsored a conference in Kansas City, Missouri, to determine how all agencies that were interested in publishing fire service training material could coordinate their efforts. Four states were represented at this conference and it was decided that, since the representatives from Oklahoma had done some pioneering in fire training manual development, other interested states should join forces with them. This merger made it possible to develop nationally recognized training material which was broader in scope than material published by an individual state agency. This merger further made possible a reduction in publication costs, since it enabled each state to benefit from the economy of relatively large printing orders. These savings would not be possible if each individual state developed and published its own training material.

From the original four states, the adoption list has grown to forty-four American States; six Canadian Provinces; the British Territory of Bermuda; the Australian State of Queensland; the International Civil Aviation Organization Training Centre in Beirut, Lebanon; the Department of National Defence of Canada; the Department of the Army of the United States; the Department of the Navy of the United States; the United States Air Force; the United States Bureau of Indian Affairs; The United States General Services Administration; and the National Aeronautics and Space Administration (NASA). Representatives with various areas of expertise are invited to serve as a voluntary group of individuals who recommend procedures, and validate material before it is published. Most of the representatives are members of other international fire protection organizations and this meeting brings together individuals from several related and allied fields, such as:

> ... key fire department executives and drillmasters,
> ... educators from colleges and universities,
> ... representatives from governmental agencies,
> ... delegates of firefighter associations and organizations, and
> ... engineers from the fire insurance industry.

This unique feature provides a close relationship between the International Fire Service Training Association and other fire protection agencies, which helps to correlate the efforts of all concerned.

The publications of the International Fire Service Training Association are compatible with the National Fire Protection Association's Standard 1001, "Fire Fighter Professional Qualifications (1981)," and the International Association of Fire Fighters/International Association of Fire Chiefs "National Apprenticeship and Training Standards for the Fire Fighter." The standards are an effort to attain professional status through progressive training. The NFPA and IAFF/IAFC Standards were prepared in cooperation with the Joint Council of National Fire Service Organizations of which IFSTA is a member.

The International Fire Service Training Association meets each July at Oklahoma State University, Stillwater, Oklahoma. Fire Protection Publications at Oklahoma State University publishes all IFSTA training manuals and texts.

Prologue

It is fitting that this manual **Essentials of Fire Fighting,** second edition, is being published in 1983, the 50th anniversary of the International Fire Service Training Association. This manual is the leading text for instructing the basics of fire fighting and meeting the performance objectives for firefighter certification to levels I and II.

The Essentials manual exemplifies the purpose of our association to provide quality training materials for international use. It is a summary of training techniques validated by fire service practitioners from throughout the United States and Canada. Since this is a compilation of practical information from other manuals, committee people, association members, and Fire Protection Publications staff, thousands of donated hours have gone into its production as with all IFSTA manuals. The manuals continue to reflect the dedicated spirit of the validation committee, the editorial staff, and all those fire service personnel who contribute to their completion.

In this anniversary year, it is appropriate to indicate the leaders that have brought the association to this point in history.

Editors	**Validation Conference Chairmen**
Fred Heisler 1934 - 1955	Fred Heisler 1934 - 1950
Everett E. Hudiburg 1955 - 1975	R. J. Douglas 1950 - 1962
John Peige 1975 - 1977	Emmett Cox 1962 - 1976
Jerry Laughlin 1978 - 1979	Howard Boyd 1976 - Present
Gene P. Carlson 1980 - Present	

The Executive Board was started in 1962 and elected Everett Hudiburg its first Executive Director in 1971. He served until 1975 and now is the Director Emeritus.

These are but a few of the individuals who have contributed to the IFSTA history. To all those who have assisted in creating the IFSTA system of providing the best fire training materials for firefighters around the world, we thank you.

Harold R. Mace

Harold R. Mace
Executive Director
Stillwater, Oklahoma 1983

Preface

This is the second edition of **Essentials of Fire Fighting**. There have been some substantial changes in the new edition that reflect the revision of NFPA Standard 1001, *Fire Fighter Professional Qualifications* (1981). Many new illustrations and photographs have been added and metric conversions have been included throughout the manual.

Acknowledgement and grateful thanks are extended to the members of the validating committee, who assisted with the final draft of this manual.

Chairman
Joseph L. Donovan
Superintendent, National Fire Academy
Emmitsburg, Maryland

Vice-Chairman
Carl McCoy, FEMA Regional Representative
United States Fire Administration
Denton, Texas

Secretary
William R. Cooper, Captain
Huntington Beach Fire Department
Huntington Beach, California

Other persons assisting on the committee during its tenure were:

Louis Amabili
Edward Bent
Gerald Brinkman
Robert Hasbrook
Eric Haussermann
John Hoglund

John Horn
Jesse Jackson
Kenneth Mitten
Junius Murray
Nick Renihan
T. R. Spencer

Special acknowledgement and thanks are extended to Fire Chief Ken Mitten and the Merced City (California) Fire Department for their cooperation and assistance in providing personnel, equipment, and facilities for the Fire Protection Publications photographers. Their ability to accurately demonstrate the proficiencies prescribed in NFPA 1001 and illustrated in this manual is greatly appreciated.

A book of this scope would be impossible to publish were it not for the assistance of many persons and organizations. To the following and the many others as noted in the captions we owe a great debt as they gave freely of advice, equipment, copyrighted material, illustrations, and photographs.

Akron Brass Company
Chicago Fire Academy staff
Elkhart Brass Manufacturing Company, Inc.
Maryland Fire and Rescue Institute
Massachusetts Firefighting Academy
    R. Goddard, J. Harrington, E. Hartin,
    J. Peltier, G. Reardon, J. Russell
Montogomery County, Maryland Fire and
  Rescue Training Academy
National Fire Protection Association
Rose Fire Company

Stillwater Fire Services

Cover Photos:
    Chicago Fire Department
    and Fire Protection Publications staff

Chapter Divider Photos:
    Chicago Fire Department
    San Clemente Fire Department
    Jim Nichols
    Fort Wayne *News-Sentinel*

We also express our thanks to the following Oklahoma State University fire protection students for their assistance in demonstrating certain evolutions described in this manual.

Kevin Johnson
John Sepaphur
Michael Mallory
Scott Stookey
Jeff Sipes

Kevin Roche
Richard A. Brenner
Jay Holton
Brad Shoefstall

Gratitude is also extended to the following members of the Fire Protection Publications staff whose contributions made the final publication of the manual possible.

William J. Vandevort — Associate Editor
Samuel O. Goldwater — Marketing Associate
Gary M. Courtney — Research Technician
Scott D. Kerwood — Research Technician
Don Davis — Coordinator, Publications Production
Karen Murphy — Phototypesetter Operator II
Desa Porter — Phototypesetter Operator I
Ann Moffat — Graphic Designer
Christi Ward — Graphic Intern
Carol Smith — Publications Validation Assistant

Gene P. Carlson
Editor

Introduction

Some years ago the fire service recognized the need for professional standards for firefighters. Representatives of all the major fire service organizations in the United States met and formulated NFPA Standard No. 1001, *Fire Fighter Professional Qualifications*.

The development of a concise standard and a workable program has inspired a professional growth attitude throughout the nation's fire services. IFSTA manuals and training aids have been developed to meet this attitude and foster professionalism in the fire service.

It is important to note that the purpose of NFPA Standard 1001 is to specify in terms of performance objectives, the minimum requirements of competence required for service as a firefighter. In many instances, this manual illustrates just one method for accomplishing each task. The methods shown throughout the text have been approved by the International Fire Service Training Association as accepted methods for accomplishing each task. However, they are *not* to be interpreted as the only methods to accomplish a given task. The specific methods for students to utilize in achieving these performance objectives should be specified by their own jurisdiction.

Scope and Purpose

The **Essentials of Fire Fighting** manual is designed to provide the firefighter recruit with the information needed to meet the performance objectives in NFPA Standard 1001, *Fire Fighter Professional Qualifications,* for levels I and II. The many methods, techniques, and in-depth explanations contained in other IFSTA manuals do not appear in this text. It is highly recommended that the student and instructor consult current editions of these publications in order to gain a more complete understanding of the many concepts and subject areas discussed in this manual.

This manual has been developed to be used in conjunction with the IFSTA manuals **Fire Service First Aid Practices** and **Fire Service Orientation and Indoctrination.** The Indoctrination manual contains the history, traditions, and organization of the fire service; fire department operations; firefighter responsibilities and duties; as well as an extensive glossary of fire service terms.

For ease of organization, the Essentials manual has been divided into chapters titled the same as in the NFPA Standard 1001. At the beginning of each chapter, those behavioral objectives have been outlined as they appear in NFPA Standard 1001. Where feasible, the content of each chapter reflects this order. It should be noted that the Standard does not require for the objectives to be mastered in the order they appear. The local and state training program should establish instructional priority and the training program content to prepare individuals to meet the performance objectives set forth by NFPA Standard 1001.

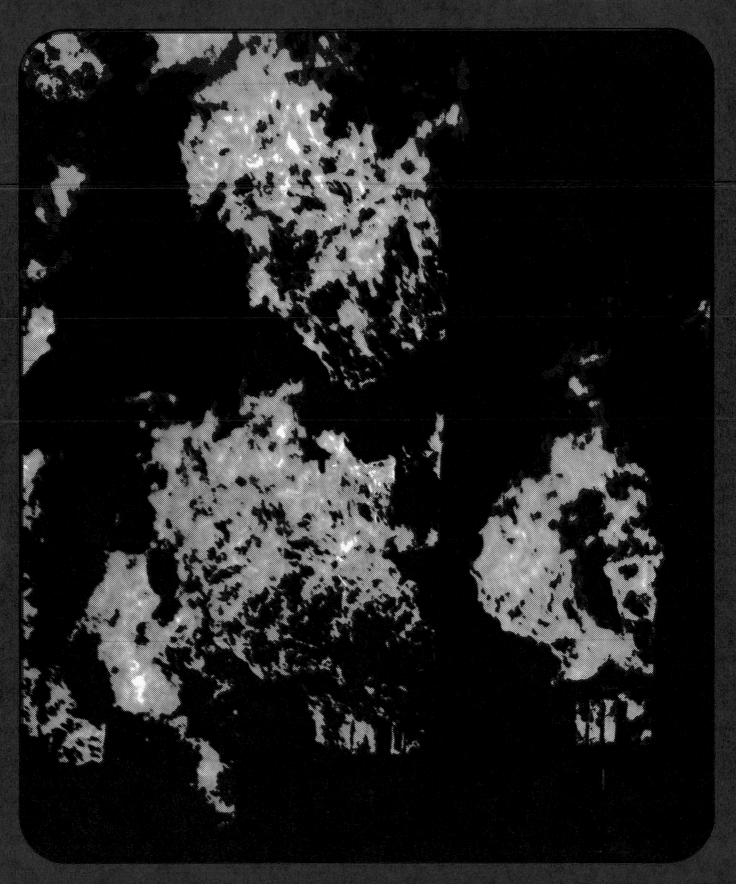

NFPA STANDARD 1001
FIRE BEHAVIOR
Fire Fighter I

3-10 Ventilation

3-10.6 The fire fighter shall define the theory of a "back draft explosion."

3-16 Fire Behavior

3-16.1 The fire fighter shall define fire.

3-16.2 The fire fighter shall define the fire triangle.

3-16.3 The fire fighter shall identify two chemical, mechanical, and electrical energy heat sources.

3-16.4 The fire fighter shall define the following potential stages of fire:

    (a) Incipient
    (b) Flame spread
    (c) Hot smoldering
    (d) Flash over
    (e) Steady state
    (f) Clear burning.

3-16.5 The fire fighter shall define the three methods of heat transfer.

3-16.6 The fire fighter shall define the three physical stages of matter in which fuels are commonly found.

3-16.7 The fire fighter shall define the hazard of finely divided fuels as they relate to the combustion process.

3-16.8 The fire fighter shall define flash point and ignition temperature.

3-16.9 The fire fighter shall define concentrations of oxygen in air as it affects combustion.

3-16.10 The fire fighter shall identify three products of combustion commonly found in structural fires which create a life hazard.*

Fire Fighter II

4-18 Fire Behavior

4-18.1 The fire fighter shall define the following units of heat measurement:

    (a) British Thermal Unit (BTU)
    (b) Fahrenheit (°F)
    (c) Celsius (°C)
    (d) Calorie (C).

4-18.2 The fire fighter shall define thermal balance and imbalance.*

*Reprinted by permission from NFPA Standard No. 1001, *Standard for Fire Fighter Professional Qualifications*. Copyright © 1981, National Fire Protection Association, Boston, MA.

IFSTA's Fire Behavior Transparencies are designed to complement this chapter.

Chapter 1
Fire Behavior

Effective fire control and extinguishment requires a basic understanding of the chemical and physical nature of fire. This includes information describing sources of heat energy, composition and characteristics of fuels, and environmental conditions necessary to sustain the combustion process.

Combustion is the the self-sustaining process of rapid oxidation of a fuel being reduced by an oxidizing agent along with the evolution of heat and light (Figure 1.1). Most fires involve a fuel that

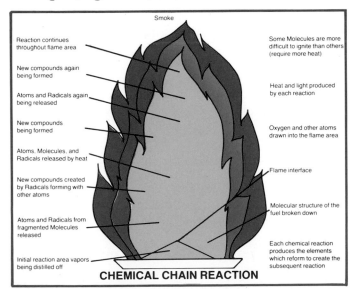

Figure 1.1 Combustion actions that are associated with the chemical chain reaction.

is chemically combined with the oxygen normally found in atmospheric air. Atmospheric air contains 21 percent oxygen, 78 percent nitrogen, and 1 percent of other gases. Substances such as chlorine gas and its compounds will also sustain combustion. Other substances, such as organic peroxides, are composed in such a way that they contain both fuel and oxidizer molecules within the compound allowing them to burn in the absence of oxygen.

Fires are defined by their physical characteristics. They may vary from very slow oxidation, as in rusting, to very fast oxidation, such as detonations and explosions. Somewhere between these extremes are the two most common reactions concerning firefighters: smoldering fires and flaming or free-burning fires.

DEFINITIONS

Following are some terms used to define and describe fire activity.

HEAT - The form of energy that raises temperature. Heat is measured by the amount of work it does.

BRITISH THERMAL UNIT (BTU) - The amount of heat needed to raise the temperature of one pound of water one degree Fahrenheit.

FAHRENHEIT - On the Fahrenheit scale, 32 degrees is the melting point of ice; 212 degrees is the boiling point of water.

CALORIE - The amount of heat needed to raise the temperature of one gram of water one degree Centigrade.

CENTIGRADE (Celsius) - On the Centigrade scale, zero is the melting point of ice; 100 degrees is the boiling point of water.

BOILING POINT - The temperature of a substance where the rate of evaporation exceeds the rate of condensation.

FLASH POINT - The minimum temperature at which a liquid fuel gives off sufficient vapors to form an ignitable mixture with the air near the surface. At this temperature, the ignited vapors will flash, but will not continue to burn.

FIRE POINT - The temperature at which a liquid fuel will produce vapors sufficient to support combustion once ignited. The fire point is usually a few degrees above the flash point.

IGNITION TEMPERATURE - The minimum temperature to which a fuel in air must be heated in order to start self-sustained combustion independent of the heating source.

ENDOTHERMIC HEAT REACTION - A chemical reaction where a substance absorbs heat energy.

EXOTHERMIC HEAT REACTION - A chemical reaction where a substance gives up heat energy.

FLAMMABLE OR EXPLOSIVE LIMITS - The percentage of a substance in air that will burn once it is ignited. Most substances have an upper (too rich) and a lower (too lean) flammable limit.

OXIDATION - The complex chemical reaction of organic materials with oxygen or other oxidizing agents in the formation of more stable compounds.

SOURCES OF HEAT ENERGY

Heat is a form of energy that may be described as a condition of "matter in motion" caused by the movement of molecules. All matter contains some heat regardless of how low the temperature is because molecules are constantly moving all the time. When a body of matter is heated, the speed of the molecules increases, and thus, the temperature also increases. Anything that sets the molecules of a material in motion produces heat in that material. There are four general categories of heat energy and they include:

- Chemical heat energy
- Electrical heat energy
- Mechanical heat energy
- Nuclear heat energy

Chemical Heat Energy

Heat of Combustion - The amount of heat generated by the combustion (oxidation) process.

Spontaneous Heating - The heating of an organic substance without the addition of external heat. Spontaneous heating occurs most frequently where sufficient air is not present to dissipate the heat produced. The speed of a heating reaction doubles with each 18°F (8°C) temperature increase.

Heat of Decomposition - The release of heat from decomposing compounds. These compounds may be unstable and release their heat very quickly or they may detonate.

Heat of Solution - The heat released by the mixture of matter in a liquid. Some acids, when dissolved, give off sufficient heat to pose exposure problems to nearby combustibles.

Electrical Heat Energy

Resistance Heating - The heat generated by passing an electrical force through a conductor such as a wire or an appliance.

Dielectric Heating - The heating that results from the action of either pulsating direct current, or alternating current at high frequency on a non-conductive material.

Induction Heating - The heating of materials resulting from an alternating current flow causing a magnetic field influence.

Leakage Current Heating - The heat resulting from imperfect or improperly insulated electrical materials. This is particularly evident where the insulation is required to handle high voltage or loads near maximum capacity.

Heat From Arcing - Heat released either as a high-temperature arc or as molten material from the conductor.

Static Electricity Heating - Heat released as an arc between oppositely charged surfaces. Static electricity can be generated by the contact and separation of charged surfaces or by fluids flowing through pipes.

Heat Generated by Lightning - The heat generated by the discharge of thousands of volts from either earth to cloud, cloud to cloud, or from cloud to ground.

Mechanical Heat Energy

Frictional Heat - The heat generated by the movement between two objects in contact with each other.

Friction Sparks - The heat generated in the form of sparks from solid objects striking each other. Most often at least one of the objects is metal.

Heat of Compression - The heat generated by the forced reduction of a gaseous volume. Diesel engines ignite fuel vapor without a spark plug by the use of this principle.

Nuclear Heat Energy

Nuclear Fission and Fusion - The heat generated by either the splitting or combining of atoms.

THE BURNING PROCESS

The initiation of combustion requires the conversion of fuel into the gaseous state by heating. Fuel may be found in any of three states of matter: solid, liquid, or gas. Fuel gases are evolved from solid fuels by pyrolysis. This is defined as the chemical decomposition of a substance through the action of heat.

TABLE 1.1
PYROLYSIS

Temperature	Reaction
392°F (200°C)	Production of water vapor, carbon dioxide, formic, and acetic acids
392° - 536°F (200° - 280°C)	Less water vapor - some carbon monoxide - still primarily an endothermic reaction. (Absorbing Heat)
536° - 932°F (280° - 500°C)	Exothermic reaction (giving off heat) with flammable vapors and particulates. Some secondary reaction from charcoal formed.
Over 932°F (500°C)	Residue primarily charcoal with notable catalytic action.

Fuel gases are evolved from liquids by vaporization. This process is the same as boiling water or evaporation of a pan of water in sunlight. In both cases, heat caused the liquid to vaporize. No heat input is required with gaseous fuels and this places considerable restraints on the control and extinguishment of gas fuel fires.

Characteristics of Fire Behavior

Solid fuels have definite shape and size. One primary consideration with solid fuels is the surface area of the material in relation to its mass. The larger the surface area for a given mass the more rapid the heating of the material and increase in the speed of pyrolysis. The physical position of a solid fuel is also of great concern to fire fighting personnel. If the solid fuel is in a vertical position, fire spread will be more rapid than if it is in a horizontal position. This is due to increased heat transfer through convection and direct flame contact in addition to conduction and radiation.

Liquid fuels have physical properties that increase the difficulty of extinguishment and hazard to personnel. Liquids will assume the shape of their container. When a spill occurs, the liquid will assume the shape of the ground (flat) and will flow and accumulate in low areas.

The density of liquids in relation to water is known as specific gravity. Water is given a value of one. Liquids with a specific gravity less than one are lighter than water while those with a specific gravity greater than one are heavier than water. It is interesting to note that most flammable liquids have a specific gravity of less than one.

The solubility of a liquid fuel in water is an important factor. Hydrocarbon liquids as a rule will not mix with water. Alcohols and polar solvents mix with water and if large volumes of water are used, they may be diluted to the point where they will not burn. Consideration must be given to which extinguishing agents are effective on hydrocarbons (insoluble) and which affect porous solvents and alcohols (soluble).

The volatility or ease with which the liquid gives off vapor influences fire control objectives. The density of gas or vapor in relation to air is of concern with volatile liquids and with gas fuels.

Gases tend to assume the shape of their container but have no specific volume. If the vapor density of a gas is such that it is less dense than air (air is given a value of one), it will rise and tend to dissipate. If a gas or vapor is heavier than air, it will tend to hug the ground and travel as directed by terrain and wind.

The second phase in initiating combustion is the mixture of the fuel vapor with an oxidizer (air). The mixture of the fuel vapor and air must be within the flammable range. The upper and lower limits of concentration of vapor in air will allow flame propagation when contacted by a source of ignition.

The flammable range varies with the fuel and with the ambient temperature. Usually the flammable range is given for temperatures of 70°F (21°C).

TABLE 1.2
EXAMPLES OF FLAMMABLE RANGES

Fuel	Lower Limit	Upper Limit
Gasoline Vapor	1.4	7.6
Methane (Natural Gas)	5.0	17
Propane	2.2	9.5
Hydrogen	4.0	75
Acetylene	2.5	100

When the proper fuel vapor/air mixture has been achieved, it must then be raised to its ignition temperature.

Fire burns in two basic modes — flaming and/ or surface combustion. The flaming mode of combustion is represented by the fire tetrahedron (a four sided figure) with the four sides representing fuel, temperature, oxygen, and the uninhibited chemical chain reaction (Figure 1.2). The surface or smoldering mode of combustion is represented by the fire triangle with the three sides representing fuel, temperature, and oxygen (Figure 1.3).

The fuel segment of both theories is any solid, liquid, or gas that can combine with oxygen in the chemical reaction known as oxidation. Temperature is the measure of the molecular activity within a substance. A fuel with a sufficiently high temperature will ignite if an oxidizing agent is present. Combustion will continue as long as enough energy or heat is present. Under most conditions

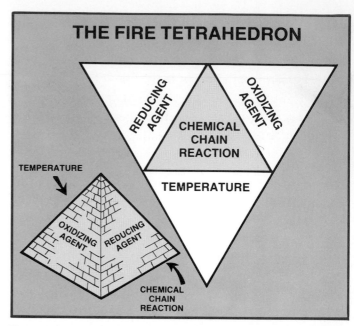

Figure 1.2 The fire tetrahedron includes the chemical chain reaction as component of burning, thus converting the fire triangle into a four-sided figure resembling a pyramid.

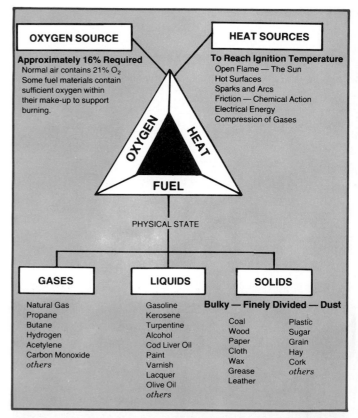

Figure 1.3 The fire triangle is used to explain the components necessary for burning to occur.

the oxidizing agent will be the oxygen in air, but the term helps explain how some material that releases its own oxygen during combustion such as sodium nitrate and potassium chlorate can burn in an oxygen-free atmosphere. While scientists only partially understand what happens in a chemical chain reaction, they do know that heating produces vapors that contain substances that will combine with oxygen and burn.

A self-sustaining combustion reaction of solids and liquids depends on radiative feedback. This is radiant heat providing energy for continued vaporization. When sufficient heat is present to maintain or increase this feedback, the fire will remain constant or will grow depending on the heat produced. When heat is fed back to the fuel, this is known as a positive heat balance. If heat is dissipated faster than it is generated, a negative heat balance is created. A positive heat balance is required to maintain combustion.

Phases of A Fire* (See page 17)

Fires may start at any time of the day or night if the hazard exists. If the fire happens when the area is occupied, the chances are that it will be discovered and controlled in the beginning phase. But if it occurs when the building is closed and deserted, the fire may go undetected until it has gained major proportions. The condition of a fire in a closed building is one of chief importance for ventilation.

When fire is confined in a building or room, a situation develops that requires carefully calculated and executed ventilation procedures if further damage is to be prevented and danger reduced. This type of fire can best be understood by an investigation of its three progressive stages.

A firefighter may be confronted by one or all of the following phases of fire at any time; therefore, a working knowledge of these phases is important for understanding of ventilation procedures.

INCIPIENT PHASE

In the first phase, the oxygen content in the air has not been significantly reduced and the fire is producing water vapor (H_2O), carbon dioxide (CO_2), perhaps a small quantity of sulfur dioxide (SO_2), carbon monoxide (CO), and other gases. Some heat is being generated, and the amount will increase with the progress of the fire. The fire may be producing a flame temperature well above 1,000°F (537°C), yet the temperature in the room at this stage may be only slightly increased (Figure 1.4).

FREE-BURNING PHASE

The second phase of burning encompasses all of the free-burning activities of the fire. During this phase, oxygen-rich air is drawn into the flame as convection (the rise of heated gases) carries the heat to the uppermost regions of the confined area. The heated gases spread out laterally from the top downward, forcing the cooler air to seek lower levels, and eventually igniting all the combustible

INCIPIENT PHASE
- **SLIGHTLY OVER 100°**
- **RISING HOT GASES**
- **RISING AIR APPROXIMATELY 20% OXYGEN**

Figure 1.4 Incipient fires are initiating heat, smoke, and flame damage.

FREE BURNING
- **APPROXIMATELY 1300°F**
- **HEAT ACCUMULATES AT UPPER AREAS**
- **REDUCED OXYGEN SUPPLY**

Figure 1.5 Free burning fires are rapidly burning using up oxygen and building up heat.

material in the upper levels of the room (Figure 1.5). This heated air is one of the reasons that firefighters are taught to keep low and use protective breathing equipment. One breath of this superheated air can sear the lungs. At this point, the temperature in the upper regions can exceed 1,300°F (700°C). As the fire progresses through the latter stages of this phase, it continues to consume the free oxygen until it reaches the point where there is insufficient oxygen to react with the fuel. The fire is then reduced to the smoldering phase and needs only a supply of oxygen to burn rapidly or explode.

SMOLDERING PHASE

In the third phase, flame may cease to exist if the area of confinement is sufficiently airtight. In this instance, burning is reduced to glowing embers. The room becomes completely filled with dense smoke and gases to the extent that it is forced from all cracks under pressure (Figure 1.6). The fire will continue to smolder, and the room will completely fill with dense smoke and gases of combustion at a temperature of well over 1,000° (537°C). The intense heat will have vaporized the lighter fuel fractions such as hydrogen and methane from the combustible material in the room. These fuel gases will be added to those produced by the fire and will further increase the hazard to the firefighter and create the possibility of a backdraft.

Backdraft

Firefighters responding to a confined fire that is late in the free-burning phase or in the smoldering phase risk causing a backdraft or smoke explosion if the science of fire is not considered in opening the structure.

SMOLDERING
- **OXYGEN BELOW 15%**
- **TEMPERATURE THROUGHOUT IS HIGH**
- **CO AND CARBON MAY CAUSE BACKDRAFT**

Figure 1.6 Smoldering fires are those with an oxygen content below 15%, little flames, but dense smoke and high heat fill the area.

In the smoldering phase of a fire, burning is incomplete because not enough oxygen is available to sustain the fire (Figure 1.7). However, the heat from the free-burning phase remains, and the unburned carbon particles and other flammable products of combustion are just waiting to burst into rapid, almost instantaneous combustion when more oxygen is supplied. Proper ventilation releases smoke and the hot unburned gases from the upper areas of the room or structure. Improper ventilation at this time supplies the dangerous missing link — oxygen. As soon as the needed oxygen rushes in, the stalled combustion resumes; and it can be devastating in its speed, truly qualifying as an explosion (Figure 1.7).

Combustion is related to oxidation, and oxidation is a chemical reaction in which oxygen combines with other elements. Carbon is a naturally abundant element present in wood, among other things. When wood burns, carbon combines with oxygen to form carbon dioxide (CO_2), or carbon monoxide (CO), depending on the availability of

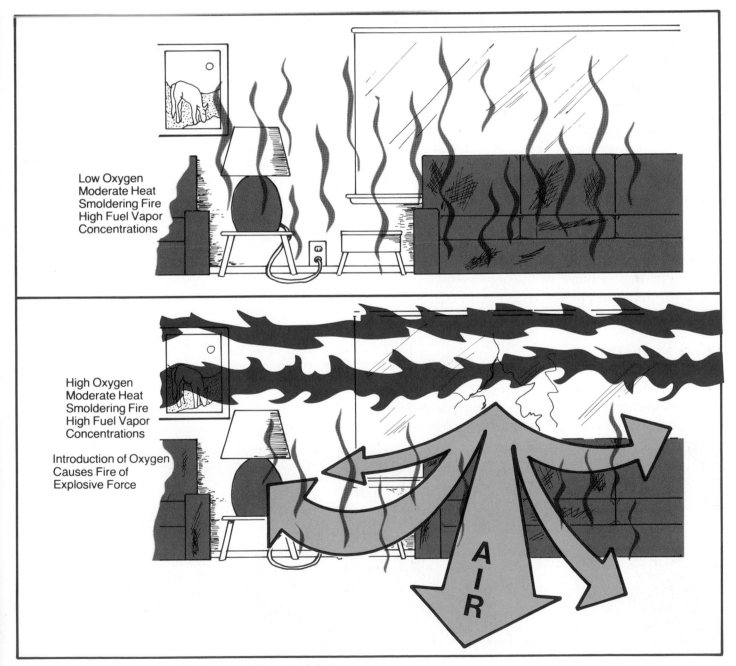

Low Oxygen
Moderate Heat
Smoldering Fire
High Fuel Vapor
Concentrations

High Oxygen
Moderate Heat
Smoldering Fire
High Fuel Vapor
Concentrations

Introduction of Oxygen
Causes Fire of
Explosive Force

AIR

Figure 1.7 With rising heat and high fuel concentrations, a backdraft can occur with explosive force when oxygen is introduced into the area.

oxygen. When oxygen is no longer available, free carbon is released in the smoke. A warning sign of possible backdraft is dense, black (carbon-filled) smoke.

The following characteristics may indicate a backdraft or smoke explosion condition.

- Smoke under pressure.
- Black smoke becoming dense gray yellow.
- Confinement and excessive heat.
- Little or no visible flame.
- Smoke leaves the building in puffs or at intervals.
- Smoke-stained windows.
- Muffled sounds.
- Sudden rapid movement of air inward when opening is made.

This type of condition can be made less dangerous by proper ventilation. If the building is opened at the highest point involved, the heated gases and smoke will be released, reducing the possibility of an explosion.

Flashover

Flashover occurs when a room or other area becomes heated to the point where flames flash over the entire surface or area. Originally, it was believed that flashover was caused by combustible gases released during the early stages of the fire. It was thought that these gases collected at the ceiling level and mixed with air until they reached their flammable range, then suddenly ignited causing flashover.

It is now believed that while this may occur, it precedes flashover. The cause of flashover is not attributed to the excessive build up of heat from the fire itself. As the fire continues to burn, all the contents of the fire area are gradually heated to their ignition temperatures. When they reach this point, simultaneous ignition occurs and the area becomes fully involved in fire.

HEAT TRANSFER

Heat can travel throughout a burning building by one or more of three methods, commonly referred to as conduction, convection, and radiation. Since the existence of heat within a substance is

Figure 1.8 Heat is transferred by conduction when an intervening medium carries the heat from floor to floor or room to room.

caused by molecular action, the greater the molecular activity, the more intense the heat. A number of natural laws of physics are involved in the transmission of heat. One is called the *Law of Heat Flow*; it specifies that heat tends to flow from a hot substance to a cold substance. The colder of two bodies in contact will absorb heat until both objects are the same temperature.

Conduction

Heat may be conducted from one body to another by direct contact of the two bodies or by an intervening heat-conducting medium (Figure 1.8). The amount of heat that will be transferred and its rate of travel depends upon the conductivity of the material through which the heat is passing. Not all materials have the same heat conductivity. Aluminum, copper, and iron are good conductors. Fibrous materials, such as felt, cloth, and paper are poor conductors.

Liquids and gases are poor conductors of heat because of the movement of their molecules. Air is a relatively poor conductor. Certain solid materials when shredded into fibers and packed into batts make good insulation because the material itself is a poor conductor and there are air pockets within the batting. Double building walls that contain an air space provide additional insulation.

Convection

Convection is the transfer of heat by the movement of air or liquid (Figure 1.9). When water is heated in a glass container, the movement within the vessel can be observed through the glass. If some sawdust is added to the water, the movement is more apparent. As the water is heated, it expands and grows lighter, hence, the upward movement. In the same manner, air becomes heated near a steam radiator by conduction. It expands, becomes lighter and moves upward. As the heated air moves upward, cooler air takes its place at the lower levels. When liquids and gases are heated, they begin to move within themselves. This movement is different from the molecular motion discussed in conduction of heat and is known as heat transfer by *convection*.

Heated air in a building will expand and rise. For this reason, fire spread by convection is mostly

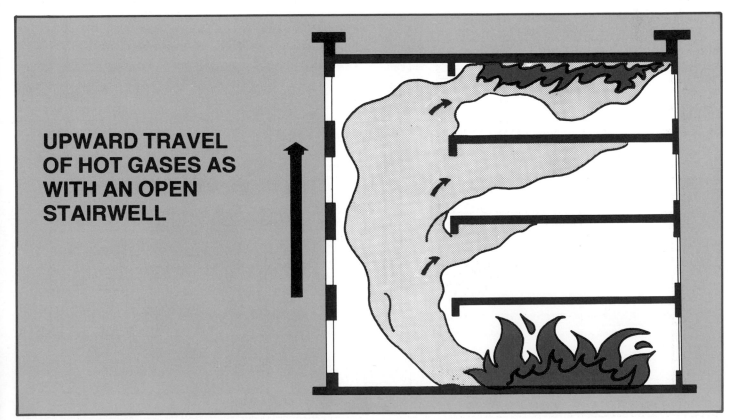

UPWARD TRAVEL OF HOT GASES AS WITH AN OPEN STAIRWELL

Figure 1.9 Heat is transferred by convection when air or water rise as they become heated.

in an upward direction, although air currents can carry heat in any direction. Convected currents are generally the cause of heat movement from floor to floor, from room to room, and from area to area. The spread of fire through corridors, up stairwells and elevator shafts, between walls, and through attics is mostly caused by the convection of heat currents and has more influence upon the positions for fire attack and ventilation than either radiation or conduction.

Another form of heat transfer by convection is direct flame contact. When a substance is heated to the point where flammable vapors are given off, these vapors may be ignited, creating a flame. As other flammable materials come in contact with the burning vapors, or flame, they may be heated to a temperature where they, too, will ignite and burn.

Radiation

The warmth of the sun is felt soon after it rises. When the sun sets, the earth begins to cool with similar rapidity. We carry an umbrella to shade our bodies from the heat of the sun. A spray of water between a firefighter and a fire will lessen the heat reaching the firefighter. Although air is a poor conductor, it is obvious that heat can travel where matter does not exist. This method of heat transmission is known as *radiation* of heat waves (Figure 1.10). Heat and light waves are similar in nature, but they differ in length per cycle. Heat

waves are longer than light waves and they are sometimes called infrared rays. Radiated heat will travel through space until it reaches an opaque object. As the object is exposed to heat radiation, it will in return radiate heat from its surface. Radiated heat is one of the major sources of fire spread, and its importance demands immediate attention at points where radiation exposure is severe.

PRODUCTS OF COMBUSTION

When a material (fuel) burns, it undergoes a chemical change. None of the elements making up the material are destroyed in the process, but all of the matter is transformed into another form or state. Although dispersed, the products of combustion equal in weight and volume that of the fuel before it was burned. When a fuel burns there are four products of combustion: fire gases, flame, heat, and smoke (Figure 1.11).

Heat is a form of energy that is measured in degrees of temperature to signify its intensity. In this sense, heat is the product of combustion that is responsible for the spread of fire. In a physiological sense, it is the direct cause of burns and other forms of personal injury. Injuries caused by heat include dehydration, heat exhaustion, and injury to the respiratory tract, in addition to burns.

Flame is the visible, luminous body of a burning gas. When a burning gas is mixed with the

Figure 1.10 Heat is transferred by radiation when heat waves hit an object and heat it.

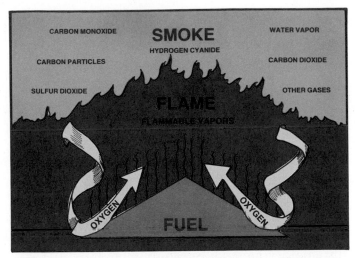

Figure 1.11 A burning fuel generates numerous new products of combustion.

proper amounts of oxygen, the flame becomes hotter and less luminous. This loss of luminosity is because of a more complete combustion of the carbon. For these reasons, flame is considered to be a product of combustion. Heat, smoke, and gas, however, can develop in certain types of smoldering fires without evidence of flame.

The smoke encountered at most fires consists of a mixture of oxygen, nitrogen, carbon dioxide, carbon monoxide, finely divided carbon particles (soot), and a miscellaneous assortment of products that have been released from the material involved.

Some materials give off more smoke than others. Liquid fuels generally give off dense black smoke. Oils, tar, paint, varnish, molasses, sugar, rubber, sulfur, and many plastics also generally give off a dense smoke in large quantities.

FIRE EXTINGUISHMENT THEORY

The extinguishment of fire is based on an interruption of one or more of the essential elements in the combustion process. With flaming combustion the fire may be extinguished by reducing temperature, eliminating fuel or oxygen, or by stopping the uninhibited chemical chain reaction (Figure 1.12). If a fire is in the smoldering mode of combustion, only three extinguishment options exist: reduction of temperature, elimination of fuel, or oxygen.

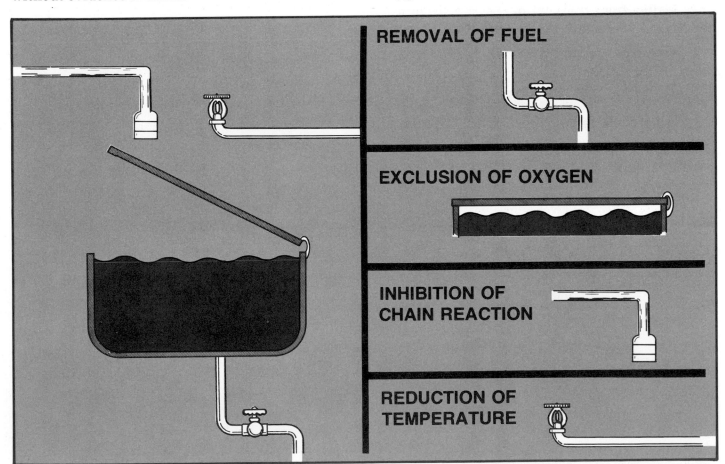

Figure 1.12 Four methods are used to extinguish fire.

Extinguishment by Temperature Reduction

One of the most common methods of extinguishment is by cooling with water. The process of extinguishment by cooling is dependent on cooling the fuel to a point where it does not produce sufficient vapor to burn. If we look at fuel types and vapor production, we find that solid fuels and liquid fuels with high flash points can be extinguished by cooling. Low flash point liquids and flammable gases cannot be extinguished by cooling with water as vapor production cannot be sufficiently reduced. Reduction of temperature is dependent on the application of an adequate flow in proper form to establish a negative heat balance.

Extinguishment by Fuel Removal

In some cases, a fire is effectively extinguished by removing the fuel source. This may be accomplished by stopping the flow of liquid or gaseous fuel or by removing solid fuel in the path of the fire. Another method of fuel removal is to allow the fire to burn until all fuel is consumed.

Extinguishment by Oxygen Dilution

The method of extinguishment by oxygen dilution is the reduction of the oxygen concentration to the fire area. This can be accomplished by introducing an inert gas into the fire or by separating the oxygen from the fuel. This method of extinguishment will not work on self-oxidizing materials or on certain metals as they are oxidized by carbon dioxide or nitrogen, the two most common extinguishing agents.

Extinguishment by Chemical Flame Inhibition

Some extinguishing agents, such as dry chemicals and halons, interrupt the flame producing chemical reaction, resulting in rapid extinguishment. This method of extinguishment is effective only on gas and liquid fuels as they cannot burn in the smoldering mode of combustion. If extinguishment of smoldering materials is desired, the addition of cooling capability is required.

CLASSIFICATION OF FIRES AND EXTINGUISHMENT METHODS

Class A Fires

Fires involving ordinary combustible materials, such as wood, cloth, paper, rubber, and many plastics (Figure 1.13).

Water is used in a cooling or quenching effect to reduce the temperature of the burning material below its ignition temperature.

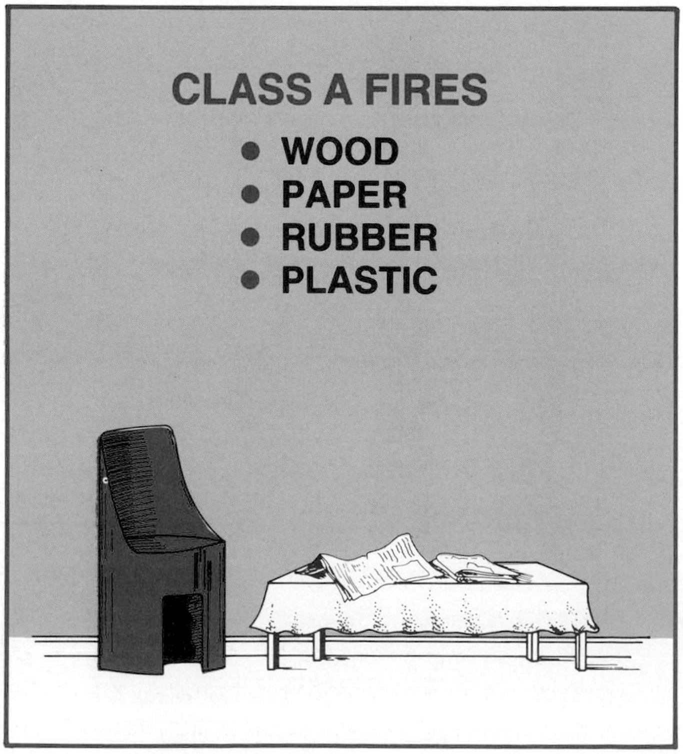

CLASS A FIRES
- **WOOD**
- **PAPER**
- **RUBBER**
- **PLASTIC**

Figure 1.13 Class A fires involve ordinary combustibles.

CLASS B FIRES
- **LIQUIDS**
- **GREASES**
- **GASES**

Figure 1.14 Class B fires involve flammable liquids, gases, and solids.

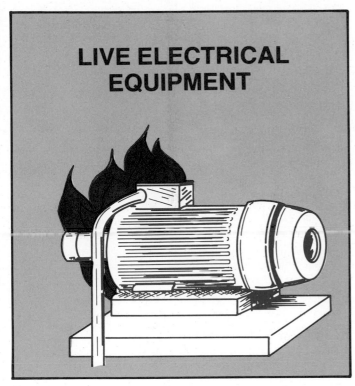

Figure 1.15 Class C fires involve energized electrical equipment.

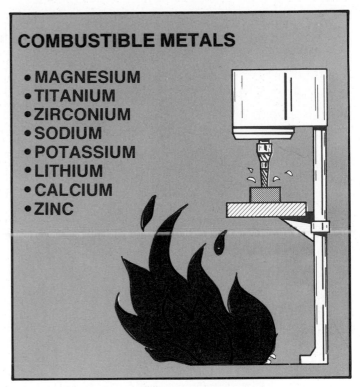

Figure 1.16 Class D fires involve combustible metals.

Class B Fires

Fires involving flammable liquids, greases, and gases (Figure 1.14).

The smothering or blanketing effect of oxygen exclusion is most effective. Other extinguishing methods include removal of fuel and temperature reduction.

Class C Fires

Fires involving energized electrical equipment (Figure 1.15).

This fire can sometimes be controlled by a non-conducting extinguishing agent. The safest procedure is always to attempt to de-energize high vol-

tage circuits and treat as a Class A or B fire depending upon the fuel involved.

Class D Fires

Fires involving combustible metals, such as magnesium, titanium, zirconium, sodium, and potassium (Figure 1.16).

The extremely high temperature of some burning metals makes water and other common extinguishing agents ineffective. There is no agent available that will effectively control fires in all combustible metals. Special extinguishing agents are available for control of fire in each of the metals and are marked specifically for that metal.

---

*Editor's Note

The approach taken to describe the stages of fire may be broken down further if a more detailed description is desired. The free burning or flame spread stage may include descriptions of steady state and clear burning.

Steady State — A fully involved fire condition with a supply of air to feed the fire and carry the products of combustion away.

Clear Burning — That phase of burning accompanied by high temperatures and complete combustion. Thermal columns will normally occur with high air speeds at the base of the fire.

NFPA STANDARD 1001
PORTABLE FIRE EXTINGUISHERS
Fire Fighter II

4-16 Portable Extinguishers

4-16.1 The fire fighter shall identify the classification of types of fire as they relate to the use of portable extinguishers.

4-16.2 The fire fighter, given a group of differing extinguishers, shall identify the appropriate extinguishers for the various classes of fire.

4-16.3 The fire fighter shall define the portable extinguisher rating system.*

*Reprinted by permission from NFPA Standard No. 1001, *Standard for Fire Fighter Professional Qualifications.* Copyright © 1981, National Fire Protection Association, Boston, MA.

IFSTA's Portable Extinguishers Transparencies are designed to complement this chapter.

Chapter 2
Portable Extinguishers

Portable fire extinguishers are classified according to their intended use on the four classes of fires (A, B, C, and D). In addition to the letter classification, extinguishers also receive a numerical rating. The number preceding the letter designates the potential size fire the extinguisher can be expected to extinguish (Figure 2.1).

EXTINGUISHER RATING SYSTEM

The rating system is based on tests conducted by the Underwriter's Laboratories, Inc. and Underwriters' Laboratories of Canada that are designed to determine the extinguishing potential for each size and type of extinguisher. These ratings consist of both a numeral and letter for extin-

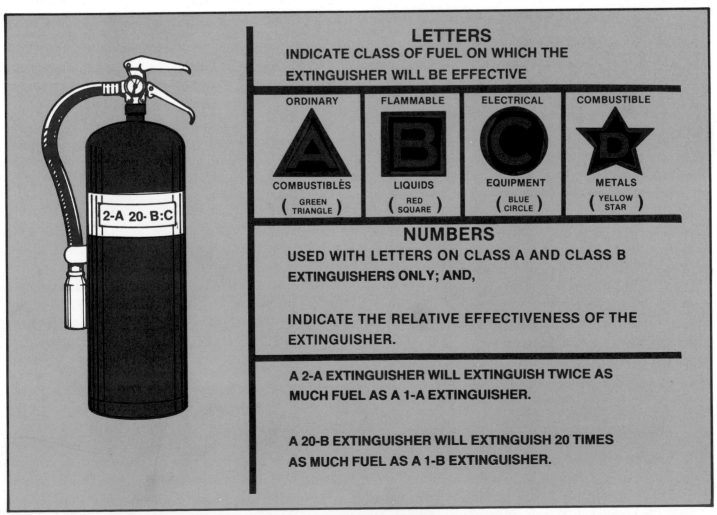

LETTERS
INDICATE CLASS OF FUEL ON WHICH THE
EXTINGUISHER WILL BE EFFECTIVE

ORDINARY	FLAMMABLE	ELECTRICAL	COMBUSTIBLE
A	B	C	D
COMBUSTIBLES	LIQUIDS	EQUIPMENT	METALS
(GREEN TRIANGLE)	(RED SQUARE)	(BLUE CIRCLE)	(YELLOW STAR)

NUMBERS
USED WITH LETTERS ON CLASS A AND CLASS B
EXTINGUISHERS ONLY; AND,

INDICATE THE RELATIVE EFFECTIVENESS OF THE
EXTINGUISHER.

A 2-A EXTINGUISHER WILL EXTINGUISH TWICE AS
MUCH FUEL AS A 1-A EXTINGUISHER.

A 20-B EXTINGUISHER WILL EXTINGUISH 20 TIMES
AS MUCH FUEL AS A 1-B EXTINGUISHER.

2-A 20-B:C

Figure 2.1 Extinguishers are marked for the class of fire and relative extinguishing capability.

guishers intended for use on Class A and Class B fires. Extinguishers for use on Class C fires only receive the letter rating because there is no readily measurable quantity for Class C fires, which are essentially Class A or B fires involving energized electrical equipment. Class D extinguishers, likewise, do not contain a numerical rating. The effectiveness of the extinguisher on Class D metals is detailed on the faceplate. Multiple letters or numeral-letter ratings are used on extinguishers which are effective on more than one class of fire (Figure 2.2).

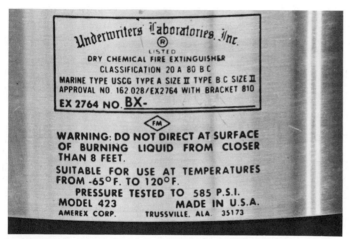

Figure 2.2 An example of a multi-use extinguisher marking.

Ratings from 1-A through 40-A are designated for units capable of extinguishing Class A fires. The numerical rating reflects the relative quantity of fire that can be extinguished by a certain size fire extinguisher. One and one-fourth gallons (5 liters) of water are required for a 1-A fire. A 2-A classification requires 2½-gallons (10 liters) or twice the 1-A capacity. Therefore, a dry chemical extinguisher rated 10-A is equivalent to four 2½-gallon (10 liters) water extinguishers. As the tests are conducted, the size of fire that can be extinguished by a non-expert or expert operator does not differ greatly for Class A fires. Technique and application are not as important in a Class A fire as they are on a Class B fire, according to the rating board. Therefore, even though expert operators are used to conduct the tests, no differential in rating is made.

There are three types of Class A fires that must be extinguished for an extinguisher to receive a 1-A through 6-A rating. These three types of fires are a wood crib, a wood panel, and an excel-

sior fire. Each presents a greatly different type of burning although each is a Class A combustible.

Extinguishers suitable for use on Class B fires are classified with numerical ratings ranging from 1-B through 640-B. The tests used by UL to determine the rating of Class B extinguishers consist of burning the flammable liquid, n-Heptane, in square steel pans.

Underwriters' Laboratories always use an expert operator in conducting the tests. However, the numerical rating of the extinguisher is applied assuming usage by a non-expert or untrained operator. To do this, a working rating is determined to be only 40 percent of the fire area that the expert operator can consistently extinguish in the tests. For example, a unit rated 60-B can extinguish a flammable liquid fire in a pan with 150 square feet (45 square meters) when used by the expert operator. The non-expert operator can be expected to extinguish 40 percent of the 150, or 60 square feet (45, or 18 square meters) of fire. Consequently, the rating of 60-B is applied to the unit.

The numerical rating on Class B fire extinguishers serves to indicate the relative fire extinguishing potential of various sizes and types of extinguishers suitable for Class B fires. It also gives an approximate indication of the square-foot area of flammable liquid fire that a non-expert operator can extinguish.

There are no fire tests specifically conducted for Class C ratings. In assigning a Class C designation, the extinguishing agent is tested only for electrical non-conductivity. If the agent meets the test requirements, it is then assigned a Class C letter rating providing the extinguisher also has an established numerical Class A, B, or both, rating. The Class C rating is provided in conjunction with a rating previously established for Class A and/or Class B fires.

Test fires for establishing Class D ratings vary for the type of combustible metal being tested. Several factors are considered during each test, including: reactions between metal and agent; toxicity of agent; toxicity of fumes produced and products of combustion; and possible burnout of metal instead of extinguishment. When an extinguishing agent is determined to be safe and effective for use on a metal, the details of instruction are included on the facepiece of the extinguisher, although no numerical rating is applied.

Multiple Markings

Extinguishers suitable for more than one class of fire should be identified by multiples of the symbols previously described. Most present-day extinguishers have these markings on them when they

are purchased. If a new extinguisher is not properly marked, the seller should be requested to supply the proper decals.

The "picture-symbol" labeling system now in use is designed to make the selection of fire extinguishers more effective and safe to use through the use of less confusing pictorial labels. The system also emphasizes when *not* to use an extinguisher on certain types of fires. Examples of this labeling system are shown in Figures 2.3-2.4.

Figure 2.3 These are the basic symbols for the classes of fire.

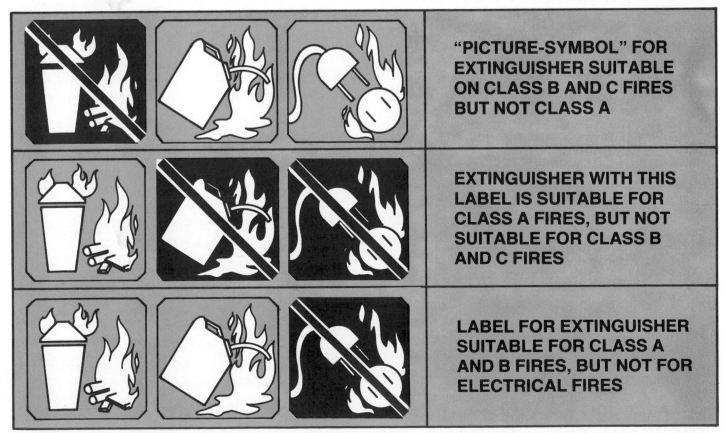

Figure 2.4 Samples of the "picture symbol" system.

Examples of Extinguisher Rating Labels
FOAM EXTINGUISHER RATED 4-A, 6-B

This extinguisher should extinguish a larger fire than a 2-A extinguisher. It will also extinguish approximately six times as much Class B fire as a 1-B extinguisher. Also, this extinguisher should extinguish a fire in a deep-layer flammable liquid, such as a dip tank having a surface area of six square feet (2 square meters).

DRY CHEMICAL EXTINGUISHER RATED 5-10-B:C

This extinguisher should extinguish approximately five to ten times as much Class B fire as a 1-B unit and should successfully extinguish a deep-layer flammable liquid fire of a 5 to 10 square foot (2 to 3 square meter) area. It is also safe to use on fires involving energized electrical equipment.

The term, dry chemical, is used for several similar but different types of extinguishing agents. Each should be studied separately to insure correct usage. Dry powder is very effective on certain combustible metals but the application of the proper type should be only after individual investigation and testing has been accomplished. It is most important NOT to confuse dry chemical extinguishers with combustible metal (Class D) dry powder extinguishers.

MULTIPURPOSE EXTINGUISHER RATED 4-A, 20-B:C

This extinguisher should extinguish a certain size Class A fire and approximately twenty times as much Class B fire as a 1-B extinguisher, and a deep-layer flammable liquid fire of a 20 square foot (6 square meter) area. It is also safe to use on fires involving energized electrical equipment.

SELECTION OF EXTINGUISHERS

The selection of a proper fire extinguisher will depend upon numerous factors including hazards to be protected, severity of the fire, atmospheric conditions, personnel available, ease of handling extinguishers, and any life hazard or operational concerns.

Portable extinguishers come in many shapes, sizes, and types. While the operating procedures of each type of extinguisher are similar, operators should become familiar with the detailed instructions found on the label of the extinguisher. In an emergency every second is of great importance, therefore everyone should be acquainted with the following general instructions applicable to most portable fire extinguishers.

The general operating instructions follow the letters P - P - P - S.

P — Pull the pin at the top of the extinguisher that keeps the handle from being pressed. Break the plastic or thin wire inspection band.

P — Point the nozzle or outlet toward the fire. Some hose assemblies are clipped to the extinguisher body. Release it and point.

P — Press the handle above the carrying handle to discharge the agent inside. The handle can be released to stop the discharge at any time.

S — Sweep the nozzle back and forth at the base of the flames to disperse the extinguishing agent. After the fire is out, probe for remaining smoldering hot spots or possible reflash of flammable liquids. Make sure the fire is out.

Pull pin — Point nozzle — Press handle — Sweep agent back and forth.

Modern extinguishers are designed to be carried to the fire in an upright position. When instructing the general public in the use of extinguishers, it should be strongly emphasized that they are operated in an upright position. Only obsolete soda-acid, foam, and cartridge operated water extinguishers are designed to be turned upside down. Avoid them. Do not attempt to activate the extinguisher until close enough to the fire to be within the reach of the stream of that particular extinguishing agent. Smaller extinguishers will require closer approach to the fire.

FIRE EXTINGUISHER OPERATION
Pump Tank Water Extinguishers (Figure 2.5)

Sizes: 2½ to 5 gallons (9½ to 19 liters)

Applicable To: Class A fires

Stream Reach Under Normal Conditions: 30 to 40 feet (9 to 12 meters)

Discharge Time Under Normal Conditions: 45 seconds to 3 minutes other for a broken spray stream. The filler cap is provided with a tiny vent which must be kept clear so that air may replace the water as it is discharged from the tank. In some cases, it may be desirable to place one finger into the discharge stream at the nozzle to break the solid stream up into a broken stream.

Figure 2.5 A hand-operated pump tank type water extinguisher for Class A fires.

Protect From Freezing: Yes

Operating Principle: Hand-pump operated

Servicing: Service in accordance with instructions displayed on face of the extinguisher and in compliance with the recommendations contained in NFPA Standard No. 10.

Method Of Operation: There are several kinds of pump tank extinguishers. These are generally equipped with a double action pump that delivers a continuous stream of water. To operate, carry the extinguisher by the handle to the fire. **CAUTION:** Do not run to the fire. Operate the pump with an up and down (or side to side) stroke. Short continuous strokes of the pump handle may provide a better stream than long strokes. Some nozzles are provided with two tips, one for straight stream and the

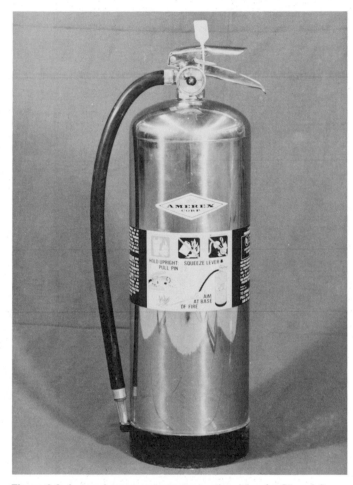

Figure 2.6 A stored-pressure water type extinguisher for Class A fires.

Stored-Pressure Water Extinguishers (Figure 2.6)

Size: 2½ gallons (9½ liters)

Applicable To: Class A fires

Stream Reach Under Normal Conditions: 30 to 40 feet (9 to 12 meters)

Discharge Time Under Normal Conditions: 60 seconds

Protect From Freezing: Yes, unless containing antifreeze solution

Operating Principle: The water is ejected by compressed air that is stored in the same shell or chamber with the water. When the shutoff device is released, a stream of water is expelled through the hose. With this type, the pressure is ready to release the extinguishing agent at any time.

Servicing: Service in accordance with instructions displayed on the face of the extinguisher and in compliance with the recommendations contained in NFPA Standard No. 10.

Method Of Operation: Stored-pressure water extinguishers are designed to be carried to the fire in an upright position. The hose is held in the operator's hand at all times, ready for use. Direct the stream at the base of the fire.

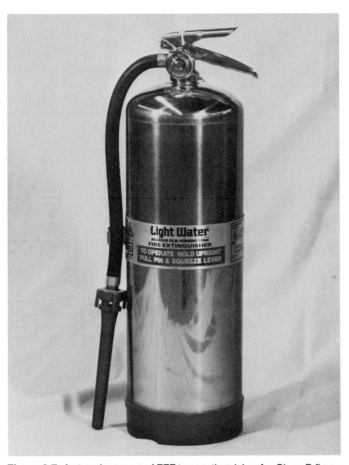

Figure 2.7 A stored-pressure AFFF type extinguisher for Class B fires.

Aqueous Film Forming Foam Extinguishers (Figure 2.7)

Size: 2½ gallon (9½ liter)

Applicable To: Class A and B fires

Stream Reach Under Normal Conditions: 30 feet (9 meters)

Discharge Time Under Normal Conditions: 60-65 seconds

Protect From Freezing: Yes, unless antifreeze solution added

Operating Principle: The water and aqueous film forming foam (AFFF) solution is expelled by compressed air stored in the same shell or chamber with the solution. When the shutoff device is released, the solution is expelled through the hose and aspirator, where air mixes with the solution and forms the foam. With this type, the pressure is ready to expel the extinguishing agent at any time. **NOTE:** Although AFFF concentrate is intended for use in an air aspirating nozzle, it may be used with some decrease in performance with non-air aspirating nozzles (Figure 2.8).

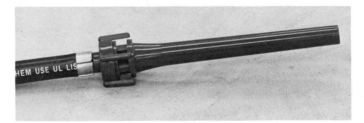

Figure 2.8 The nozzle for most AFFF extinguishers is designed to introduce air into the stream.

Servicing: The standard 2½ gallon (9½ liter) extinguisher should only be recharged with a special kit by experienced personnel to insure the proper mixtures in the solution. Recharging should be done in accordance with manufacturers' recommendations and NFPA Standard No. 10.

Method Of Operation: This extinguisher is designed to be carried to the fire in an upright position. Operation includes pulling the pin and squeezing the lever. A special aspirator nozzle creates the foam from the 6 percent AFFF solution. Use side to side sweeping motions across the entire width of the fire but avoid splashing liquid fuels. This special foam has the ability to make water float on fuels that are lighter than water. The vapor seal that is created extinguishes the flame and prevents reignition. The foam also has good wetting and penetrating properties on Class A fires.

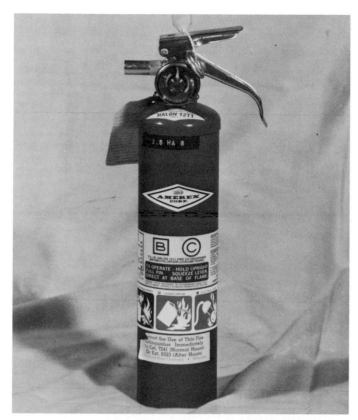

Figure 2.9 A stored-pressure vaporizing liquid extinguisher containing Halon for Class B:C fires.

Halon 1211 Extinguishers (Figure 2.9)

Sizes: 2½ to 25 pounds (1 to 11 kg)

Applicable To: Class B and C fires

Stream Reach Under Normal Conditions: 4 to 10 feet (1 to 3 meters)

Discharge Time Under Normal Conditions: 8 to 25 seconds

Protect From Freezing: No

Operating Principle: Halon 1211 (Bromochlorodifluoromethane) extinguishers operate on the principle of liquefied compressed gas, which will not support combustion, being discharged to smother and terminate flame propagation. Halon 1211 is stored under pressure and ready to release the extinguishing agent at any time.

Servicing: Service in accordance with instructions displayed on the face of the extinguisher and with the recommendations contained in NFPA Standard No. 10.

Method Of Operation: Halon 1211 extinguishers are designed to be carried by the top handle. They are intended primarily for use on Class C fires but are effective on Class B fires as well. They have a limited range and the discharge of the extinguishing agent may be affected by draft and wind. The initial application should be made close to the fire and the discharge should be directed at the base of the flames. The discharge should be applied to the burned surface even after the flames are extinguished. Best results will be obtained on flammable liquid fires if the discharge is directed to sweep the flame from the burning surface. Apply the discharge first at the near edge of the fire and gradually progress forward while moving from side to side.

Figure 2.10 A liquefied-gas CO_2 extinguisher for Class B:C fires.

Carbon Dioxide Extinguishers (Figure 2.10)

Sizes: 2½ - 5 pounds (1.5 to 2 kg), 10 - 15 pounds (4.5 to 7 kg), 20 pounds (9 kg)

Applicable To: Class B and C fires

Stream Reach Under Normal Conditions: 3 to 18 feet (1 to 5 meters)

Discharge Time Under Normal Conditions: 8 to 30 seconds

Protect From Freezing: No

Operating Principle: Carbon dioxide extinguishers operate on the principle of an inert gas, carbon dioxide (CO_2), which will not support combustion, being discharged thereby smothering the fire. The carbon dioxide is stored under its own pressure and ready for release at any time.

Servicing: Service in accordance with instructions displayed on the extinguisher and in compliance with the recommendations contained in NFPA Standard No. 10.

Method Of Operation: Carbon dioxide extinguishers are designed to be carried to the fire by the top handle. The discharge expels a cloud of carbon dioxide gas through the nozzle horn **CAUTION:** Often a "frost" residue will form on the nozzle horn and contact with the skin could result in frostbite. The discharge horn should be pointed at the base of the fire and application should continue even after the flames are extinguished to prevent a possible reflash of the fire. On flammable liquid fires, best results are obtained when the discharge from the extinguisher is employed to sweep the flame off the burning surface, applying the discharge first at the near edge of the fire and gradually progressing forward, moving the discharge cone very slowly from side to side.

Carbon Dioxide Wheeled Units (Figure 2.11)

Sizes: 50 (23 kg), 75 (34 kg), and 100 pounds (45 kg)

Applicable To: Class B and C fires

Stream Reach Under Normal Conditions: 8 to 20 feet (2 to 6 meters)

Discharge Time Under Normal Conditions: 30 to 80 seconds

Protect From Freezing: No

Hose Length: 15 feet (50 meters)

Method Of Operation: The principle of operation is the same for these larger units as for

Figure 2.11 Large CO_2 extinguishers are used in industrial applications.

the smaller CO_2 extinguishers. They are designed to be wheeled to the fire, and operated in accordance with the instructions on the extinguisher.

Dry Chemical Extinguishers — Ordinary Base (Figure 2.12)

Sizes: 2½ to 30 pounds (1½ to 11 kg)

Applicable To: Class B and C fires

Stream Reach Under Normal Conditions: 5 to 20 feet (2 to 6 meters)

Discharge Time Under Normal Conditions: 10 to 25 seconds

Protect From Freezing: No

Operating Principle: The chemical compound in dry chemical extinguishers consists principally of sodium bicarbonate, potassium bicarbonate, ammonium phosphate, or potassium chloride which has been chemically processed to make it moisture resistant and free flowing. This compound is discharged under pressure and directed at the fire. The extinguisher may

Figure 2.12 Dry chemical extinguishers may be either stored pressure or exterior gas cartridge operated.

contain a cartridge of carbon dioxide or nitrogen gas, either inside or alongside the main container, to expel the dry chemical. When the pressure is allowed to enter the main cylinder, the dry chemical may be expelled by opening the shutoff nozzle. Some extinguishers are pressurized with inert gas or dry air and do not have cartridges.

Method Of Operation: These extinguishers are designed to be carried to the fire. In the case of cartridge operated extinguishers, the cartridge must be activated to release the gas that pressurizes the dry chemical chamber and expels the dry chemical. With a pressurized dry chemical extinguisher, both the dry chemical and the expellant are stored in a single chamber. In either case, operation expels a cloud of dry chemical and the discharge is controlled by a shutoff valve. Best results are obtained by attacking the near edge of the fire and progressing forward while moving the nozzle with a side to side sweeping motion. The discharge should be applied to the burning surface even after the flames are extin-

guished to prevent possible reflash by coating the hot surfaces and any glowing materials present. When used on electrical fires, care should be taken around open electrical contacts to prevent costly cleaning. It may be better to use other agents in such cases.

Figure 2.13 Dry chemical extinguishers are available in larger capacities as wheeled units.

Dry Chemical Wheeled Units (Figure 2.13)

Sizes: 75 to 350 pounds (34 to 160 kg)

Applicable To: Class B and C fires

Stream Discharge Under Normal Conditions: 10 to 45 feet (3 to 14 meters)

Discharge Time Under Normal Conditions: 20 seconds to 120 seconds

Protect From Freezing: No

Hose Length: 50 to 100 feet (15 to 30 meters)

Method Of Operation: The principle of operation is the same for these larger units as for the smaller dry chemical units. They are designed to be wheeled to the fire and operated in accordance with the instructions on the extinguisher.

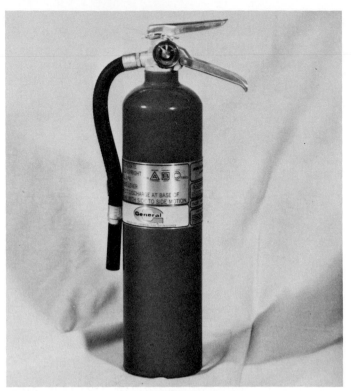

Figure 2.14 Multi-purpose dry chemical extinguishers are used extensively to reduce the total number of extinguishers required.

Dry Chemical Extinguishers - Multi-purpose Base (Figure 2.14)

Multi-purpose dry chemical (monammonium phosphate and barium sulphate) is a mixture of chemicals in powder form found suitable for use on incipient fires in ordinary combustible materials (Class A) as well as on Class B and C fires. They are not recommended for use on metal (Class D) fires, unless stated so on the nameplate. The extinguishers are available in a wide range of sizes and are identical to other dry chemical extinguishers. The method of operation is the same as with the ordinary base dry chemical units. CARE SHOULD BE TAKEN TO AVOID CONTAMINATION OF ANY TYPE DRY CHEMICAL WITH THE MULTI-PURPOSE TYPE.

On Class A fires, the discharge should be directed at the burning surfaces to cover them with chemical. When the flames have been extinguished, the chemical discharge should be intermittently directed on any glowing areas. Careful watch should be maintained for hot spots that may develop and additional agent applied to those surfaces as required to adequately coat them with the extinguishing agent.

Extinguishers and Powder Extinguishing Agents for Metal Fires (Figure 2.15)

Normal extinguishing agents generally should not be used on metal fires (Class D). Specialized techniques and extinguishing agents have been developed to control and extinguish fires of this type. A given agent does not, however, necessarily control or extinguish all metal fires. Some agents are valuable in working with several metals; others are useful for combating only one type of metal fire. Some of the agents are intended to be applied by means of a hand shovel or scoop, others by means of portable fire extinguishers designed for use with dry powders. The application of the agents should be of sufficient depth to adequately cover the fire area and provide a smothering blanket. The agent should be applied gently on metal fires to avoid breaking any crust which may have formed over the burning metal. If the crust is broken, the fire may flare up and expose more raw material to combustion.

Additional applications may be necessary to cover any hot spots that may develop. The material

Figure 2.15 Dry powder extinguishers are gas cartridge operated and contain an agent for particular metals.

should be left undisturbed until the mass has cooled before disposal is attempted. Care should be taken to avoid scattering the burning metal. If the burning metal is on a combustible surface, the fire should first be covered with powder, then a one-or two-inch (25 to 50 mm) layer of powder spread out nearby and the burning metal shoveled onto this layer with more powder added as needed. Reference should be made to the manufacturer's recommendations for use and special techniques for extinguishing fires in various combustible metals.

DAMAGED EXTINGUISHERS

Leaking, corroded, or otherwise damaged extinguisher shells or cylinders should be discarded or returned to the manufacturer for repair. **CAUTION:** NEVER TRY TO REPAIR THE SHELL OR CYLINDER SUBJECTED TO PRESSURE. If an extinguisher shows only slight damage or corrosion, and it is questionable whether it is safe to use, it should be given a hydrostatic test by the manufacturer or a qualified testing agency. Leaking hose, gaskets, nozzles, and inner chambers can be replaced by firefighters.

OBSOLETE EXTINGUISHERS

In 1969, American manufacturers stopped making inverting-type extinguishers, including soda-acid, foam, and cartridge-operated water and loaded stream extinguishers (Figure 2.16). Nevertheless, it is estimated that there are still several million in use. Some of their disadvantages are:

- They cannot be turned off once activated.
- The agent is more corrosive than water.
- They are potentially dangerous. If the discharge hose is blocked, these extinguishers can build up pressures in excess of 300 psi (2000 kPa) and explode, causing serious injury or death.

Also discontinued are extinguishers made of copper or brass joined by soft solder or rivets.

The soda-acid extinguisher is the most common obsolete type. When this extinguisher is inverted, acid from a bottle mixes with a soda-and-water solution and produces a gas that expels the

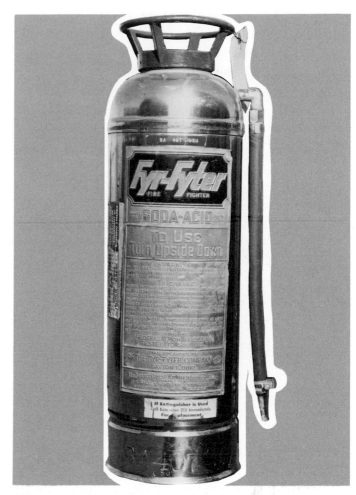

Figure 2.16 Extinguishers that require inverting to operate are now obsolete.

liquid. The pressure on an acid-corroded shell has exploded many soda-acid extinguishers.

Inverting foam extinguishers look like soda-acid extinguishers. Inverting mixes solutions from two chambers, forming a foam that expands at a 1:89 ratio and a gas that expels the foam to fight Class B and C fires. These units have been replaced by stored-pressure units.

Users of cartridge-operated water extinguishers have to invert and bump the units to puncture a CO_2 cylinder. The pressure of the gas released from the cartridge expels the water.

Some people still have liquid carbon tetrachloride extinguishers, obsolete since the 1960's. When carbon tetrachloride comes in contact with heat it releases a highly toxic phosgene gas.

Owners of obsolete extinguishers should replace them.

INSPECTION OF FIRE EXTINGUISHERS

Firefighters are frequently required to inspect fire extinguishers. Two main factors that determine a fire extinguisher's worth and which will justify its purchase and installation are its serviceability and its accessibility. An inspector should look for the following items:

- Check accessibility and proper location (Figure 2.17).

- Check tag for date of last recharge or inspection (Figure 2.18).

- Check nozzle for obstructions and operation (Figure 2.19).

- Examine for corrosion (leaks at seams) or mechanical damage.

- Check lockpin and seal (Figures 2.20, 2.21).

- Determine if full (water level, pressure gauge, or weight) (Figures 2.22 - 2.24).

- Examine condition of hose and hose coupling.

- Check horns for cracks, dirt, or grease accumulations.

- Date of this inspection and initials of inspector should be placed on tag.

Figure 2.17 Hidden extinguishers cost valuable time in an emergency. They should be hung four to five feet (1 to 2 m) above the floor.

Figure 2.18 An extinguisher not inspected for several years may be unsafe or may fail to operate. Every check or recharge should be recorded.

Figure 2.19 When checking the extinguisher body or recharging, check the nozzle for obstructions.

Figure 2.20 Make sure the extinguisher has not been tampered with since the last inspection. Is the leaded wire seal in place?

Figure 2.21 If the seal is plastic, is it in place through the lockpin?

Figure 2.22 If a gauge is provided, note the amount of pressure in the cylinder.

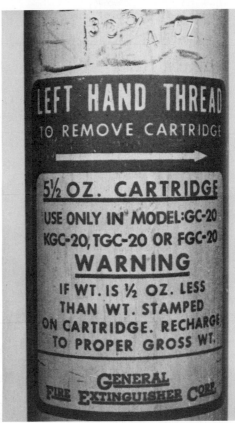

Figure 2.23 If a gas cartridge is used, it must be weighed to determine if it is full. Note the total weight of the gas and the cartridge is 30¾ ounces (861 g).

Figure 2.24 Carbon dioxide extinguishers are also weighed. If the bottle is full it weighs 39½ pounds (18 kg). The gas weight alone is 15 pounds (7 kg).

TABLE 2-1
CHARACTERISTICS OF EXTINGUISHERS

Extinguishing Agent	Method of Operation	Capacity	UL or ULC Classification
Water	Stored Pressure	2½ gal	2-A
Water	Pump Tank	1½ gal	1-A
	Pump Tank	2½ gal	2-A
	Pump Tank	4 gal	3-A
	Pump Tank	5 gal	4-A
Water (Antifreeze Calcium Chloride)	Cartridge or Stored Pressure	1¼, 1½ gal	1-A
	Cartidge or Stored Pressure	2½ gal	2-A
	Cylinder	33 gal	20 A
Water (Wetting Agent)	Stored Pressure	1½ gal	2-A
	Carbon Dioxide Cylinder	25 gal (wheeled)	10-A
	Carbon Dioxide Cylinder	45 gal (wheeled)	30-A
	Carbon Dioxide Cylinder	60 gal (wheeled)	40-A
Water (Soda Acid)	Chemically Generated Expellant	1¼, 1½ gal	1-A
	Chemically Generated Expellant	2½ gal	2-A
	Chemically Generated Expellant	17 gal (wheeled)	10-A
	Chemically Generated Expellant	33 gal (wheeled)	20-A
Water (Loaded Stream)	Stored Pressure	2½ gal	3-A
	Cartridge or Stored Pressure	33 gal (wheeled)	20-A
AFFF	Stored Pressure	2½ gal	3-A 20 B
	Nitrogen Cylinder	33 gal (wheeled)	20-A 160 B
Carbon Dioxide	Self-Expellant	2 to 5 lb	1 to 5-B:C
	Self-Expellant	10 to 15 lb	2 to 10-B:C
	Self-Expellant	20 lb	10-B:C
	Self-Expellant	50 to 100 lb (wheeled)	10 to 20-B:C
Dry Chemical (Sodium Bicarbonate)	Stored Pressure	1 lb	1 to 2-B:C
	Stored Pressure	1½ to 2½ lb	2 to 10-B:C
	Cartridge or Stored Pressure	2¾ to 5 lb	5 to 20-B:C
	Cartridge or Stored Pressure	6 to 30 lb	10 to 160-B:C
	Nitrogen Cylinder or Stored Pressure	75 to 350 lb (wheeled)	40 to 320-B:C
Dry Chemical (Potassium Bicarbonate)	Stored Pressure	1 to 2 lb	1 to 5-B:C
	Cartridge or Stored Pressure	2¼ to 5 lb	5 to 20-B:C
	Cartridge or Stored Pressure	5½ to 10 lb	10 to 80-B:C
	Cartridge or Stored Pressure	16 to 30 lb	40 to 120-B:C
	Cartridge	48 lb	120-B:C
	Nitrogen Cylinder or Stored Pressure	125 to 315 lb (wheeled)	80 to 640-B:C
Dry Chemical (Potassium Chloride)	Stored Pressure	2½ to 8½ lb	5 to 10-B:C
	Stored Pressure	5 to 9 lb	20 to 40-B:C
	Stored Pressure	10 to 20 lb	20 to 40-B:C
	Stored Pressure	135 lb	160-B:C
Dry Chemical (Ammonium Phosphate)	Stored Pressure	1 to 5 lb	1 to 2-A and 2 to 10-B:C
	Stored Pressure or Cartridge	2½ to 8½ lb	1 to 4-A and 10 to 40-B:C
	Stored Pressure or Cartridge	9 to 17 lb	2 to 20-A and 10 to 80-B:C
	Stored Pressure or Cartridge	17 to 30 lb	3 to 20-A and 30 to 80-B:C
	Cartridge	45 lb	20-A and 80-B:C
	Nitrogen Cylinder or Stored Pressure	110 to 315 lb (wheeled)	20 to 40-A and 60 to 320-B:C
Dry Chemical (Foam Compatible)	Cartridge or Stored Pressure	4¾ to 9 lb	10 to 20-B:C
	Cartridge or Stored Pressure	9 to 27 lb	20 to 30-B:C
	Cartridge or Stored Pressure	18 to 30 lb	40 to 60-B:C
	Nitrogen Cylinder or Stored Pressure	150 to 350 lb	80 to 240-B:C
Dry Chemical (Potassium Chloride)	Cartridge or Stored Pressure	2½ to 5 lb	10 to 20-B:C
	Cartridge or Stored Pressure	9½ to 20 lb	40 to 60-B:C
	Cartridge or Stored Pressure	19½ to 30 lb	60 to 80-B:C
	Stored Pressure	125 to 200 lb (wheeled)	160-B:C
Dry Chemical (Potassium Bicarbonate Urea Base)	Stored Pressure	5 to 11 lb	40 to 80-B:C
	Stored Pressure	9 to 23 lb	60 to 160-B:C
	Stored Pressure	175 lb	480-B:C
Bromotrifluoromethane	Stored Pressure	2½ lb	2-B:C
Bromochlorodifluoromethane	Stored Pressure	2 to 4 lb	2 to 5-B:C
	Stored Pressure	5½ to 9 lb	1-A and 10-B:C
	Stored Pressure	16 to 22 lb	1 to 4-A and 20 to 80-B:C

This table with Metric equivalents is printed in the Appendix.

NFPA STANDARD 1001
ROPES AND KNOTS
Fire Fighter I

3-5 Ropes

3-5.1 The fire fighter, when given the name, picture, or actual knot shall identify it and describe the purpose for which it would be used.

3-5.2 The fire fighter, when given the proper size and amount of rope, shall demonstrate tying a bowline knot, a clove hitch, and a becket or sheet bend.

3-5.3 The fire fighter, given the proper rope, shall demonstrate the bight, loop, round turn, and half hitch as used in tying knots and hitches.

3-5.4 The fire fighter, using an approved knot, shall hoist any selected forcible entry tool, ground ladder, or appliance to a height of at least 20 ft.

3-5.5 The fire fighter shall demonstrate the techniques of inspecting, cleaning, and maintaining rope.*

Fire Fighter II

4-5 Ropes

4-5.1 The fire fighter, when given a simulated fire fighting or rescue task, shall select the appropriate size, strength, and length of rope for the task.

4-5.2 The fire fighter shall select and tie a rope between two objects at least 15 ft apart, that will support the weight of a fire fighter on the rope.

4-5.3 The fire fighter shall use a rope to tie ladders, hose, and other equipment so as to secure them to immovable objects.*

---

*Reprinted by permission from NFPA Standard No. 1001, *Standard for Fire Fighter Professional Qualifications.* Copyright © 1981, National Fire Protection Association, Boston, MA.

IFSTA's Ropes and Knots Transparencies are designed to complement this chapter.

Chapter 3
Ropes and Knots

Rope practices in this chapter are limited to the basic knots and hitches most commonly used in the fire service. Local fire department policies may involve the use of knots other than those discussed, and such policies are encouraged. Only one approved method of tying these basic knots is shown, but if a firefighter has learned to tie them another way, it is suggested that the technique not be changed. Some departments encourage the practice of tying knots while blindfolded to better develop the ability of the firefighter.

Throughout the following description of how to tie knots, the terms "standing part" and "running part" are used. In order to completely understand these terms, the following definitions are offered.

STANDING PART — The part of the rope that is to be used for work, such as hoisting, pulling, snubbing, and the like.

RUNNING PART — The part of the rope that is to be used in forming the knot (commonly referred to as "the loose end.")

ELEMENTS OF A KNOT

Knots weaken a rope because the rope is bent in order to form the knot. The fibers on the outside of the bend are stretched and the fibers on the inside of the bend are crushed. It can be seen then that a knot with sharp bends will weaken a rope more than a knot with easy bends. The bends that a rope undergoes in the formation of a knot or hitch are known as the bight, the loop, and the round turn. Each of these formations is shown in Figure 3.1. The bight is formed by simply bending the rope

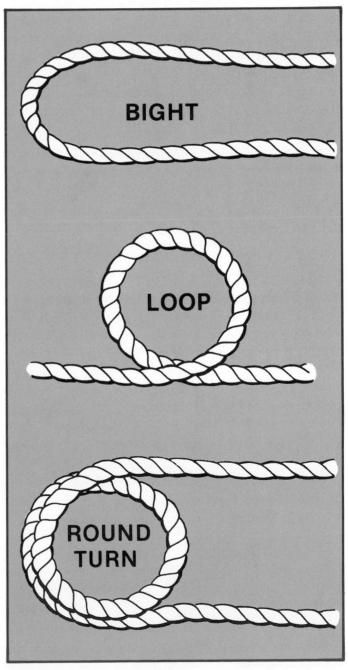

Figure 3.1 Elements of a knot or hitch.

while keeping the sides parallel. The loop is made by crossing the side of a bight. The round turn consists of further bending one side of a loop. Knots and hitches are formed by combining these elements in different ways so that the tight part of the rope bears on the free end to hold it in place. Knots and hitches for fire service use should be those which may be rapidly tied, easily untied, not subject to slippage, and have a minimum of abrupt bends.

THE BECKET OR SHEET BEND

The becket or sheet bend is used for joining two ropes and is particularly well-suited for joining ropes of unequal diameters. It is also unlikely to slip when the rope is wet. These advantages make it useful and dependable in fire service rope work. The becket bend is tied as follows and illustrated in Figures 3.2 - 3.5.

Step 1: Form a bight in one of the ends to be tied (if two ropes of unequal diameter are being tied, the bight always goes in the larger of the two) and pass the other end through the bight (Figure 3.2).

Step 2: Bring the loose end around both parts of the bight as shown in Figure 3.3.

Step 3: Tuck this end under its own standing part and over the bight standing part as shown in Figure 3.4.

Step 4: Draw the knot down snug as shown in Figure 3.5.

Figure 3.3 Step 2.

Figure 3.4 Step 3.

Figure 3.2 Step 1. Becket or sheet bend.

Figure 3.5 Step 4.

THE BOWLINE KNOT

The bowline is a good knot for forming a loop that will not slip under strain and it may be easily untied. Its use in the fire service is extensive and all firefighters should be able to tie the bowline in the open as well as around an object. The following method, as illustrated in Figures 3.6 - 3.9, is one good way of tying the bowline, although other methods may be just as effective.

Step 1: Measure off sufficient rope to form the size of the knot desired and form a loop in the standing part as shown in Figure 3.6.

Step 2: Pass the running part upward through the loop as shown in Figure 3.7.

Step 3: Pass the running part over the top of the loop under the standing part and bring the end of the running part completely around the standing part and down through the loop as shown in Figure 3.8.

Step 4: Pull the knot snugly into place, forming an "inside" bowline with the running part of the inside of the loop as shown in Figure 3.9.

NOTE: The bowline may be tied with the running part outside the loop. This is known as an outside bowline. The outside bowline is just as strong as the inside bowline.

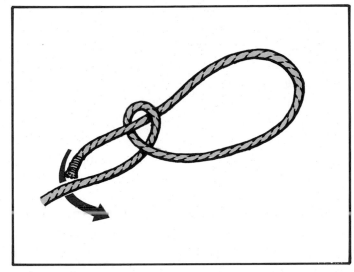

Figure 3.7 Step 2.

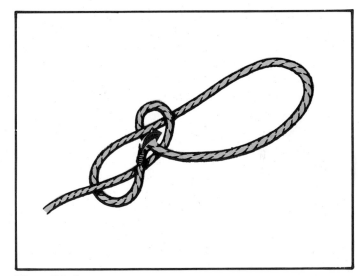

Figure 3.8 Step 3.

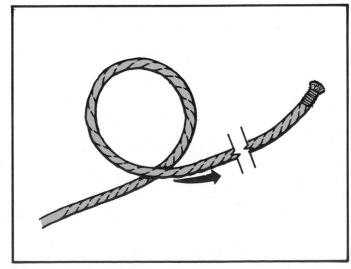

Figure 3.6 Step 1. Bowline.

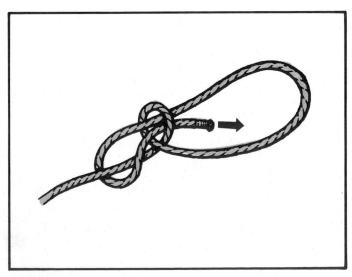

Figure 3.9 Step 4.

THE CLOVE HITCH

The clove hitch may be formed by several methods. It consists essentially of two half hitches. Its principal use is to attach a rope to an object such as a pole, post, or hose. The clove hitch may be formed anywhere in the rope from one end to the middle. When properly set, it will stand a pull in either direction without slipping. How to form a clove hitch in the open is as follows and is illustrated in Figures 3.10-3.13.

Step 1: Form a loop in the left hand with the running part to the right crossing under the standing part, as shown in Figure 3.10.

Step 2: Form another loop in the right hand again with the running part crossing under the standing part as shown in Figure 3.11.

Step 3: Slide the right hand loop on top of the left hand loop as shown in Figure 3.12. (This is the important step in forming the clove hitch.)

Step 4: Hold the two loops together at the rope and thus form the clove hitch as shown in Figure 3.13. Slip these loops over the object that the knot is to be tied around. Pull the ends in opposite directions to tighten.

Figure 3.10 Step 1. Clove hitch anywhere in the rope.

Figure 3.11 Step 2.

Figure 3.12 Step 3.

Figure 3.13 Step 4. **NOTE:** The clove hitch is used to anchor the rope to the object and the subsequently applied half-hitches are used to carry the weight of the object being hoisted.

The clove hitch, as formed by the method just described, can obviously not be placed over an object which has no free end (such as the center of a hoseline). It is, therefore, necessary to know how to tie the clove hitch around an object as illustrated in Figures 3.14 - 3.16.

Step 1: Make one complete loop around the object and bring the running part below the standing part as shown in Figure 3.14.

Step 2: Cross the running part over the standing part and complete the "round turn" about the object just above the first loop as shown in Figure 3.15.

Step 3: Pass the running part end under the upper wrap just above the cross and by pulling, properly set the hitch (Figure 3.16).

In order to insure that the clove hitch does not loosen during use, a safety hitch should be applied. This is done by taking a half hitch around the standing part of the clove hitch, with the loose end.

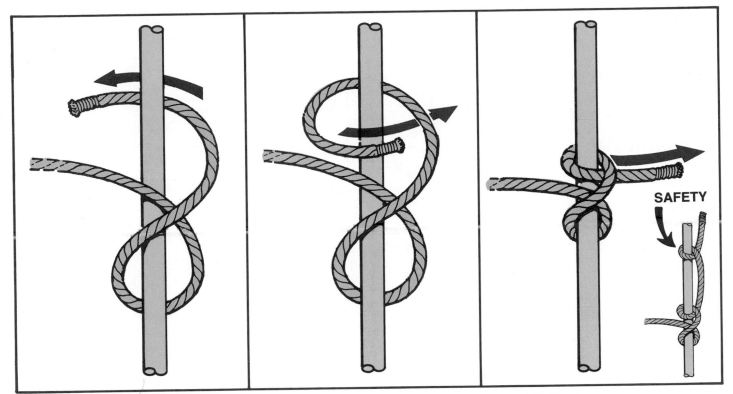

Figure 3.14 Step 1. Clove hitch around an object.

Figure 3.15 Step 2.

Figure 3.16 Step 3.

THE HALF SHEEP SHANK WITH A SAFETY

The half sheep shank consists of a half hitch formed around a bight and its principal use is in connection with the tightening of a rope between two objects such as would be done in roping off an area. The forming of a half sheep shank is as follows and is illustrated in Figures 3.17 - 3.22.

Step 1: With the line secured to one object, place the running part around another object and bring it back parallel to the standing part as shown in Figure 3.17.

Step 2: Face the standing part of the line which is secured to the first object, drop the running part across the standing part and then form a bight about 12 inches (30 cm) long just behind where the running part crosses the standing part as shown in Figure 3.18.

Step 3: Reach ahead of the cross-over point and form a loop in the standing part as shown in Figure 3.19.

Step 4: Place the loop just formed well back over the end of the bight to form a half hitch as

shown in Figure 3.20. (This forms the true half sheep shank, but for safety's sake Step No. 5 should be followed.)

Step 5: Place another half hitch over the end of the bight as shown in Figure 3.21.

Step 6: Pull the running part to tighten the half sheep shank and make it secure. The complete hitch is shown in Figure 3.22.

Figure 3.17 Step 1. Half sheep shank with a safety.

Figure 3.18 Step 2. Half sheep shank with a safety.

Figure 3.19 Step 3.

Figure 3.20 Step 4.

Figure 3.21 Step 5.

Figure 3.22 Step 6.

THE CHIMNEY HITCH

The chimney hitch is formed by several half hitches and a bight which are formed around the standing part of a rope. It is used to take up slack in a rope during an operation. Its ability to be easily untied makes it a desired hitch for anchoring a rope around a chimney. The usage gave it its name. The forming of a chimney hitch is as follows and is illustrated in Figures 3.23 - 3.27.

Step 1: With one end of the line secured and the running part around a stationary object and parallel to the standing part, form a half hitch around the standing part as shown in Figure 3.23.

Step 2: Pass the running part through the parallel ropes and form another half hitch or loop over the half hitch previously formed as shown in Figure 3.24.

Step 3: Pull the loop snugly to the side of the half hitch and wedge the half hitch against the standing part as shown in Figure 3.25.

Step 4: Place one or more half hitches back of the chimney hitch on the standing part for safety as shown in Figure 3.26. (This forms a hitch that will not slip under tension.)

Step 5: Hold the chimney hitch with one hand and pull on the running part with the other to tighten the rope and slip the chimney hitch back on the standing part to hold the rope taut as shown in Figure 3.27.

Figure 3.23 Step 1. Chimney hitch.

Figure 3.24 Step 2.

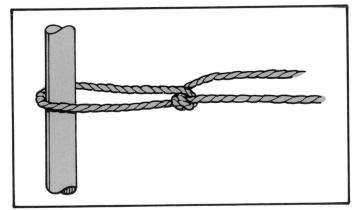

Figure 3.25 Step 3.

Figure 3.26 Step 4.

Figure 3.27 Step 5.

TIGHTENING A ROPE BETWEEN OBJECTS

It is often desirable to rope off an area of ground to keep the public back and away from fire fighting activities or a danger zone. To accomplish this task, so that the rope can be tightened, the half sheep shank and the chimney hitch may be used together. In order to avoid having too much rope to make the tie, the firefighter should first stretch the rope between two stationary objects and leave the unused rope opposite the point of the tie. The half sheep shank must be tied first, and then the chimney hitch is applied behind the half sheep shank to hold the rope taut. A rope that has been tightened between two objects by using these two hitches is shown in Figures 3.28- 3.29. The safety behind the chimney hitch has been loosened and pulled away from the chimney hitch for picture clarity.

Figure 3.28 A chimney hitch and half sheep shank can be used to secure a rope between objects.

Figure 3.29 Tightening the rope between the objects will create a barrier to keep the public away from a danger zone.

HOISTING TOOLS AND EQUIPMENT

Hoisting tools and equipment at fires involves the application of one or more of the basic knots and hitches. Some of the more common hoisting techniques are illustrated in Figures 3.30 - 3.36. These should be studied and practiced to develop efficiency. The basic knots and hitches may be applied to almost any object. Other tools and equipment may be hoisted in a similar manner by adapting the knots and hitches best suited for that purpose.

Figure 3.30 To tie a ladder for hoisting with a handline form a large loop secured by a bowline and place the loop between the rungs of the ladder about one-third of the distance from the hoisting end.

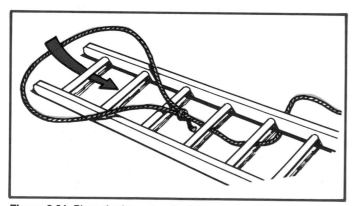

Figure 3.31 Place the loop over the beam ends and the standing line under the rungs between the beams to the end of the ladder.

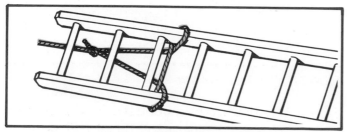

Figure 3.32 The ladder tie is completed by pulling on the standing part of the rope and tightening the loops around the beams.

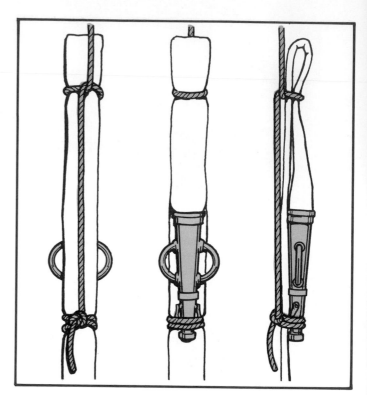

Figure 3.33 A dry hoseline is tied for hoisting by placing a clove hitch with a safety and a half hitch around a doubled section. The rope runs up the back of the hose to keep it from turning during hoisting and to protect the nozzle or couplings on the outside.

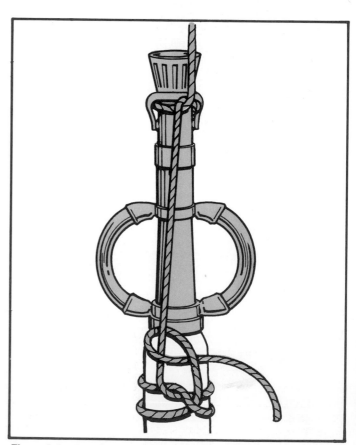

Figure 3.34 The clove hitch with safety and half hitch are used to raise or lower a charged hoseline.

Figure 3.35 The clove hitch and one or two half hitches are tied to a pike pole for hoisting. Hoist with the pike up.

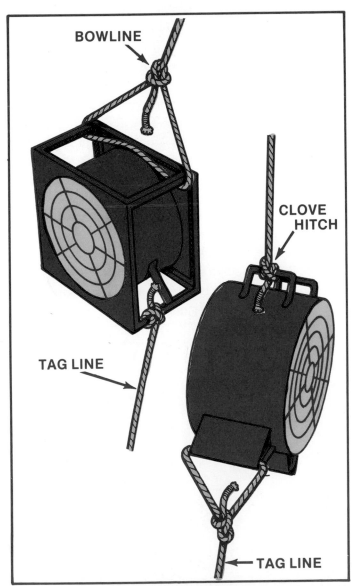

BOWLINE

CLOVE HITCH

TAG LINE

TAG LINE

Figure 3.36 Heavier objects such as a smoke ejector can be hoisted with a bowline or the clove hitch, half hitch combination. A tag line is recommended to keep the object from striking the building.

COILING A ROPE FOR SERVICE

Coiling a rope so that it may be placed into use with a minimum of delay is very essential in the fire service. An improperly coiled handline or lifeline may result in failure of an evolution. A method of coiling a rope which may be done according to the following steps is illustrated in Figures 3.37 - 3.40.

Step 1: Select enough rope at the loop end to make a tie around the coil when completed. This amount is usually about three times the distance between standards (Figure 3.37).

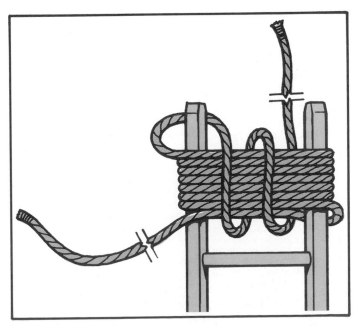

Figure 3.37 Step 1. Coiling a rope.

Step 2: Wrap the rope around the standards until sufficient width is developed. It may be necessary to coil the rope in two layers to use a sufficient amount of rope. (This amount can be determined by trial.) (Figure 3.38).

Step 3: Coil the remainder of the rope around the loops and fasten the end securely with a clove hitch (Figure 3.39).

Step 4: Make the tie illustrated so the finished coil is secure (Figure 3.40).

To uncoil the rope, release the tie, grasp the inside end, pull out two or three loops to loosen the coil, and drop the coil from the top of a building or window, as the case may be.

Figure 3.40 Step 4.

Some firefighters prefer to coil their handlines in a different manner. Two main objections to the above described coil are the numerous bends and the lack of air circulation should the line become wet. The use of hand rope lines in the fire service requires them to be ready for use at all times in a manner in which they are free from tangles and kinks. In order to fulfill this requirement, it is sometimes necessary to sacrifice some of the fundamental principles of rope care and maintenance by having the rope in a compact bundle that will pay out evenly when needed. Another method that has been satisfactorily used by some departments is to carry the line within a canvas bag, as illustrated in Figure 3.41. The eye splice end may be left out for a hand hold when the bag is dropped from the roof or upper floors. As the bag descends, the rope pays out

Figure 3.38 Step 2.

Figure 3.39 Step 3.

Figure 3.41 A rope may also be stored in a bag or plastic bucket. The rope may or may not be laid in an orderly coil inside.

in an orderly manner. The bag may have a draw string and shoulder straps for carrying. A common size is 12 inches (30 cm) in diameter and 32 inches (81 cm) in height which will provide adequate capacity for a 150-foot (45 m) length of ¾-inch (16 mm) manila hemp rope. Burlap type bags can be used in place of canvas.

INSPECTION AND CARE OF ROPES

Rope should be examined periodically, particularly after each use, because the life and the safety of firefighters may depend on it. A surface examination of each rope over its entire length for cuts, severe abrasions, and spots that indicate chemical damage should be made. If rope has musty odors or shows brownish spots, untwist it to enable a close examination between the strands. If the inner fibers break easily, show separation, or have lost their luster, discard the rope. Synthetic ropes with outer covers should be examined for exterior damage, wear, and overall condition. Interior damage to kernmantle ropes can be felt as

bunching or thinness under the cover. Severe strains on colored ropes may show up as blurring of colors on the cover or transfer of dye to contacted surfaces. Rope inspection tags should be attached directly to the rope or its container and the results of inspections recorded. Any rope that is subject to excessive loads should be removed from service as a lifeline, although it may continue to be used as a utility line. Rope that is replaced at the proper time is an investment in safety.

Rope should be properly selected, used, and cared for to insure safety and satisfactory service. The needs of the fire department and the situations in which the rope may be used will determine, in part, the type of rope selected for use. Different rope designs and the materials from which they are made determine the load capacity, inherent safety factor, stretch and rebound capability, and care requirements (Table 3.1). Static kernmantle rope, for instance, is usually found to be more desirable for fire department use than dynamic ropes used by mountain climbers.

TABLE 3.1
ROPE CHARACTERISTICS

	MANILA	SISAL	POLY-P	POLY-E	NYLON	DACRON (Polyester)
Moisture Regain	Up to 60%	Up to 60%	0%	0%	to 9%	Less than 1%
⅝" Dia. Strength	4,400 lbs.	3,520 lbs.	6,200 lbs.	5,600 lbs	10,400 lbs.	10,000 lbs.
Elongation break in rope	13%	13%	24%	22%	35%	20%
Change of strength:	Up to +20%	Up to +20%	No change	No change	Less than -10%	No change
Floatability	No	No	Yes	Yes	No	No
Resistance to rot, mildew and attack by marine organisms	Poor	Very Poor	100% Resistant	100% Resistant	100% Resistant	100% Resistant
Resistance to Surface abrasion	Good	Fair	Good	Good	Very Good	Excellent
Acids	Very Poor	Very Poor	Excellent	Excellent	Fair	Very Good - Excellent
Alkalies	Very Poor	Very Poor	Good	Good	Excellent	Very Good
Solvents	Good	Good	Good	Good	Excellent	

This Table with Metric weights is printed in the Appendix.

The static line has better abilities to withstand heavy shocks such as a rescuer and victim falling rather than the stretching ability of the dynamic ropes. Because they are made from living plants, sisal or manila ropes begin to deteriorate from the moment that they are cut from the parent plant. When selecting ropes, the manufacturers' information on load capacity, shock loading, and stretch and rebound should be examined. Some ropes may be guaranteed for a number of "falls" or only for a limited time after they are manufactured (Table 3.2).

Rope is subject to damage from direct chemical contact and from vapors of chemicals. Water run-off often contains harmful chemicals in suspension which may damage rope fibers. Alkalies, like acids, are injurious to rope fibers and may be seen as spots or discoloration on the rope. Rope that is dragged along the ground or carelessly laid on the ground will suffer damage to the fibers from dirt, glass, and other abrasives.

Rope can be damaged, even before it is used, by improper removal from the coil. Follow instructions for uncoiling that are given on tags attached to coils of rope. Kinking can also occur after uncoiling when a rope is wet. A strain should not be placed on a rope that is kinked or snarled. New rope should be uncoiled from the center of the coil in a counterclockwise direction. If it starts to un-wind in a clockwise direction, the coil should be turned over and the end of the rope pulled up through the center. Uncoiling rope in the wrong direction causes it to twist and kink. Kinks in large ropes are hard to straighten and they may impair the strength of the rope.

Sharp bends should be avoided when using ropes. Ties should be made around a smooth surface of sufficient size when possible. Rope should never be dragged along the ground, stepped on, or pulled over rough or sharp surfaces. Rope, new or used, should be kept in a cool place, with a good circulation of air, and not exposed to direct sunlight. Ropes stored in apparatus compartments should be separated from gasoline fueled power tools and batteries. Additionally, contact with iron should not be permitted, since iron rust is very harmful to rope fibers. Never place used rope in storage until it has dried well.

When ropes have been used they should be inspected for damage and their condition recorded on the rope inspection tag. In addition, the purpose for which the rope was last used, the estimated or actual load that was placed on the rope, and the duration of the load should be recorded. Ropes that are dirty or that have been exposed to chemicals should be cleaned only with clear water and permitted to dry completely. Another inspection of the rope should be performed after drying and before returning the rope to service.

TABLE 3.2
SAFE WORKING CAPACITIES FOR ROPES

Diameter (inches)	Manila (new)	(used)[1]	Sisal (new)	(used)[1]	Nylon (new)	(used)[1]	Dacron (new)	(used)[1]	Polypropylene (new)	(used)[1]	Braided Nylon Cover Nylon Core (new)	(used)[1]
3/8"	270	135	216	108	363	181	287	143	408	204	840	420
1/2"	530	265	424	212	726	363	492	246	731	365	1,500	750
5/8"	880	440	704	352	1,122	561	803	401	1,122	561	2,400	1,200
3/4"	1,080	540	864	432	1,485	742	1,100	550	1,644	822	3,400	1,700
7/8"	1,540	770	1,232	616	2,145	1,072	1,496	748	1,921	960	4,740	2,370
1"	1,800	900	1,440	720	2,640	1,320	1,980	990	2,856	1,428	5,700	2,850
1 1/4"	2,700	1,350	2,160	1,080	3,960	1,980	2,695	1,347	3,859	1,929	8,800	4,400
1 1/2"	3,700	1,850	2,960	1,480	5,610	2,805	3,795	1,897	5,525	2,762	13,000	6,500
1 3/4"	5,300	2,650	4,240	2,120	8,250	4,125	5,610	2,805	7,684	3,842	19,200	9,600
2"	6,200	3,100	4,960	2,480	9,845	4,927	6,710	3,355	9,197	4,598	21,000	10,500
	(20%)[2]	(10%)[2]	(20%)[2]	(10%)[2]	(11%)[2]	(5.6%)[2]	(11%)[2]	(5.6%)[2]	(17%)[2]	(8.5%)[2]	(20%)[2]	(10%)[2]

[1]According to manufacturers' information, manila and sisal rope are considered "used" rope after they have been in service for six months, providing the rope has had proper usage, care, and storage. Persons using synthetic rope should consult the manufacturer for their interpretation of "used" rope.

[2]Based on manufacturer's recommendations for new rope.

This Table with Metric weights is printed in the Appendix.

Chapter 4
Self-Contained Breathing Apparatus

NFPA STANDARD 1001
PROTECTIVE BREATHING APPARATUS
Fire Fighter I

3-3 Protective Breathing Apparatus

3-3.1   The fire fighter shall identify at least four hazardous respiratory environments encountered in fire fighting.

3-3.2   The fire fighter shall demonstrate the use of all types of protective breathing apparatus in a dense smoke environment.

3-3.3   The fire fighter shall identify the physical requirements of the wearer, the limitations of the protective breathing apparatus, and the safety features of all types of protective breathing apparatus.

3-3.4   The fire fighter shall demonstrate donning breathing apparatus while wearing protective clothing.

3-3.5   The fire fighter shall demonstrate that the protective breathing apparatus is in a safe condition for immediate use.

3-3.6   The fire fighter shall identify the procedure for cleaning and sanitizing protective breathing apparatus for future use.*

Fire Fighter II

4-3 Protective Breathing Apparatus

4-3.1   The fire fighter shall identify the procedure for daily inspection and maintenance of breathing apparatus.

4-3.2   The fire fighter, given each type of breathing apparatus, shall demonstrate the correct procedure for recharging.

4-3.3   The fire fighter shall demonstrate the following emergency techniques using breathing apparatus to:
    (a) Assist other fire fighters,
    (b) Conserve air,
    (c) Restricted use of by-pass valves.*

IFSTA's Protective Breathing Apparatus Transparencies are designed to complement this chapter.

Chapter 4
Self-Contained Breathing Apparatus

Special attention should be devoted to protective breathing equipment. The lungs and respiratory tract are probably more vulnerable to injury than any other body area, and the gases encountered in association with fires are, for the most part, dangerous in one way or another. It should be a fundamental rule in fire fighting that no one be permitted to enter a building that is charged with smoke and gas unless equipped with self-contained breathing apparatus. Failure to use this equipment may incapacitate firefighters and could lead to the failure of rescue attempts.

RESPIRATORY HAZARDS

The intent of this section is to examine the four most common atmospheres found at fires (Figure 4.1).

- Oxygen Deficiency
- Elevated Temperatures
- Smoke
- Toxic gases

Oxygen Deficiency

The combustion process consumes oxygen

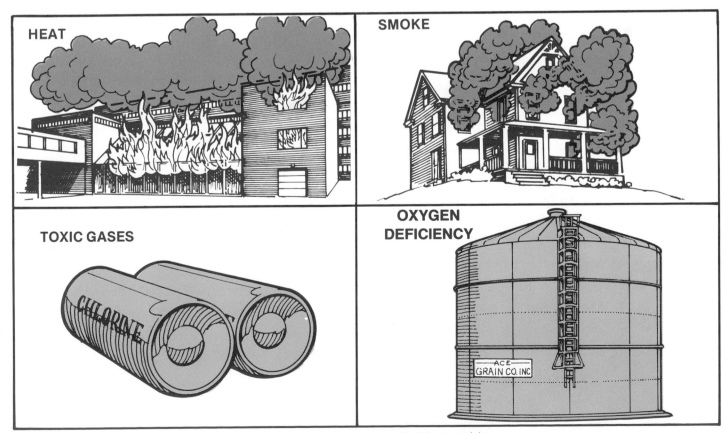

Figure 4.1 Breathing apparatus are necessary due to the hazardous atmospheres firefighters work in.

while producing toxic gases that either physically displace oxygen or dilute its concentration. When oxygen concentrations are below 18% the human body responds by increasing the respiratory rate. Symptoms of oxygen deficiency by percentage of available oxygen are shown in Table 4.1.

TABLE 4.1
Physiological Effects of Reduced Oxygen

Percent of Oxygen in Air	Symptoms
21	None — normal condition.
17	Some impairment of muscular coordination; increase in respiratory rate to compensate for lower oxygen content.
12	Dizziness, headache, rapid fatigue.
9	Unconsciousness.
6	Death within a few minutes from respiratory failure and concurrent heart failure.

NOTE: The data cannot be considered absolute because it does not account for differences in breathing rate or length of time exposed.

These symptoms occur only from reduced oxygen. If the atmosphere is contaminated with toxic gases, other symptoms may develop.

Elevated Temperatures

Exposure to heated air can damage the respiratory tract, and if the air is moist, the damage can be much worse. Excessive heat, temperatures exceeding 120° F (49° C) to 130° (54° C), taken quickly enough into the lungs can cause a serious decrease in blood pressure and failure of the circulatory system. Inhaling heated gases can cause edema (fluid collection) in the lungs, which can cause death from asphyxiation. The tissue damage from inhaling hot air is not immediately reversible by introducing fresh, cool air.

Smoke

Most of the smoke at a fire is a suspension of small particles of carbon and tar, but there is also some ordinary dust floating in a combination of heated gases. The particles provide a means for the condensation of some of the gaseous products of combustion, especially aldehydes and organic acids formed from carbon. Some of the suspended particles in smoke are merely irritating, but others may be lethal. The size of the particle will determine how deeply into the unprotected lungs they will be inhaled.

Fire-Related Toxic Gases

The firefighter should remember that a fire means exposure to combinations of the irritants and toxicants that cannot be predicted accurately beforehand. In fact, the combination can have a synergistic effect in which the combined effect of two or more substances is more toxic or more irritating than the total effect would be if each were inhaled separately (Figure 4.2).

Inhaled toxic gases may have several harmful effects on the human body. Some of the gases directly cause disease of the lung tissue and impair its function. Other gases have no directly harmful effect on the lungs but pass into the bloodstream and to other parts of the body and impair the oxygen-carrying capacity of the red blood cells.

The particular toxic gases given off at a fire vary according to four factors:

- Nature of the combustible
- Rate of heating
- Temperature of the evolved gases
- Oxygen concentration

CARBON MONOXIDE

More fire deaths occur from carbon monoxide (CO) than from any other toxic product of combustion. This colorless, odorless gas is present with every fire, and the poorer the ventilation and the more inefficient the burning, the greater the quantity of carbon monoxide formed. A rule of thumb, although subject to much variation, is that darker smoke means higher carbon monoxide levels. Black smoke is high in particulate carbon and carbon monoxide because of incomplete combustion.

The blood's hemoglobin combines with and carries oxygen in a loose chemical combination called oxyhemoglobin. The most significant char-

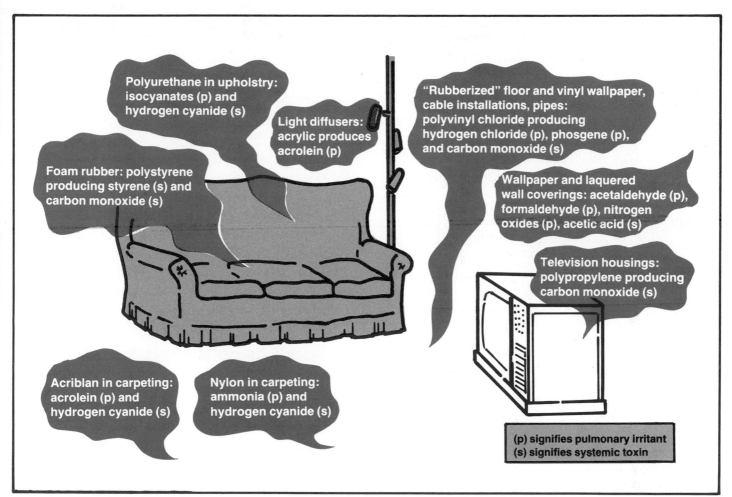

Figure 4.2 The common products in today's household release many toxicants.

acteristic of carbon monoxide is that it combines with the blood's hemoglobin so readily that the available oxygen is excluded. The loose combination of oxyhemoglobin becomes a stronger combination called carboxyhemoglobin (COHb). In fact, carbon monoxide combines with hemoglobin about 200 times more readily than does oxygen. The carbon monoxide does not act on the body, but crowds oxygen from the blood and leads to eventual hypoxia of the brain and tissues, followed by death if the process is not reversed.

Concentrations of carbon monoxide in air above five hundredths of one percent (0.05 percent) can be dangerous. When the level is more than one percent there is no sensory warning in time to allow escape. At lower levels there is headache and dizziness before incapacitation, so sufficient warning is possible. Table 4.2 shows the toxic effect of different levels of carbon monoxide in air, although it is not absolute, because it does not show the variations in breathing rate or length of exposure that

would cause the toxic effect to appear faster. The characteristic cherry red skin color of carbon monoxide poisoning is not always a reliable sign, particularly in long exposures to low concentrations.

Measurements of carbon monoxide concentrations in air are not the best way to predict rapid physiological effects, because the actual reaction is from the concentration of carboxyhemoglobin in the blood, causing oxygen starvation. High oxygen users such as the heart and brain are damaged early. The combination of carbon monoxide with the blood will be greater when the concentration in air is greater. An individual's general physical condition, age, degree of physical activity, and length of exposure all affect the actual carboxyhemoglobin level in the blood.

Experiments have provided some comparison relating air and blood concentrations to carbon monoxide. A one percent concentration of carbon

TABLE 4.2
Toxic Effects of Carbon Monoxide

CO (Parts per Million)	Percent CO in Air	Symptoms
100	0.01	No symptoms — no damage.
200	0.02	Mild headache; few other symptoms.
400	0.04	Headache after 1 to 2 hours.
800	0.08	Headache after 45 minutes; nausea, collapse, and unconsciousness after 2 hours.
1,000	0.10	Dangerous — unconsciousness after 1 hour.
1,600	0.16	Headache, dizziness, nausea after 20 minutes.
3,200	0.32	Headache, dizziness, nausea after 5 to 10 minutes; unconsciousness after 30 minutes.
6,400	0.64	Headache, dizziness after 1 to 2 minutes; unconsciousness after 10 to 15 minutes.
12,800	1.28	Immediate unconsciousness, danger of death in 1 to 3 minutes.

monoxide in a room will cause a 50 percent level of carboxyhemoglobin in the bloodstream in two and one-half to seven minutes. A five percent concentration can elevate the carboxyhemoglobin level to 50 percent in only 30 to 90 seconds. Because the newly formed carboxyhemoglobin may be traveling through the body, a person previously exposed to a high level of carbon monoxide may react later in a safer atmosphere. A person so exposed should not be allowed to use breathing apparatus or resume fire control activities until the danger of toxic reaction has passed. Even with protection a toxic condition could be endangering consciousness.

A hardworking firefighter may be incapacitated by a one percent concentration of carbon monoxide. The stable combination of carbon monoxide with the blood is only slowly eliminated by normal breathing. Administering pure oxygen is the most important element in first aid care.

After an uneventful convalescence from a severe exposure, signs of nerve or brain injury may appear anytime within three weeks. Again, this is a reason why an overcome firefighter who quickly revives still should not be allowed to reenter a smoky atmosphere.

HYDROGEN CHLORIDE

Hydrogen chloride (HCl) is colorless but is easily detected by its pungent odor and intense irritation of the eyes and respiratory tract. Although not a general poison, hydrogen chloride causes swelling and obstruction of the upper respiratory tract. Breathing is labored and suffocation can result. This gas is more commonly present in fires because of the increase in plastics such as polyvinyl chloride (PVC) containing chlorine.

In addition to the usual presence of plastics in homes, firefighters can expect to encounter plastics containing chlorine in drug, toy, and general merchandise stores. The overhaul stage is especially dangerous because breathing apparatus is often removed, and toxic fumes linger in a room. Heated concrete can remain hot enough to decompose the plastics in telephone or electrical cables and release more hydrogen chloride.

The other gases given off when those plastics are heated are carbon monoxide and carbon dioxide. One investigator studying firefighters exposed to hydrogen chloride began his survey after a relatively small, smoky fire in an office photocopier killed one firefighter and sent others to the hospital. He discovered that hydrogen chloride acts as an irritant to the heart muscle and causes irregular rhythms.

HYDROGEN CYANIDE

Hydrogen cyanide (HCN) interferes with respiration at the cellular and tissue level. The proper exchange of oxygen and carbon dioxide is hampered, so hydrogen cyanide is classified as a chemical asphyxiant. The gas inhibits the enzymes by which the tissues take up and use oxygen. Hydrogen cyanide also may be absorbed through the skin.

Materials that give off hydrogen cyanide include wool, nylon, polyurethane foam, rubber, and paper. Unusually hazardous atmospheres might be found at fires in clothing stores or rug shops. Exposure to this colorless gas that has a noticeable almond odor might cause gasping respirations, muscle spasms, and increased heart rate, possibly up to 100 beats per minute. Collapse is often sudden. An atmosphere containing 135 parts per million (1.0123 percent) is fatal within 30 minutes; a concentration of 270 ppm is fatal. Nearly all materials tested in an experiment with aircraft interior materials yielded some hydrogen cyanide.

Businesses with vermin problems sometimes use hydrogen cyanide as a fumigant. Owners should be instructed to notify the fire department whenever the building is being fumigated.

Cyanide asphyxia is one of the most rapid killers at a fire. Death is quick and painless, noted authorities say.

CARBON DIOXIDE

Carbon dioxide (CO_2) must be considered because it is an end product of the complete combustion of carboniferous materials. Carbon dioxide is nonflammable, colorless, and odorless. Freeburning fires should generally form more carbon dioxide than do smoldering fires. Normally its presence in air and its exchange from the bloodstream into the lungs stimulates the respiratory center of the brain. Air normally contains about 0.03 percent carbon dioxide. At a 5 percent concentration in air there is a marked increase in respiration, along with headache, dizziness, sweating, and mental excitement. Concentrations of 10 to 12 percent cause death within a few minutes from paralysis of the brain's respiratory center. Unfortunately, increased breathing increases the inhalation of other toxic gases. As the gas increases, the initially stimulated breathing rate becomes depressed before total paralysis takes place.

Firefighters should anticipate high carbon dioxide levels when a carbon dioxide total-flooding system has been activated. These systems are designed to extinguish a fire by excluding the oxygen, and they will have the same effect on a firefighter. According to the American Conference of Industrial Hygienists, exposure for even a short time to carbon dioxide concentrations greater than 15,000 ppm should be avoided.

NITROGEN OXIDES

There are two dangerous oxides of nitrogen: nitrogen dioxide and nitric oxide. Nitrogen dioxide is the most significant because nitric oxide readily converts to nitrogen dioxide in the presence of oxygen and moisture. Nitrogen dioxide is a pulmonary irritant that has a reddish brown color. When inhaled in sufficient concentrations it causes pulmonary edema that blocks the body's natural respiration processes and leads to death by suffocation.

Additionally, all oxides of nitrogen are soluble in water and react in the presence of oxygen to form nitric and nitrous acids. These acids are neutralized by the alkalis in the body tissues and form nitrites and nitrates. These substances chemically attach to the blood and can lead to collapse and coma. Nitrates and nitrites can also cause arterial dilation, variation in blood pressure, headaches, and dizziness. The effects of nitrites and nitrates are secondary to the irritant effects of nitrogen dioxide but can become important under certain circumstances and cause delayed physical reactions.

Nitrogen dioxide is an insidious gas because its irritating effects in the nose and throat can be tolerated even though a lethal dose is being inhaled. Therefore, its hazardous effects from its pulmonary irritation action or chemical reaction may not become apparent for several hours after exposure.

PHOSGENE

Phosgene ($COCl_2$) is colorless, tasteless gas with a disagreeable odor. It may be produced when refrigerants such as freon, contact flame. It is a strong lung irritant, the full poisonous effect of which is not evident for several hours after exposure. The musty-hay odor of phosgene is perceptible at 6 ppm, although lesser amounts cause coughing and eye irritation. Twenty-five ppm is deadly. When phosgene contacts water it decomposes into hydrochloric acid. Since the lungs and bronchial tubes are always moist, phosgene forms hydrochloric acid in the lungs when inhaled.

Toxic Atmospheres not Associated with Fire

Hazardous atmospheres can be found in numerous situations in which fire is not involved

Figure 4.3 When toxic gases are released at a stationary location or during a transportation incident breathing apparatus is imperative. *Courtesy of Andy Levy.*

(Figure 4.3). Many industrial processes use extremely dangerous chemicals to make ordinary items. For example, quantities of carbon dioxide would be stored at a facility where wood alcohol, ethylene, dry ice, or carbonated soft drinks are manufactured. Any other specific chemical could be traced to numerous wide-ranging, common products.

Many refrigerants are toxic and may be accidentally released, causing a rescue situation to which firefighters may be called. Ammonia and sulfur dioxide are two dangerous refrigerants that irritate the respiratory tract and eyes. Sulfur dioxide reacts with moisture in the lungs to form sulfuric acid. Other gases also form strong acids or alkalies on the delicate surfaces of the alveoli.

Chlorine gas leaks obviously can be encountered at manufacturing plants or, not so obviously, at swimming pools. At either place incapacitating concentrations can be found. Chlorine is also used in manufacturing plastics, foam, rubber, and synthetic textiles, and is commonly found at water and sewage treatment plants.

Sometimes the leak is not at the manufacturing plant but during transportation of the chemical. Train derailments have resulted in container failures, exposing the public to toxic chemicals and gases. The large quantities involved can travel long distances.

Rescues in sewers, caves, trenches, storage tanks, tank cars, bins, silos, manholes, pits, and other confined places require the use of self-contained breathing apparatus because some toxic gas is usually present or there is an oxygen deficiency to cause a rescue need in the first place. Workers have also been overcome by harmful gases in large tanks during cleaning or repairs. Unfortunately, personnel attempting a rescue while unprotected often have also been overcome. In addition, the atmosphere in many of these areas is oxygen deficient and will not support life even though there may be no toxic gas.

The smallest town, even without a chemical processing plant or without any manufacturing plant using dangerous chemicals, is susceptible to hazardous conditions from accidents involving dangerous chemicals in transit by rail or truck. Many of the chemicals are especially damaging when inhaled. The need to properly use self-contained breathing apparatus is just as important in these situations even when there is no fire.

TYPES OF BREATHING APPARATUS
Demand Apparatus

The demand regulator type of self-contained breathing equipment provides face and respiratory protection for the user, but it is limited to the amount of air or oxygen that is carried in the supply cylinder. This equipment consists essentially of a full facemask, corrugated flexible breathing tube, demand regulator, air or oxygen supply cylinder, and harness (Figure 4.4).

A regulator pressure gauge should be in view of the user at all times. This gauge indicates cylinder pressure and provides an indication of the reserve supply. During normal operation, the emergency bypass valve should be fully closed (turn clockwise for closed), and the regulator control valve to the mainline should be fully open (turn counterclockwise for open) and locked in position by the locking device. This valve is provided for shutting off the automatic demand regulator in the event of its failure or damage. It should be closed only after the emergency bypass has been opened. *Once the valves are set in this position, they should not be changed unless the emergency bypass valve is needed.* The air or oxygen supply is controlled by a main valve on the cylinder. (Open and close these valves with the fingers; do not use force.)

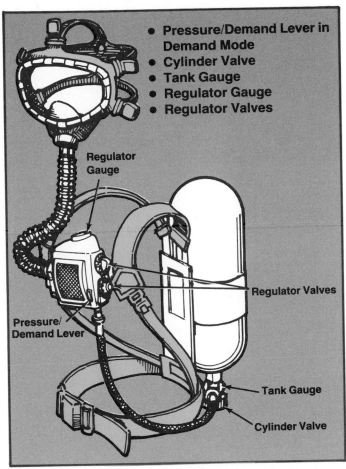

- Pressure/Demand Lever in Demand Mode
- Cylinder Valve
- Tank Gauge
- Regulator Gauge
- Regulator Valves

Regulator Gauge

Regulator Valves

Pressure/Demand Lever

Tank Gauge

Cylinder Valve

Figure 4.4 Self-contained breathing apparatus has five major parts: facemask, breathing tube, regulator, air cylinder, and harness.

NOTE: The operations herein described are intended for one particular brand of demand regulator equipment. Operations for other types or brands are very similar, but the location of some of their parts may be different. This manual suggests these operations as a pattern from which a procedure can be developed for all demand regulator equipment.

Positive-Pressure Apparatus

Most positive-pressure units look almost exactly like the standard demand units. The cylinder and backpack assemblies are similar. The major difference is that in the positive-pressure unit the diaphragm in the regulator is held open to create a slight pressure in the low-pressure hose and facepiece. This pressure is held in the facepiece by a spring-loaded exhalation valve so the pressure inside the facepiece is slightly higher than atmospheric pressure (Figure 4.5), preventing the entry of smoke particles and toxic gases. The insignificant amount of extra breathing air expended is well worth the extra safety to the user. If the seal of the facepiece against the face is not good, however, there is still the possibility that toxic substances will be drawn into the facepiece if the user is

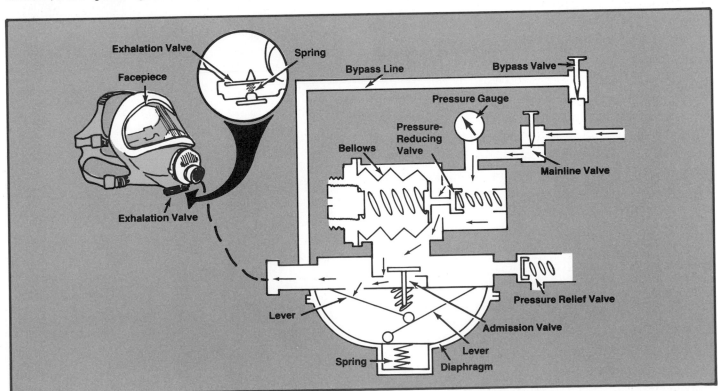

Figure 4.5 Schematic drawing of how a positive-pressure regulator works. *Courtesy of International Fire Chief.*

breathing heavily. Therefore, the seal of the positive-pressure facepiece is as important as it is for the demand facepiece.

Training is clearly necessary for firefighters to use the positive-pressure units efficiently. Some positive-pressure units can be converted to demand units by flicking a switch, but an unswitchable unit requires the user to consciously be sure the cylinder valve or mainline valve is turned off until the facepiece is donned unless there is another shutoff valve. Firefighters may prefer to keep the mainline valve closed when the unit is not in use. When the cylinder valve is turned on, air will flow only as far as the regulator, but will be ready when needed. There may be confusion if the firefighter uses both demand and unswitchable positive-pressure units. With demand units always keep the mainline valve open; with the unswitchable positive-pressure units keep the mainline valve closed until ready to don the facepiece before entering the hazardous area. Refer to manufacturer's instructions and department policy for specific units.

Closed-Circuit Positive-Pressure Compressed-Oxygen Apparatus

Closed-circuit apparatus, also known as re-breathers, recycle the user's exhaled breath after removing carbon dioxide and moisture and adding supplemental oxygen as needed. None of the oxygen used in this system and none of the exhaled waste gas is released outside the facepiece. For all practical purposes all gases in the system stay in the system, traveling in a closed circuit. The oxygen within the system comes from a cylinder of compressed oxygen. The compressed oxygen in the system is supplied at a rate greater than that needed for breathing alone. The extra breathing gas increases the pressure in the facepiece during inhalation and exhalation. The slight positive pressure is maintained mechanically in the breathing chamber by a device that exerts a force on the breathing diaphragm. Reuse of the exhaled air results in longer duration and lower unit weight. The carbon dioxide scrubber must be changed and the oxygen cylinder recharged or replaced after each use. For additional information see IFSTA's **Self-Contained Breathing Apparatus.**

MOUNTING PROTECTIVE BREATHING APPARATUS

Methods of storing self-contained, demand breathing apparatus vary from department to department. Each department should use the method more appropriate to facilitate quick and easy donning. Types of storage include the seat mount, side mount, and compartment mount; two of these are shown in Figure 4.6.

Figure 4.6. A variety of mountings are in use to expedite donning of breathing apparatus. Jump seat and compartment mountings are illustrated.

DONNING BREATHING APPARATUS

Several methods can be used to don self-contained breathing apparatus, depending on how the apparatus is stored. The methods used in the fire service include the over-the-head method, the coat method, and donning from a seat, or compartment mount. The steps needed to get the apparatus onto the body differ with each method, but once the apparatus is on the body the method of securing the unit will be the same for any one model (there are different steps for securing different makes and models).

Over-the-Head Method

All self-contained breathing apparatus must be stored ready to don, with the backpack harness straps arranged so they will not be in the way of grasping the cylinder. Be sure to turn up the turnout coat collar before donning the self-contained breathing apparatus so the shoulder straps will not keep the collar held down.

Step 1: Crouch or kneel at the end opposite the cylinder valve. Check the cylinder gauge to make sure the air cylinder is full.

Step 2: Open the cylinder valve all the way (Figure 4.7) and check the regulator gauge (Figure 4.8). Both gauges should register about the same pressure. (If the unit is positive pressure [other than an un-

Figure 4.8 Step 2.

switchable Scott 2.2 or 4.5] leave the cylinder valve open and be sure the mainline valve is closed after this gauge check. If it is an unswitchable positive-pressure unit, leave the cylinder valve closed unless the unit has a special valve at the facepiece.)

Step 3: Grasp the backplate or cylinder with both hands, one at each side (Figure 4.9). There should be no straps between the hands.

Figure 4.7 Step 1. Over-the-head method.

Figure 4.9 Step 3.

Step 4: Raise the cylinder overhead and let the elbows find their respective loosened harness shoulder strap loops (Figures 4.10 and 4.11). Keep the elbows close to the body and let the straps fall easily into place.

Step 5: Lean slightly forward to balance the cylinder on the back, then pull down on the two underarm straps (Figure 4.12).

Step 6: Fasten and adjust the lower waist strap so the unit fits snugly (Figure 4.13).

NOTE: Some departments, with the mistaken intent of facility, have removed waist straps from self-contained breathing apparatus. Without a waist strap fastened, however, the self-contained breathing apparatus wearer is caused undue stress from side-to-side shifting of the unit and from improper weight distribution of the unit.

Step 7: Don the facepiece (see pages 63-66).

Figure 4.10 Step 4.

Figure 4.11 Step 4. The straps fall into place.

NOTE: When donning a self-contained breathing apparatus on a slippery or ice coated surface, it is safer to kneel beside the case while using either method.

Figure 4.12a Step 5.

Figure 4.12b Step 5.

Figure 4.13 Step 6.

Coat Method

The backpack breathing apparatus may be donned like a coat, putting one arm at a time through the shoulder strap loops. The equipment should be arranged in the case so a shoulder strap can be grasped for lifting.

Step 1: Check the cylinder gauge to determine whether the cylinder is full.

Step 2: Open the cylinder valve fully (Figure 4.14) and listen for the audible alarm as the system pressurizes (Figure 4.15). If the audible alarm does not sound, use another unit.

Figure 4.14 Step 1. Coat method.

Figure 4.15 Step 2.

Figure 4.16 Step 3.

Step 3: Grasp with the right hand the shoulder strap that will be worn over the right shoulder (Figure 4.16). (Alternative: left hand, left strap.) (Note that one reaches across the apparatus to do this.)

Step 4: Bring the unit up so the strap rests on the shoulder. During this move, the elbow of this arm should slip between the shoulder strap and the backpack frame (Figures 4.17 and 4.18). As the unit swings across the back, the opposite arm should be inserted through its strap opening.

Figure 4.17 Step 4.

Figure 4.18 Step 4. Note protection afforded the regulator during donning.

Step 5: Fasten or adjust the shoulder and chest straps as recommended by the manufacturer (Figure 4.19).

Step 6: Fasten and adjust the waist strap (Figure 4.20).

Step 7: Don the facepiece (see pages 63-66).

Figure 4.21 Breathing apparatus conveniently located in the jump seat.

Figure 4.19 Step 5.   **Figure 4.20** Step 6.

Seat Mount Method

Valuable time can be gained if the breathing equipment is mounted on the back of the firefighter's seat (Figure 4.21). By having a seat mount the firefighter can don the self-contained breathing apparatus while enroute to the emergency.

Donning enroute is done by inserting the arms through the straps while sitting, then adjusting the straps for a snug fit (Figure 4.22 and 4.23). At no time should the firefighter stand up while donning the self-contained breathing apparatus when the vehicle is moving. The cylinder's position should match the proper wearing position for the firefighter. The visible seat-mounted self-contained breathing apparatus reminds and even encourages personnel to check the equipment more frequently and also keeps it exposed so the checks are handled more conveniently.

Figure 4.22 Place the arms through the straps and adjust them for the proper fit.

Figure 4.23 Fasten and adjust the waist strap.

Compartment Mount Method

Although not allowing donning enroute, compartment mounted self-contained breathing apparatus may be desirable (Figure 4.24). Time savings are possible because the steps needed to re-

Figure 4.24 Many pumpers have breathing apparatus mounted in convenient compartments.

Figure 4.25 The donning method will depend on the mounting of the breathing apparatus.

move the equipment case from the apparatus, place the case on the ground, open the case, and pick up the unit are eliminated. The method used to don the breathing apparatus will depend on how it is mounted (Figure 4.25).

Donning the Facepiece

The facepieces for most self-contained breathing apparatus are donned similarly. One important difference in facepieces is the number of straps used to tighten the head harness. Different models of the same manufacturer may have a different number of straps. The shape and size of the lens may also differ, but the uses and donning are not changed.

NOTE: Interchanging facepieces, or any other part, from one manufacturer's equipment to that of another without express permission from the manufacturer makes any warranty and certification void.

The facepiece should not be worn loosely or it will not seal against the face properly and will allow toxic gases to enter and be inhaled. Firefighters should not wear long hair, sideburns, or beards, because those will prevent the outer edges of the facepiece from making contact and a good seal with the skin. A facepiece tightened too much, however, will be uncomfortable or may cut off circulation.

The facepiece may be packed in a case or stored in a bag or coat pouch. Whichever, the straps should be left fully extended for donning ease.

Step 1: If the firefighter is using a Nomex hood, the hood should be donned before the turnout coat and the self-contained breathing apparatus facepiece are donned. Put the hood over the head and pull the hood back and down so the face opening is around the neck.

Step 2: If the facepiece harness is stored over the front of the facepiece, pull it to the rear.

NOTE: Some facepiece harnesses, notably those of MSA's Clearvue and Ultravue facepieces, should not be stored in front of the facepiece, because the facepiece seal will be damaged.

Step 3: Grasp the head harness, with the thumbs through the straps from the inside, and spread it (Figure 4.26).

Step 4: Push the harness top up the forehead to brush hair from the facial seal area and continue up and over the head (Figure 4.27) until the harness is centered at the rear of the head and the chin is in the facepiece chin cup.

Step 5: Tighten the bottom straps by pulling them evenly and simultaneously to the rear (Figure 4.28).

NOTE: Pulling the straps outward, to the sides, will damage them.

Figure 4.27 Step 4.

Inset photos courtesy of Dennis Sargent.

Figure 4.26 Step 3. Donning the facepiece.

Figure 4.28 Step 5.

Step 6: Tighten the temple straps (Figure 4.29).

Step 7: Tighten the top strap or straps (Figure 4.30).

> **NOTE:** Do not overtighten the top straps. Circulation will be impeded and the facepiece fit might be altered.

Figure 4.29 Step 6. For positive pressure units place helmet chin strap under the chin. If regulator is in place, inhale to check facepiece seal, readjust if needed.

Figure 4.30 Step 7. For positive pressure units, if needed, insert regulator, check gasket, place purge valve upright in line with nose.

Step 8: Check the facepiece seal (Figure 4.31). Exhale deeply, seal the end of the low-pressure hose with a hand, and inhale deeply and *slowly* (quick sucking will seal any leak and will give a false sense of security). If there is evidence of leaking, adjust or redon the facepiece.

Step 9: Check the exhalation valve: Inhale, seal the end of the low-pressure hose, and exhale. If the exhalation does not go through the exhalation valve, keep the low-pressure hose sealed, press the sides of the facepiece against the face and temples, and blow vigorously to free the valve.

Step 10: If wearing a Nomex hood, pull the hood into place, making sure all exposed skin is covered and that vision is unobscured.

Step 11: Don the helmet, first inserting the low-pressure hose through the helmet's chin strap. Be sure to put the helmet strap under the chin.

> **NOTE:** Do not put the helmet strap around the lens, the exhalation valve, or any part of the facepiece. A blow to the helmet would be likely to dislodge the facepiece, letting toxic atmosphere inside.

Figure 4.31 Step 8. On positive pressure, open cylinder valve, check audible alarm, inhale and hold the breath. There should be no audible flow of air into the facepiece.

Figure 4.32 Step 12. If needed in the positive pressure, turn the regulator to the left a quarter turn to secure it in place. Check facepiece seal.

Step 12: Connect the low-pressure hose to the regulator (Figure 4.32). If the unit is demand, open the mainline valve fully. If the unit is positive pressure, open the cylinder valve. Be sure the mainline valve is fully open and the bypass valve is fully closed on either demand or positive-pressure apparatus.

To remove the facepiece, disconnect the low-pressure hose from the regulator. A switchable demand/positive-pressure unit must be switched to "demand" before the low-pressure hose is disconnected. If the unit is positive pressure without a switch, close the cylinder valve. Close the mainline valve, grasp the bottom of the facepiece at the chin, and pull the facepiece away from the face and over the head.

INSPECTION AND CARE
Daily Inspections

Self-contained breathing apparatus requires proper care and inspection before and after each use in order for it to provide the protection for which it was designed. This can best be done by making a daily inspection conducted as soon as possible after reporting for duty (Figure 4.33).

- Full cylinder?
- Gauges work?
- Alarm work?
- Hose connections ok?
- Facepiece ok?
- Shoulder harness ok?
- Bypass and mainline valve operational?
- Bypass valve fully closed?

Breathing apparatus should be cleaned and sanitized immediately after each use. Moving parts that are not clean may malfunction. A facepiece that has not been cleaned and sanitized may contain an unpleasant odor and can spread germs throughout a department. An air cylinder with less air than prescribed by the manufacturer renders the apparatus inefficient, if not useless.

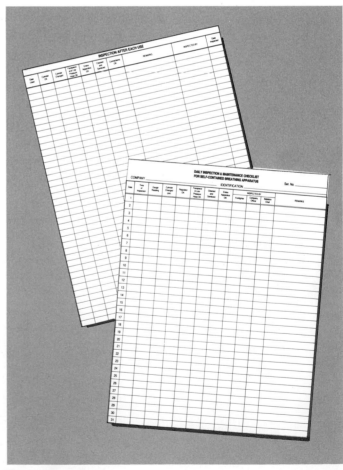

Figure 4.33 Daily and after use inspections should be performed on all breathing apparatus.

The facepiece should be thoroughly washed with warm water containing any mild commercial disinfectant and then rinsed with clear warm water. Special care should be given to the exhalation valve to insure proper operation. The air hose should be inspected for cracks or tears. Then the facepiece should be dried with a lint free cloth or air dried (Figure 4.34). **CAUTION:** Do not use paper towel to dry the lens as the paper towel will scratch the plastic lens.

Periodic Inspection and Care

After each three month period, it is advisable to remove the equipment from service and check valves, pressure regulators, gauges, harness, and facepiece. The following functional test and inspection should then be made: check the facepiece, hose, and exhalation valve by inhaling slowly with the thumb over the end of the hose connection. Make the hose connection and check the performance of the regulator. Inhale deeply and quickly. The regulator should supply a full flow to give the user all the air that is needed. If, on slow inhalation, a "honking" sound is heard in the regulator, it can usually be stopped by inhaling faster. The sound is caused by the bellows vibrating and in no way affects the performance or safety of the regulator. If the bellows vibrate continuously or excessively, the regulator should be overhauled by competent technicians recommended by the manufacturer. If the demand valve sticks open slightly (this may be caused by the diaphragm being cold), the breathing gas will continue to flow when the wearer is not inhaling. This condition can usually be corrected by "blowing back" on the regulator. Operate the regulator several minutes to exercise the diaphragm and valves before condemning the regulator. With the hose out of the connection, close off the cylinder valve. With 1,980 psi (13,650 kPa) indicated on the regulator gauge, the regulator and the regulator hose assembly should hold the trapped-in pressure.

After two and one-half years, the regulator with regulator hose should be returned to the fac-

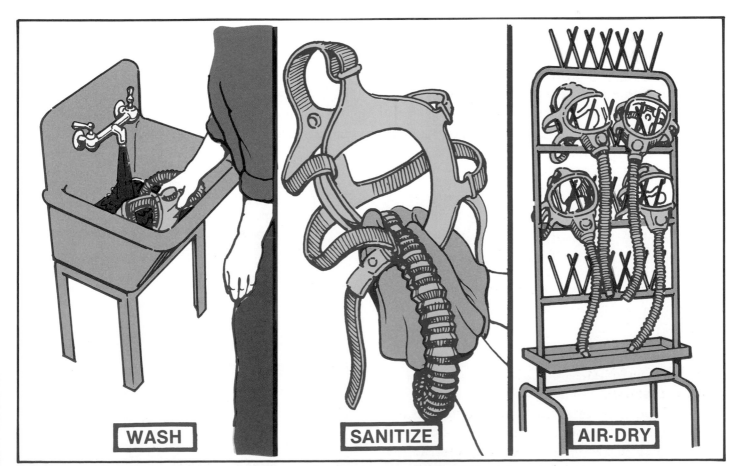

WASH   **SANITIZE**   **AIR-DRY**

Figure 4.34 Breathing apparatus facepieces should be cleaned, sanitized, and properly dried for the next use.

tory or to their representative for test and/or repair. After each five year period, the cylinders should be hydrostatically tested. Each cylinder is stamped with the month and the year of manufacture and the date of the last test. This procedure is necessary to meet requirements of the United States Department of Transportation (formerly Interstate Commerce Commission). Always empty cylinders before returning them for service and test.

TESTING BYPASS VALVE FOR LEAKS

With the regulator shut-off valve and bypass valve closed (cylinder valve open), place a soap bubble solution across the hose connection fitting on the regulator. If the bypass valve is leaking, the bubble will expand and break. Soap bubbles that are derived from ordinary soaps or detergents may be so heavy or dry that they will not detect small leaks. It has been found that those specially prepared bubble solutions, which are for children to use in bubble pipes are best suited to detect leaks which might not otherwise be found.

TESTING REGULATOR SHUT-OFF VALVE FOR LEAKS

With the regulator shut-off valve and bypass valve closed (cylinder valve open), draw (inhale) from the equipment until regulator gauge reads "0." Then watch the gauge to see if pressure will build up. If gauge pointer raises indicating pressure, the regulator shut-off valve is leaking. Do not use force in closing either regulator shut-off valve or bypass valve; "finger tight" is sufficient.

With some SCBA units it may be necessary to use the following procedure. With the mainline and bypass valve closed and the cylinder valve open, inhale from the equipment to reduce pressure in the pressure reduction valve, then close the tank valve. Watch the gauge to see if pressure drops. A drop in pressure will indicate a leak in the system.

Regulator shut-off and bypass valve stem seals may be checked for leakage by the following method: remove both valve knobs, turn them over and reinstall on their respective valve stems so that the valve stem will be uncovered. Open the cylinder valve and operate the two valves while applying soap and water solution to valve stems and packing nuts. Bubbles will be formed if the stem seals are leaking.

TESTING CYLINDER VALVE FOR LEAKS

Inspect the place where the cylinder pressure gauge connects to the valve body and the safety plugs with a soap solution. With the regulator hose and regulator attached to the cylinder valve, open the cylinder valve. If bubbles appear around the valve stem and packing gland nut when making soap solution test, the packing nut should be tightened or the gland packings should be replaced.

With the regulator hose disconnected, close the cylinder valve. If bubbles form at the regulator hose connection when a soap solution is applied, the valve seat is leaking. Open and close the valve quickly several times and allow pressure to blow through quickly. This procedure may clear the valve seat of dirt and correct the trouble. If the leak continues, the cylinder should be returned to the factory for test and repair.

RECHARGING CYLINDERS

Recharging of air cylinders is usually done from a bank of large air cylinders. In some cases, the large cylinders are connected to an air compressor designed specifically for air breathing systems. This system is called a "Cascade System" (Figure 4.35). Steps for recharging cylinders include:

Step 1: Inspect cylinder for damage and hydrostatic test date.

Step 2: Place cylinder in charging station, connect charging hose.

Step 3: Open cylinder valve.

Step 4: Slowly open valve of cascade cylinder with lower pressure.

Step 5: When SCBA and cascade cylinder pressure is equal, close cascade cylinder valve and open valve on cascade cylinder with next highest pressure.

Step 6: Repeat step 5 until SCBA cylinder is charged.

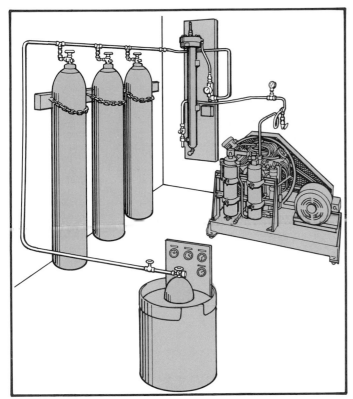

Figure 4.35 A cascade system is used for recharging small cylinders. The cascade of large bottles may or may not be connected to an air compressor and purification system.

SAFETY PRECAUTIONS

Any fire fighting is strenuous, demanding activity, so firefighters need to be in good physical condition. Their protective gear can work against them while at the same time protecting them. The basic required turnout coat can be a virtual sweat box which builds up body heat and hinders movement to increase firefighter exhaustion. This should be stressed even more when self-contained breathing equipment is used under emergency conditions. The difference between the weight of ordinary street clothes and fire fighting gear plus the mask unit has been measured at an extra 47 pounds (21 kg); the breathing unit alone has been weighed at from 33 to 43 pounds (15 - 30 kg) depending on size and type.

In exercise tests by the New York Fire Department, first in street clothes and then in full protective gear with mask, levels of fatigue were recorded for different men. Smaller men (5'6-5'8) showed an increase of 81% in energy consumption when wearing the gear. Larger men, usually stronger, also showed an increase, by 19%, in energy expenditure.

Fire scene operations would often require more energy use than the test.

> **Federal OSHA Standards state:**
> "Persons should not be assigned to tasks requiring use of respirators unless it has been determined that they are physically able to perform the work and use the equipment."

That applies to the private sector. More comprehensive cautions should apply to firefighters.

When using self-contained breathing apparatus, the following items should be remembered and observed for maximum safety.

- A firefighter with a respiratory ailment should not perform duties requiring the use of protective breathing apparatus.

- Demand breathing apparatus should not be used immediately after performing strenuous work.

- The so-called 30-minute breathing apparatus should be expected to last 1 minute for every 100 psi (700 kPa) as indicated on cylinder gauge. EXPECT NO MORE.

- The "Mainline" valve should always be open and locked.

- The "Bypass" valve is for emergency only.

- When "Bypass" must be used, "Mainline" valve must be closed.

- Once entering a contaminated area, breathing apparatus should not be removed until you have left contaminated area. Because visibility improves does not insure the area is free of contamination.

- When working in breathing apparatus, work in pairs.

Emergency Situations

The emergencies created by the malfunction of demand type breathing apparatus can be overcome in several ways. In all of these emergencies, the conservation of air is of the utmost importance.

- Don't panic!
  — Causes rapid breathing using more valuable air.
- Stop and Think!
  — How did you get to where you are? Downstairs? Upstairs? Left turns?
- Different methods to use to find a way out.
  — Follow hoseline out if possible (male couplings toward fire, female away from fire).
  — Crawl in straight line (hands flat to floor, move knee to hand).
  — Once in contact with wall, crawl in one direction (all left hand turns, all right hand turns).
  — Control breathing while crawling.
  — Call for directions, call out or make noise for other firefighters to assist you.

When it is necessary to use the "Bypass" valve due to malfunction of the regulator, the following procedures should be followed:

Step 1: Open "Bypass" valve slowly just to a point where you can breathe comfortably.

Step 2: Close "Main-Line" valve.

Step 3: To conserve air, take a deep breath and hold it. Shut "Bypass" valve off until you need another breath. Continue procedure in Steps 1-3 until you are out of contaminated area.

Another emergency situation is when the facepiece is damaged during fire fighting operations such as: broken harness, torn facepiece or breathing tube, or faulty exhalation valve. In these cases proceed as follows:

Step 1: Disconnect breathing tube from regulator.

Step 2: Remove facemask.

Step 3: Loosen body harness.

Step 4: Lift regulator to your mouth and breathe directly from the threaded connection on the regulator.

Running out of air while in a contaminated area is the most frequent emergency encountered. This situation can be overcome by a procedure called "Buddy Breathing."

"Buddy Breathing" is performed by two firefighters. When one firefighter runs out of air (buddy breather), the second (air carrier) shares air until both are out of the contaminated area.

For firefighters with other than facepiece mounted regulators the following procedure is used. The buddy breather, after signaling the air carrier, can disconnect the low-pressure hose from the regulator and give it to the air carrier. This firefighter cracks the bypass valve, then pushes the hose inside the facepiece between the cheekbone and the jaw, just far enough so the low-pressure hose coupling will not excessively interfere with the facepiece seal. Then the air carrier should pinch off the opening in the facepiece around the inserted hose, making sure not to pinch off the hose in the process. Wearing gloves will help plug the gaps between the facepiece and the low-pressure hose. This and the positive pressure should keep out contaminants as the firefighters proceed to a safe area.

**NFPA STANDARD 1001
LADDERS
Fire Fighter I**

3-9 Ladders

3-9.1   The fire fighter shall identify each type of ladder and define its use.

3-9.2   The fire fighter, operating as an individual and as a member of a team, shall demonstrate the following ladder carries:
  (a) One person carry
  (b) Two person carry
  (c) Three person carry
  (d) Four person carry
  (e) Five person carry
  (f) Six person carry.

3-9.3   The fire fighter, operating as an individual and as a member of a team, shall raise each type and size of ground ladder using several different raises for each ladder.

3-9.4   The fire fighter shall climb the full length of every type of ground and aerial ladder.

3-9.5   The fire fighter shall climb the full length of each type of ground and aerial ladder carrying fire fighting tools or equipment while ascending and descending.

3-9.6   The fire fighter shall climb the full length of each type of ground and aerial ladder and bring an "injured person" down the ladders.

3-9.7   The fire fighter shall demonstrate the techniques of working from ground or aerial ladders with tools and appliances, with and without a life belt.

3-9.8   The fire fighter shall demonstrate the techniques of cleaning ladders.*

Fire Fighter II

4-9 Ladders

4-9.1   The fire fighter shall identify the materials used in ladder construction.

4-9.2   The fire fighter shall identify the load safety features of all ground and aerial ladders.

4-9.3   The fire fighter shall demonstrate inspection and maintenance techniques for different types of ground and aerial ladders.*

*Reprinted by permission from NFPA Standard No. 1001, *Standard for Fire Fighter Professional Qualifications*. Copyright © 1981, National Fire Protection Association, Boston, MA.

IFSTA's Handling Ground Ladders and Ladder Carries and Raises Transparencies are designed to complement this chapter.

Chapter 5
Ladders

Although the many subdivisions of firefighting activities require firefighters to use a variety of tools and devices, fire service ladders are essential in the performance of both major functions: saving lives and extinguishing fires. Fire service ladders are similar to any other ladder in shape, design, and purpose, but the way in which they are used requires them to be constructed under rigid specifications. Their use under adverse conditions further requires them to provide a margin of safety not usually expected of commercial ladders. The National Fire Protection Association, Standard 1931, *Standard for Fire Department Ground Ladders* provides recommended specifications of fire department ground ladders. Fire service ground ladders, like fire hose, are considerably more than just a tool or appliance, since special training for individual skill and team performance is required for their efficient use at fires. Mechanical power has made possible the use of power-operated aerial ladders and elevating platforms, but hand-operated ground ladders will always be relied upon by firefighters to gain access to areas which cannot be reached by normal means.

LADDER TERMS

Fire service ground ladders are designed to perform varying functions. The specific task at hand will determine which types of ladders are to be used. Although these functions are usually similar in different parts of the country, the nomenclature and trade terms of the various types of ladders may vary considerably. Because of these variances, it is sometimes difficult to discuss ladder terms without an understanding of their meaning.

Firefighters should become familiar with the generally-accepted terms used in this chapter. To help the firefighter identify and use these terms, the following explanations are offered to help make ladder terms more easily understood. See Figure 5.1 on next page.

ANGLE OF INCLINATION — Refers to the angle of a ladder in place in relation to horizontal.

BASE SECTION — (Also called bed section or main section.) The lower section of an extension ladder.

BEAM — The side rail of a ladder.

BEAM BOLTS — Bolts which pass through both rails at a truss block of a wood ladder to tie the two truss rails together.

BUTT — The bottom of ground end of a ladder.

BUTT SPURS — Metal safety plates attached to the butt end of ground ladder beams.

DOGS — See Pawls.

EXTENSION LADDER — A term to identify a ladder with two or more sections.

FLY — The upper section or top sections of an extension ladder.

GROUND LADDER — A term to designate the difference between ladders raised on the ground and those raised from the apparatus.

GUIDES — Wood or metal strips on an extension ladder which guide the fly section while being raised. (Sometimes in the form of slots or channels.)

HALYARD — A rope or cable used for hoisting fly sections.

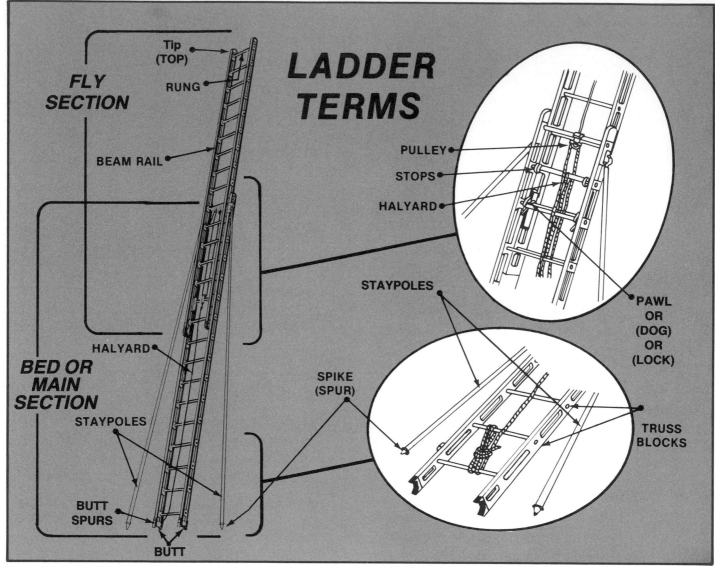

Figure 5.1 Ladder terminology.

HEAT SENSOR LABEL — A label affixed to the ladder beam near the tip used to provide a warning that the ladder has been subjected to excessive heat.

HOOKS — A pair of sharp curved devices which fold outward from each beam at the top of a roof ladder.

LOCKS — See Pawls.

PAWLS (Dogs, Locks) — Devices attached to the inside of the beams on fly sections used to hold the fly section in place after it has been extended.

PROTECTION PLATES — Plates fastened to a ladder to prevent wear at points where it comes in contact with mounting brackets.

PULLEY — A small grooved wheel through which the halyard is drawn.

RAILS — The two lengthwise members of a trussed ladder beam which are separated by truss or separation blocks.

RUNGS — Cross members (usually round) between the beams on which people climb.

SAFETY SHOES — Rubber or neoprene spike plates, usually of the swivel type, attached to the heel of a ground ladder.

SPURS — Metal points at the lower end of tormentor poles.

STAYPOLES — The poles which are attached to long extension ladders to assist in raising and

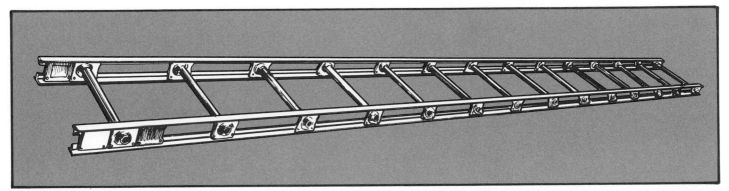

Figure 5.2 Single ladder.

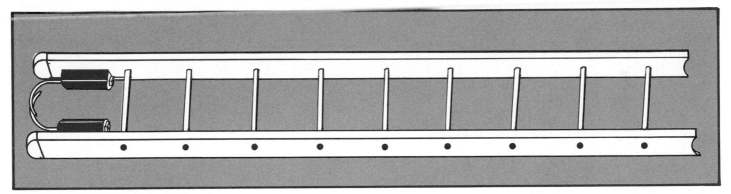

Figure 5.3 Roof ladder with hooks closed.

steadying the ladder. Some poles are permanently attached and some are removable.

STOPS — Wood or metal pieces which prevent the fly section from being extended too far.

SINGLE LADDER — Term used to identify a ladder consisting of one section.

TIE RODS — Metal rods running from one beam to the other.

TOGGLE — A device by which a staypole is attached to a ladder.

TOP OR TIP — The extreme top of a ladder.

TRUSS BLOCK — Separation pieces between the rails of a trussed ladder. Sometimes used to support rungs.

LADDER TYPES

All of the various types of fire service ladders have a purpose. Many of them, however, are more adaptable to a specific function than they are to general use. Their identifying name is often significant regarding the use to which they are applied, and firefighters frequently make reference to them by association. The definitions which follow are offered to more clearly identify fire service ladders.

Single Ladders

A single ladder is nonadjustable in length and consists of only one section (Figure 5.2). Its size is designated by the overall length of the side rails. The single ladder, sometimes called a wall ladder, is used for quick access to windows and roofs on one and two-story buildings. Single ladders must be constructed to have a maximum strength and minimum weight and may be of the trussed type in order to reduce their weight. Single ladders are generally used in lengths of 12, 14, 16, 18, and 24 feet (4, 4.3, 5, 5.5, and 7 m) but some longer single ladders do exist.

Roof Ladders

Roof ladders are single ladders equipped with folding hooks at the top end which provide a means of anchoring the ladder over the roof ridge or other roof part (Figure 5.3). Roof ladders are generally required to lie flat on the roof surface so that the firefighters may stand on the ladder for roof work and the ladder will distribute the firefighters' weight and will help to prevent slipping. Roof ladders may also be used as single wall ladders. Their lengths range from 12 to 20 feet (4 to 6 m).

Folding Ladders

Folding ladders are single ladders that have hinged rungs allowing them to be folded up so that one beam rests against the other. This allows them to be carried in narrow passageways and for use in attic scuttle holes and small room or closet work. Folding ladders are commonly found in 10 feet (3 m) lengths since they are only required to reach a short distance. All folding ladders should be equipped with safety shoes to prevent slipping (Figure 5.4).

Extension Ladders

An extension ladder is adjustable in length. It consists of two or more sections which travel in guides or brackets to permit length adjustment (Figure 5.5). Its size is designated by the length of the sections, measured along the side rails, when fully extended. Extension ladders provide access to windows and roofs within the limits of their extendable length. Most of the longer extension ladders are of the trussed type. Extension ladders are heavier than single ladders and more personnel are needed to safely handle them. Extension ladders generally range from 24 to 50 feet (7 to 15 m) in length.

Pole Ladders

Pole ladders are extension ladders that have staypoles for added stability. Lengths vary from 35 to 65 feet (11 to 20 m). However, most do not exceed 50 feet (15 m). They are of truss construction and have one to three fly sections. (Figure 5.6).

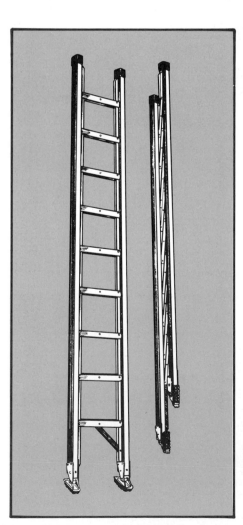

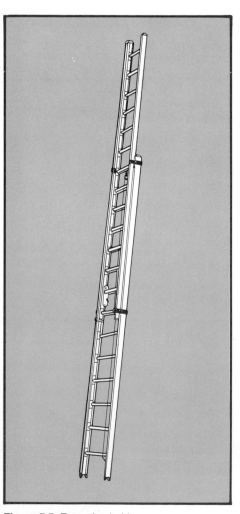

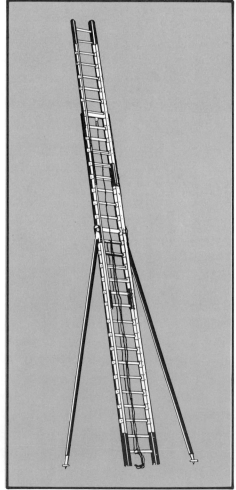

Figure 5.4 Folding ladder in closed and open positions.

Figure 5.5 Extension ladder.

Figure 5.6 Three-section pole-type extension ladder.

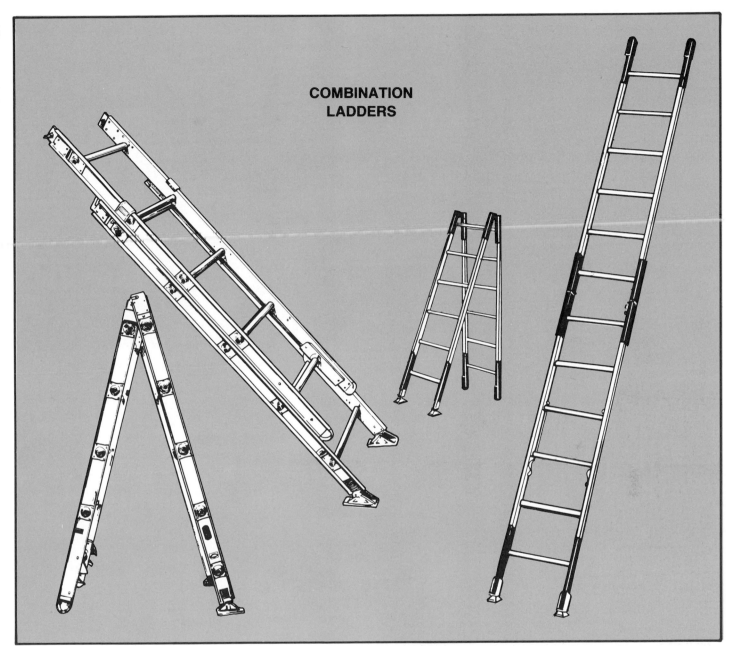

COMBINATION LADDERS

Figure 5.7 Combination ladders can be used in three ways.

Combination Ladders

Combination ladders are designed so that they may be used as either a single, extension, or "A" frame ladder. Lengths range from 8 to 14 feet (2 to 4.3 m) (Figure 5.7).

Pompier Ladders

The pompier ladder is a single beam ladder with rungs projecting from both sides. It has a large metal "gooseneck" projecting at the top for inserting into windows or other openings (Figure 5.8). Lengths vary from 10 to 20 feet (3 m to 6 m).

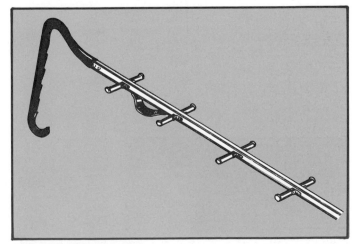

Figure 5.8 Pompier ladder.

Figure 5.9 Aerial ladder.

Aerial Ladders

Aerial ladder apparatus consists of a power-operated metal ladder mounted on a specially built chassis. The aerial ladder is hydraulically powered and may be mounted on a single chassis or a three-axle tractor-drawn vehicle (Figure 5.9). Aerial apparatus consists of a continuous walkway for the purpose of fire and rescue operations ranging in lengths from 65 to 135 feet (20 to 41 m).

Aerial-Ladder Towers

Aerial-ladder tower apparatus combine some of the features of both aerial ladders and telescoping booms. They provide a continuous walkway in the form of an aerial ladder in addition to an enclosed platform attached to the end of the ladder (Figure 5.10). Ladder towers are equipped with built-in piping and nozzles for providing elevated streams and range in length from 30 to 100 feet (9 to 30 m).

Telescoping-Aerial Towers

A telescoping-aerial tower has a telescoping boom with an enclosed platform attached to the end of the boom (Figure 5.11). These towers have a ladder mounted on the boom; however, the ladder

Figure 5.10 Aerial-ladder tower. *Courtesy of Ladder Towers Incorporated.*

is not constructed nor designed for the same use as an aerial ladder. They are equipped with built in piping and nozzles for providing elevated streams and range in length from 50 to 100 feet (15 to 30 m).

Articulating-Aerial Towers

Articulating-aerial tower apparatus consists of two or more booms or sections with an enclosed platform attached to the top boom (Figure 5.12). The platform provides a stable base to carry out fire and rescue operations. They are equipped with built in piping and nozzles for providing elevated streams and range in length from 50 to 150 feet (15 to 46 m).

CONSTRUCTION AND MAINTENANCE

Fire service ladders must be able to take considerable abuse such as sudden overloading, exposure to temperature extremes, falling debris, and for use other than what they are designed for.

Because of the importance of eliminating, or at least reducing, any structural defects and design weaknesses, ladder specifications should be written to meet NFPA Standard 1931, *Standard For*

Figure 5.11 Telescoping-Aerial Tower. *Courtesy of Mack Trucks, Incorporated.*

Figure 5.12 Articulating-Aerial Tower.

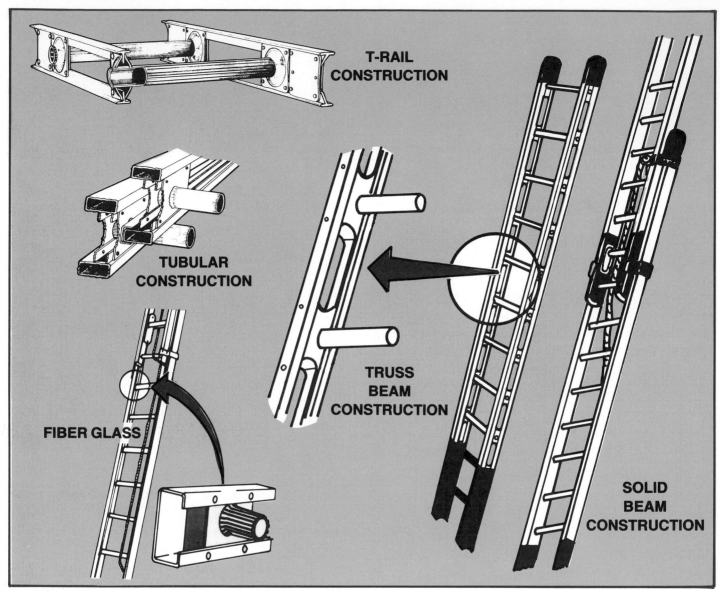

Figure 5.13 Ladders are manufactured with solid or truss beam construction.

Fire Department Ground Ladders. All ladders meeting NFPA 1931 are required to have a certification label affixed to the ladder by the manufacturer indicating that they meet the standard.

Construction Features

The major components of ladder construction are the beams and the rungs. Solid beam and truss beam construction are the two designs that are used (Figure 5.13). However, both types must meet the same specifications.

For lengths of 24 feet (8m) or less, solid beam construction provides a lighter weight ladder. For intermediate lengths of 25 to 35 feet (8 to 11 m), solid beam metal ladders will usually still be light-

est in weight, although with wood construction the opposite is true. For longer ladders with lengths over 35 feet (11 m) truss construction provides the lightest weight ladder.

Ladder rungs must not be less than 1¼-inches (3 cm) in diameter and spaced on 14-inch (36 cm) centers. When metal rungs are used, they must be constructed of heavy duty corrugated, serrated, knurled, or dimpled material. As an alternative the rungs may be coated with a skid resistant material.

Fire service personnel should be alert to the maximum loading capacities for ladders. The load capacity is the total weight on the ladder including

TABLE 5.1
Maximum Ladder Loading

Type of Ladder	Load
Folding Ladder	300 lbs. (136 kg)
Roof Ladder Hanging from Hooks	500 lbs. (227 kg)
Roof Ladder when Resting on Roof	500 lbs. (227 kg)
Single Ladder	500 lbs. (227 kg)
Extension Ladder 26 ft. (7.9 m) or Less	500 lbs (227 kg)
Extension and Pole Ladders 27 to 45 ft. (8.2 to 13.7 m)	600 lbs. (272 kg)
Pole Ladders Over 45 ft. (13.7 m)	700 lbs. (318 kg)

all persons, their equipment, and any other weight such as charged hoselines (Table 5.1).

Fire service ladders may be constructed of metal, wood, or fiber glass. Regardless of the type of material used in ladder construction it must be of high quality. The variety of materials used is probably a result of material availability and a constant effort to improve the construction of ladders.

METAL LADDERS

Metal fire service ladders are usually built with heat-treated aluminum alloy. Three principal reasons for using aluminum are:

- It is light in weight.

- Has adequate strength.

- Permits a reliable visual inspection of all ladder parts.

Aluminum ladders usually require less maintenance than wood because an artificial coating to protect their exterior is unnecessary. Also, it is sometimes difficult for a person to perform a visual maintenance inspection of wood ladders, since pitch pockets, knots, nodes, and other defects may sometimes be concealed. It is relatively simple process to keep an aluminum ladder clean, and other maintenance requirements are seldom necessary. There are, however, some disadvantages to using aluminum for fire department ladders. They are a good conductor of electricity and caution must be exercised whenever metal ladders are used near electrical power sources. Another disadvantage is the increased possibility of freezing to

the ladder in extremely cold weather. Aluminum can become very cold in winter and noticeably hot in summer because of its good conductive qualities.

WOOD LADDERS

Although there are relatively few wood fire department ladders manufactured today, many are still in service. Douglas fir has long been a favored wood for ladder beams because it is relatively free from knots and pitch pockets. White ash or hickory is used for rungs. A two-year drying period is usually required before ladder stock can qualify with a moisture content between 9% and 12%. Ladders constructed from wood with excessive moisture content are likely to shrink, resulting in warping or loose rungs. Wood shrinkage is particularly noticeable after the ladder has been subjected to low humidity and artificial heat. This slow drying process is a primary cause of loose rungs and cracked rails in wood ladders.

Some ladders are manufactured as composites of wood and aluminum to take advantage of lightness and strength and resistance to electricity. These ladders have wood rails and aluminum rungs.

FIBER GLASS LADDERS

Ladders made of fiber glass are relatively new to the fire service. These ladders are not totally constructed of fiber glass but a combination of fiber glass beams and metal rungs. The major advantage of fiber glass is that it is a nonconductor of electricity. However, in order to meet the specifications of NFPA Standard 1931 *Standard for Fire Department Ground Ladders,* these types of ladders become relatively heavy. This is a result of the dense qualities of fiber glass and the amount of material needed to meet the strength requirements of the standard.

Maintenance

Regular and proper cleaning of ladders is more than a matter of appearance. Unremoved dirt or debris from a fire may collect and harden to the point where ladder sections are no longer operable. They should be cleaned after every use.

Remove dirt with a brush and running water. Use solvent cleaners to remove any oily or greasy residues. After rinsing, or anytime a ladder is wet, wipe it dry. During each cleaning period, firefighters should look for defects in the ladder.

Although many ground ladders are constructed from materials other than wood, it is possible for all ladders to have or develop defects and to deteriorate because of improper maintenance. Damaged or weakened portions of a ladder can best be found by regular and systematic inspections. The inspection should cover all parts of a ladder, and when a part shows excessive wear, the cause should be determined (Figure 5.14).

Metal ladders are not subject to some of the problems of moisture and climate conditions which affect wood ladders. All braces, slides, stops, locks, halyards, rivets, pulleys, and other movable parts should be examined. The movable parts should be lubricated at least every six months with a waterproof grease. Remove old grease with a solvent.

Wood ladders require a much closer inspection than metal ladders, for their trusses and beams are subject to cracks and splinters. Wood rungs are susceptible to damage at the point where they come in contact with the locks, and wood staypoles have similar damage characteristics. Fire department ladders should be inspected at regular intervals, after each use or major repair, and tested once each year. Appropriate records should be kept for each ladder.

It is also important to thoroughly inspect ladders before they are subjected to any physical test. Rungs should be checked for tightness; bolts, rivets, and welds should be checked for looseness; and beams, trusses, and truss blocks for evidence of compression failure. Compression failure in wood is not easily seen but it can sometimes be detected by a wavy condition of the wood grain. Exposure of metal ladders to temperatures over 300°F (149°C) should be cause for testing. An indication of high

- **Use Brush, Soap, And Water**

- **Inspect Thoroughly For:**

  ✔ SLIVERS
  ✔ FRAYED ROPE
  ✔ BENT BEAMS OR RUNGS
  ✔ LOOSE PARTS
  ✔ CRACKS
  ✔ DRY ROT
  ✔ UNUSUAL WEAR

Figure 5.14 Regular inspections and maintenance are necessary for all ladders.

temperature would be if water boils when sprayed on the ladder or the manufacturer's heat indicator label has changed colors.

During the inspection of ladders, the inspector should mark all defects with chalk or some other suitable marker. Legible marks permit repairs to be made without the chance of missing some defect that was previously found. Major repairs that must be made on metal ladders may often require special tools which are usually not available to most fire departments. Repairs to ladders should be made in accordance with manufacturer's recommendations.

Varnish may be used to preserve wood ladders, since it seals in the natural oils and resins and keeps the moisture out. When the varnish finish becomes worn or scratched, it should be replaced without delay. Varnish prevents dry rot and attack from fungus growth. Paint is not recommended on fire department ladders except to identify the ladder ends, balance point, or length. Paint used as an outer cover of a wood ladder makes it practically impossible to detect fungus growth, dry rot, or cracks during inspection.

HANDLING LADDERS
Methods of Mounting Ground Ladders on Apparatus

The method by which ground ladders are mounted on fire apparatus varies with departmental requirements, the type of apparatus, the type of ladder, and manufacturer's policies. There are no established standards for mounting ground ladders on fire apparatus and a training sequence for ladder removal is a local procedure. General construction features and ladder location are, however, common with most fire apparatus.

Specifications for *Automotive Fire Apparatus* (NFPA Standard 1901) requires fire department pumpers to be equipped with an extension ladder not less than 24 feet (7 m) long and a roof ladder not less than 14 feet (4 m) long. These ladders are usually mounted on edge at or above shoulder level on one side of the apparatus. The type of bracket provided and the method used to hold the ladder are many and varied (Figure 5.15).

Figure 5.15 Ladders carried on pumpers are usually vertically mounted on the side.

National specifications also require ladder apparatus to carry a complement of ground ladders of various lengths. Ladders may be mounted in several different ways:

● Loaded from the rear lying flat in tiers usually two rows wide and two rows high. This type mounting requires the ladder to fit into runners or troughs and they are locked into position at each tier by a bar that is controlled either by a manually-operated lever or a power-operated bar activated by electric solenoid. When ladders are mounted in this manner room must be left clear at the rear of the apparatus to allow for removal and loading (Figure 5.16).

Figure 5.16 Aerial apparatus may carry ground ladders in a flat position at the rear.

Figure 5.17 Aerial apparatus may carry ground ladders vertically or flat for removal from the side.

- Nested vertically (on edge) or in flat tiers on the sides of the apparatus. This arrangement eliminates the problem of another piece of apparatus stopping too close to the rear and preventing ladder removal (Figure 5.17).

- Nested vertically (on edge) on each side. These ladders are arranged so they may be loaded and unloaded from the rear by sliding them in and out of the troughs requiring that room be left at the rear of the apparatus for removal and loading.

When ladders must extend beyond the rear end of the apparatus to the extent that they create an accident hazard, some form of guard should be placed over the heel plates. Short pieces of fire hose or a brightly painted protector are sometimes used for this purpose. Before firefighters are drilled in removing ladders from the apparatus, each firefighter should be able to answer the following questions concerning the equipment.

- Are the ladders mounted on the right or left side of the pumper?

- Are the butt plates toward the front or the rear of the apparatus?

- Can the roof ladder be removed leaving the extension ladder securely in place on the pumper?

- Is the fly section of the extension ladder on the inside or outside when the ladder is on the pumper?

- How are the ladders secured in place?

- Are the ladders provided with protection plates?

- Where two or more ladders are mounted flat, one inside the other, will all the ladders need to be removed from the tier to get only one ladder?

Selecting the Ladder for the Job

The designated length of ground ladders is derived from a measurement on the basis of their maximum usable length. These lengths have been established according to the places where they will normally reach. The 50-foot (15 m) extension ladder will normally reach fourth floor windows and, under some circumstances, the roof of a four-story building. Increasing the length of an extension ladder beyond 50 feet (15 m) results in excessive weight, reduced safety, and impractical application. The 35-foot (11 m) extension ladder is probably the most versatile of all extension ladders. It will reach the roofs of some three-story buildings. Shorter ground ladders are applicable within the limits of their reach.

Selecting a ladder to do a specific job requires firefighters to be a good judge of distance, stresses, and strength. Roughly speaking, a residential story will average 8 to 10 feet (2 to 3 m) from floor to floor, with a 3-foot (1 m) distance from the floor to the windowsill. Stories of commercial buildings will average 12 feet (4 m) from floor to floor, with a 4-foot (1 m) distance from the floor to the windowsill. In general, the following table can be used for selecting ladders.

TABLE 5.2 Ladder Selection Guide	
First story roof	16 to 20 feet (4.9 to 6.0 m)
Second story window	20 to 28 feet (6.0 to 8.5 m)
Second story roof	28 to 35 feet (8.5 to 10.7 m)
Third story window or roof	40 to 50 feet (12.2 to 15.2 m)
Fourth story roof	over 50 feet (15.2 m)

Ladders should extend a few feet (preferably three rungs) beyond the windowsill or roof edge to provide a footing and hand hold for persons step-

Figure 5.18 Place unused ladders in safe position when removing others.

ping on and off the ladder. When rescue from a window opening is to be performed, the top of the ladder should be placed just below the windowsill.

Removing Ladders from Apparatus

Ladders carried on pumpers are sometimes mounted so it is necessary to remove the complete assembly of roof and extension ladders before they are separated. Unless the roof and extension ladders have individual brackets and holding devices, one ladder may fall from the apparatus when the other is removed. Consideration should be given to providing a separate holding device for each ladder for safety reasons. Unused ladders should be put back on apparatus or kept in a safe and available location. Each department should develop a procedure for placing unused ladders in such a manner that personnel will not trip or vehicles run over them (Figure 5.18).

If two firefighters are used to remove the ladder or ladders from a pumper, they should position themselves at each locking device. After the locks have been released, each person should grasp a convenient group of rungs on each side of the locks and lift the entire assembly from the apparatus. If a single firefighter is to remove a ladder, a position at the center of the ladder near the balance point should be taken (Figure 5.19). The ladder can then

Figure 5.19 A single firefighter should take a position at the center of the ladder near the balance point before removing the ladder.

Figure 5.20 Multiple firefighters should take positions at the locking devices before removing the ladder.

be lifted over the retaining brackets. Once removed from the pumper, the ladder can then be prepared for the appropriate carry.

Some departments mount a three-section, 35-foot (11 m) extension ladder on the side of the pumper. It can be removed by two or three firefighters as shown in Figure 5.20 following the same procedure as for the two-firefighter method. They would then stay in the same position on one side of the ladder when making the desired carry. On apparatus with vertical mounted ladders, such as elevated platforms, removal should be done by two to six firefighters depending on ladder size.

When laddders are carried flat as with overhead racking on some pumpers and on aerial apparatus, the procedure begins with the release of the locking device. After the ladder locks have been released, the ladder should be pulled straight back from the ladder bed on a level plane. This can be accomplished by two to six firefighters depending on size of the ladder and the number of firefighters available to remove and carry the ladder. As the ladder is pulled from the bed, the firefighters should evenly space themselves along the length of the ladder as they grasp the beam of the ladder. The person closest to the apparatus should give a signal to stop when the ladder is about ready to clear the ladder bed (Figure 5.21). At this point, all lift the ladder from the apparatus for the desired carry.

Figure 5.21 When removing ladders carried flat, firefighters should be evenly spaced along the length of the ladder.

Proper Lifting and Lowering Methods

When lifting any object from the ground, the lifting force should come from the legs and NOT the back. To lift a ladder from the ground use the the following procedure:

● Obtain adequate personnel for the job.

● Place firefighters parallel to the ladder at ends and in the middle if necessary.

● Bend knees, keeping back as straight as possible and lift with the legs, (NOT BACK OR ARMS) on command of one of the firefighters at the rear who can see the operation. The lifting should be done in unison and as a team. If one firefighter is not ready that firefighter should make it known (Figure 5.22).

Figure 5.22 When lifting a ladder from the ground, lift in unison, using the legs.

● When setting a ladder down be sure to lower it with leg muscles and not back muscles. Also be sure to keep body and toes parallel to the ladder so that when it is placed down it does not injure any toes.

● When it is necessary to place a ladder on the ground prior to raising the reverse of the procedure for lifting is used.

Ladder Carries
ONE-FIREFIGHTER LOW-SHOULDER METHOD

Short, light ladders may be carried on the shoulder by one person. Select a balance point near the center of the ladder, face toward the butt, and insert one arm between the beams (some departments mark the center for convenience). Rest the upper beam on the shoulder and steady the ladder with both hands (Figure 5.23). Lower the butt end slightly for better balance and vision during the carry (Figure 5.24).

Figure 5.23 For the low-shoulder carry, place one arm between the beams and rest the upper beam on the shoulder.

Figure 5.24 Lower the butt end while carrying.

To pick the ladder up for the low-shoulder carry when the ladder is on the ground face toward the tip, crouch beside the ladder, and grasp the middle rung with the hand nearest the ladder (Figure 5.25). Stand, using the leg muscles and keeping the back straight and vertical. As the ladder is brought up, pivot into the ladder and insert the other arm through the rungs so the upper beam rests on the shoulder (Figure 5.26). With the firefighter now facing toward the butt, steady the

Figure 5.27 Rest the ladder on the shoulder, steady it, and lower the butt before proceeding.

ladder, lower the butt end slightly for better balance and visibility (Figure 5.27).

ONE-FIREFIGHTER HIGH-SHOULDER METHOD

The high-shoulder carry is good to use when the ladder will be raised on the beam. Face the butt and at the balance point place the palm of one hand under the bottom beam and the palm of the other hand on the top beam. Lift the ladder and rest the lower beam on the shoulder, with the butt end lowered slightly. Hold the upper beam with one hand, palm down (Figure 5.28).

Figure 5.25 For the low-shoulder carry from the ground, kneel beside the ladder facing the tip and grasp the middle rung with the palm forward.

Figure 5.26 Stand and pivot into the ladder, keeping the back straight, while inserting the arm through the rungs.

Figure 5.28 For the high-shoulder carry, place a hand under each beam and rest the lower beam on the shoulder.

CARRYING ROOF LADDERS

Normally the roof ladder is carried to its place of use with the hooks closed. If the hooks are to be utilized, they should be opened prior to taking the ladder onto the roof. When one firefighter is to carry a roof ladder up another ladder and utilize the hooks for placing it on a sloping roof, it is necessary that the ladder be carried with the tip end forward (Figure 5.29).

Figure 5.29 Carry the roof ladder with the hooks closed, top forward, and lowered. The roof ladder can be carried high or low shoulder.

TWO-FIREFIGHTER LOW-SHOULDER METHOD (From Vertical Racking)

Step 1: The two firefighters stand facing the ladder, one is positioned near the tip end and the other is positioned near the butt end; each firefighter uses both hands to grasp the ladder and remove it from the rack (Figure 5.30).

Step 2: As soon as the ladder clears the rack the firefighters continue to grasp the ladder with the hand nearest the butt end while they place the other arm between two rungs, pivot, and bring the upper beam on the shoulder (Figure 5.31).

Step 3: The forward firefighter is in a position to use one hand to push persons out of the way to prevent them being struck by the butt spur (Figure 5.32).

Figure 5.30 A firefighter is positioned at each end of the ladder (near the locking devices).

Figure 5.31 Support the ladder while pivoting and placing an arm between convenient rungs. Then bring the ladder to the shoulder.

Figure 5.32 The forward firefighters should clear the way to prevent injuries.

TWO-FIREFIGHTER LOW-SHOULDER METHOD (From Flat Racking)

Step 1: As the ladder clears the rack each firefighter grasps two rungs holding the ladder flat (Figure 5.33).

Step 2: When the ladder is clear of the rack the outside beam is lowered and the inside beam raised to the shoulder simultaneously with the firefighter pivoting and placing the arm farthest from the butt end between two rungs. The upper beam of the ladder should now be on the firefighters' shoulders and the firefighters are facing the butt end (Figure 5.34).

Figure 5.33 As the ladder leaves the apparatus, each firefighter grasps the last two rungs at the end and holds the ladder flat.

Figure 5.34 The outside beam is lowered as the firefighters pivot placing an arm between two rungs and resting the ladder on the shoulder.

OTHER TWO-FIREFIGHTER CARRIES

Two methods commonly used for carrying ladders with two firefighters are the hip or underarm method (Figure 5.35) and the arms-length method (Figure 5.36).

To pick the ladder up for the low-shoulder method with two firefighters, when the ladder is on the ground, the following procedure is used.

Step 1: The two firefighters position themselves on the same side of the ladder, one near the butt end, the other near the tip end. They then kneel next to the ladder facing the tip end and grasp a convenient rung with their near hand, palm forward (Figure 5.37).

Step 2: The firefighter at the heel gives the command to "shoulder the ladder." Both firefighters should stand up using their leg muscles to lift the ladder (Figure 5.38).

Step 3: As the ladder and the firefighters rise the far beam is tilted upward, the firefighters pivot and place the free arm between two rungs. The upper beam is placed on the shoulders with the firefighters facing the butt end (Figure 5.39).

NOTE: The lift should be smooth and continuous.

Figure 5.35 Two firefighters can use the hip or underarm carry for single ladders.

Figure 5.36 The arms-length method is also used for single ladders.

Figure 5.37 For the low-shoulder carry from the ground, firefighters kneel aside the ladder at each end facing the tip, then grasp a convenient rung with the palm forward.

Figure 5.38 On command, stand, lifting the ladder with the leg muscles.

Figure 5.39 The firefighters bring the ladder to vertical and pivot placing an arm through the rungs. Rest the upper beam on the shoulder.

THREE-FIREFIGHTER FLAT-SHOULDER METHOD (From Flat Racking)

As the ladder is pulled from the rack, two firefighters position themselves on one side of the ladder near each end and the third firefighter takes a position at midpoint on the opposite side (Figure 5.40). All three firefighters grasp the rungs and beam, lift the ladder clear of the apparatus, and prepare to pivot toward the butt end (Figure 5.41). The ladder is then raised to shoulder height as the firefighters pivot, and the beams are placed upon the shoulder (Figure 5.42).

THREE-FIREFIGHTER FLAT-SHOULDER METHOD (From the Ground)

To pick the ladder up for the flat-shoulder carry when the ladder is on the ground, the firefighters take the same positions as they would if the ladder were being removed from the apparatus. Each firefighter kneels so that the back is straight and the knee closest to the ladder is the one touching the ground and grasp the beam with the hand

Figure 5.41 Grasp convenient rungs with palms to the rear, and prepare to pivot under the ladder.

Figure 5.40 Position two firefighters on the same side at each end and one on the opposite side at the midpoint.

Figure 5.42 Lift, pivot, and place the beams on the shoulders.

closest to the ladder (Figure 5.43). The firefighter at the heel gives the command to "shoulder the ladder." All firefighters stand up using their leg muscles to lift the ladder and pivot toward the butt end (Figure 5.44). The beam is placed on the shoulder with the firefighter facing toward the butt end (Figure 5.45).

THREE-FIREFIGHTER FLAT-ARMS-LENGTH METHOD

(From the Ground)

Step 1: Two firefighters position themselves on one side of the ladder near each end and the third firefighter takes a position at midpoint on the opposite side (Figure 5.46).

Step 2: All firefighters kneel beside the ladder with the near knee touching the ground and grasp the beam. The firefighter at the heel gives the command to "pick up the ladder" and all stand up using their leg muscles to lift the ladder to the arms-length position (Figure 5.47).

Figure 5.45 The beams are lowered to the shoulders for carrying.

Figure 5.46 Position two firefighters on the same side at each end and one on the opposite side at the midpoint.

Figure 5.43 Kneel aside the ladder on the knee next to the ladder, and grasp the beam.

Figure 5.44 On command, the firefighters stand, lifting with their legs, and pivot toward the butt under the ladder.

Figure 5.47 All firefighters kneel, grasp the beam, and lift with the legs to the arms-length carry.

THREE-FIREFIGHTER LOW-SHOULDER OR HIP METHOD

The procedure for three firefighters to use for the low-shoulder method or the hip method is the same as for the two-firefighter methods, except that a third firefighter is positioned at midpoint between the other two firefighters. The completed carry is shown in Figure 5.48.

Figure 5.48 The three firefighters are positioned at each end and midpoint for the hip and low-shoulder carries.

FOUR-FIREFIGHTER CARRIES

The same methods may be used by four firefighters for carrying ladders as used by three firefighters by making a slight change in the positioning of the firefighters. For the flat-shoulder method, two firefighters are positioned at each end of the ladder opposite one another (Figure 5.49).

Figure 5.49 For four-firefighter flat-shoulder carries, a firefighter is positioned at the end of each beam.

The firefighters are positioned in the same manner for the flat-arms-length method except that the ladder is carried at arms length (Figure 5.50). When using the low-shoulder or arms-length on-edge methods, firefighters are equally spaced along one side of the ladder.

Figure 5.50 The arms-length carry with four firefighters has one at the end of each beam.

FIVE-AND SIX-FIREFIGHTER CARRIES

Five and six firefighters are normally used for carrying ladders 40 feet (12 m) or more in length. The only difference between a five-and six-firefighter carry is in the positioning of personnel. Also, because of the weight of these ladders, only the flat-shoulder method and the flat-arms-length method are recommended (Figures 5.51-5.53).

CARRYING OTHER LADDERS

Folding ladders should be carried in the closed position by the handle when provided on the ladder. This ladder is usually carried by one firefighter (Figure 5.54).

SPECIAL CARRY FOR A NARROW PASSAGE

When firefighters employing a flat carry encounter a narrow passageway that restricts the use of the normal carry, the ladder can be raised to a flat overhead position.

To place the ladder in this position, the signal is given to change position. All firefighters then raise the ladder upward and swing underneath it with arms outstretched (Figure 5.55).

Figure 5.51 For the five-firefighter carry, position two at the top (pole-men), two at the midpoint (beammen), and one at the butt (heelman).

Figure 5.52 For the six-firefighter shoulder carry, two are positioned at the top, midpoint, and butt.

Figure 5.53 For the six-firefighter arms-length carry, the firefighters use the same positions as the shoulder carry.

Figure 5.54 Folding ladders are carried by one firefighter in the collapsed position.

Figure 5.55 To go to a narrow passage carry, the firefighters on the right move forward while those on the left move behind, swinging under and carrying the ladder above the head.

Proper Ladder Placement

The placement of ladders is determined by the intended use of the ladder and the positioning of it for safe and easy climbing. Factors that have an effect on ladder placement regarding intended use include:

- If the ladder is to be used by a firefighter to effect ventilation from a window it should be placed alongside the window to windward side with the tip about even with the top of the window.

- The same position is used when firefighters desire to climb into or out of narrow windows.

- If the ladder is to be used for rescue from a window, consideration must be given to the size of the window. Normally the ladder tip is placed even with or slightly below the sill. If the sill projects out from the wall, the tip of the ladder can be wedged up under the sill for additional stability. Where the window opening is wide enough to permit placing the ladder inside the window opening and still leave room beside it to facilitate the rescue, it should be placed so that two or three rungs extend above the sill.

- The same position can be used for firefighters to climb in or out of wide window openings.

- When a ladder is to be used as a vantage point from which to direct a hose stream into a window opening and no entry is to be made, it is raised directly in front of the window with tip on the wall above the window opening.

Other factors that affect ladder placement include:

- Overhead obstructions such as wires, tree limbs, signs, cornices, and building overhangs

- Uneven terrain and soft spots

- Obstructions on the ground such as bushing and parked cars

- Main paths of travel that firefighters or evacuees may use

When ladders are raised into place they should be at an angle that is safe and easy to climb. The distance of the butt from the building establishes the angle formed by the ladder and the ground. If the butt is too close to the building, the ladder's stability is reduced and when it is climbed the tip of the ladder may be pulled away from the building. If the butt is too far from the building, the load-carrying capacity of the ladder is reduced and the ladder may also slip away from the building. An angle of 75 degrees gives adequate stability, insures safe stress on the ladder, and is easy to climb. With a ladder at this angle a climber can stand perpendicular to the ground on the rungs and climb at arm's length from the rungs. The proper angle and the position of a person when climbing is shown in Figure 5.56.

An easy way to determine the proper distance between the heel and the building is to divide the used length of the ladder by 4. For example, if 20 feet (6 m) of a 24-foot (7 m) ladder is to be used, the heel should be 5 feet (2 m) (20 ÷ 4) from the building. Exact measurements are unnecessary on the fire scene. Firefighters will soon have the experience to judge visually what will be the proper place for positioning the ladder.

Ladders placed to the roof should be extended three rungs above the roof edge to aid in climbing onto and off the ladder and so firefighters on the roof can find the ladder (Figure 5.57).

RAISING AND CLIMBING LADDERS

The process of raising a ladder when it is needed will not in itself extinguish a fire, but a well-positioned ladder becomes a means by which other operations can be performed. Teamwork, smoothness, and rhythm are very necessary when raising and lowering fire department ladders if speed and accuracy are to be developed. However, prior to learning the technique of raising ladders, firefighters should be aware of certain general procedures that affect the raising of ladders.

General Procedures

TRANSITION FROM CARRY TO RAISE

With the exception of pole ladders, it is not necessary to place the ladder flat on the ground

Figure 5.56 The proper angle can be determined by standing erect at the butt of the ladder with the toes against the beams on the bottom rung. The outstretched arms should reach the ladder.

prior to raising it; only the butt end need be placed on the ground. The transition from carry to raise should be a smooth continuous series of motion.

ELECTRICAL HAZARDS

A major concern when raising ladders is contact with live electric wires or equipment either by the ladder or by the person climbing it. The danger of metal ladders in this respect has been previously stressed; however, many firefighters do not realize that wet wooden or fiber glass ladders present the same hazard. Care must be taken before beginning a raise to be sure that this hazard is avoided.

POSITION OF THE FLY SECTION ON EXTENSION LADDERS

The question of whether the fly on an extension ladder should be in (next to the building) or out (away from the building) must be settled before starting the discussion on individual raises. This has been a matter of controversy for many years.

Figure 5.57 Ladders placed to the roof should extend three rungs above the parapet or roof edge.

Figure 5.58 The fly section may be placed on the top (fly out) or the bottom (fly in) of the bed section. The fly out method is recommended because of greater strength. **NOTE:** Ladders in the following photos will be shown fly in as this was the local fire department option at the time they were taken.

A recommended practice is that the FLY SECTION BE OUT (away from the building) as shown in Figure 5.58. This recommendation is based on information received from major fire service ground ladder manufacturers in which they advise that the ladder in the fly out position is stronger, from NFPA Standard 1931, *Fire Department Ground Ladders,* which calls for the fly to be out, and from a report of strength tests conducted on ground ladders by the National Bureau of Standards which states that extension ladders are stronger with the fly out.

PIVOTING LADDERS

There are times when it becomes necessary to pivot a ladder once it has been raised to the vertical position. This may be due to a beam raise, a flat raise parallel to a building, or the need to reposition the ladder so that the fly section is facing out. Whenever possible pivoting should be done before extending the ladder. The procedure for pivoting a ladder is as follows:

Step 1: Two firefighters face each other through the ladder and grasp the beams with both hands. Each firefighter places their foot against the beam that the ladder will pivot on (Figure 5.59).

Figure 5.59 In preparation for pivoting a ladder, grasp the beams and place the feet against the beam that will serve as the pivot point.

Another method is for one firefighter to place a foot on the bottom rung.

Step 2: The opposite beam of the ladder is tilted until it clears the ground (Figure 5.60).

Step 3: Pivot the ladder into position. Both firefighters must simultaneously adjust their position as the ladder is moved around (Figure 5.61). If three or four firefighters are used, extra positions should be taken along the outside of the beam to provide additional support (Figure 5.62).

ROLLING A LADDER

In order to place the fly out on a one-firefighter raise of an extension ladder, it becomes necessary to reposition the ladder once it has been raised.

Figure 5.60 Tilt the opposite beam until it clears the ground.

Figure 5.61 The firefighters adjust their position as the ladder moves. Note the alternate method being used in the photo.

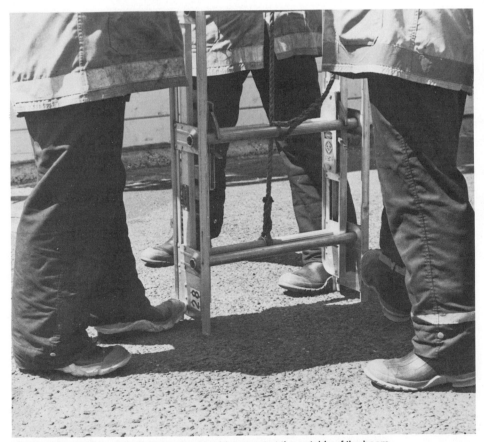

Figure 5.62 Additional firefighters should add support at the outside of the beam.

This is best done by rolling the ladder over while it is up against the building.

Step 1: Stand to one side of the raised ladder at its base, reach over and grasp the opposite beam, pull the beam up and over.

Step 2: Continue to roll the raised ladder from beam to beam until one complete turn has been made. If this procedure is not adequate to relocate the ladder in an adjoining window, repeat the operation. After the ladder has been rolled into position it may need to be straightened, adjusted, or even skidded into proper position.

LADDER RAISES

One-Firefighter Raise

There are two common methods of raising ladders with one firefighter. The type of raise used will depend upon the carrying method, weight of the ladder, and the strength of the firefighter. When raising the ladder from the high-shoulder carry, the following procedure is used.

Step 1: Lower the butt to the ground at a point determined for establishing the proper climbing angle (Figure 5.63).

Figure 5.63 Place the butt at a location that will provide a good climbing angle.

Step 2: As the ladder is brought to a vertical position the firefighter pivots the ladder 90° and takes a position facing the ladder on the side away from the building (Figure 5.64).

Figure 5.64 Pivot the ladder and bring it to a vertical position facing the building.

CAUTION: The area overhead should be visually checked for obstructions before bringing the ladder to a vertical position.

Step 3: To extend the ladder, place one foot at the butt of one beam and with the instep, knee, and leg, steady the ladder (Figure 5.65).

Figure 5.65 Place the foot, leg, and knee to steady the beam before extending the fly.

Figure 5.66 Extend the fly with a hand-over-hand motion.

Step 4: Grasp the halyard and extend the fly section with a hand-over-hand motion (Figure 5.66). When the tip is at the desired elevation, make sure the ladder locks are in place.

Step 5: To lower the ladder, place one foot against a butt spur or on the bottom rung while grasping the beams, and lower the ladder gently into the building (Figure 5.67).

Step 6: In order to position the ladder so the fly section is facing out, the ladder must be rolled over. Some repositioning may be necessary after the turn has been completed.

The major difference in using the one-firefighter raise from the low-shoulder carry is the placement of the butt. In this instance a building is used to heel the ladder and prevent the ladder butt from slipping while the ladder is brought to the vertical position. When raising the ladder from the low-shoulder carry, the following procedure is used.

Figure 5.67 Heel the ladder with one foot and lower it in by the beams.

Step 1: Place the butt end of the ladder on the ground with the butt spurs against the wall of the building (Figure 5.68).

Step 2: With the free hand, grasp a rung in front of the shoulder while removing the opposite arm from between the rungs (Figure 5.69).

Step 3: Step beneath the ladder and grasp a convenient rung with the other hand (Figure 5.70).

Step 4: Advance hand-over-hand down the rungs toward the butt until the ladder is in a vertical position (Figure 5.71).

CAUTION: The area overhead should be visually checked for obstructions before bringing the ladder to a vertical position. The terrain in front of the firefighter should also be visually checked before stepping forward.

Figure 5.69 Grasp the rung in front of the shoulder as the arm is withdrawn from between the rungs.

Figure 5.68 Place the butt of the ladder on the ground against the building wall.

Figure 5.70 Use the arm along the beam to push the ladder up into a flat position and step beneath.

Figure 5.71 Bring the ladder to a near vertical position by hand-over-hand down the rungs.

Figure 5.72 Steady the ladder against the building with the upper hand while moving the butt out with the lower hand.

Step 5: To position the ladder for climbing, push against an upper rung to keep the ladder against the building. Grasp a lower rung with the other hand and carefully move the ladder butt out from the building to the desired location (Figure 5.72).

Step 6: To extend the ladder from this position, it is necessary to first bring the ladder into a vertical position. Grasp a convenient rung with both hands, heel the ladder, and pull it away from the building into a vertical position. To complete the extension and placement procedure, follow Steps 3-6 for the raise from a high-shoulder carry as described earlier (Figure 5.73).

Figure 5.73 Bring the ladder to vertical, brace it, and extend the fly to the necessary height.

Two-Firefighter Raises

Whether a ladder is raised parallel with or perpendicular to a building makes little difference. If raised parallel with the building, the ladder can always be pivoted after it is in the vertical position. Whenever two or more firefighters are involved in raising a ladder, the firefighter at the butt is responsible for placing the butt at the desired distance from the building and determining whether the ladder will be raised parallel with or perpendicular to the building. The procedure for raising ladders with two firefighters is as follows.

FLAT RAISE

Step 1: When the desired location for the raise has been reached, the heelman places the ladder butt on the ground while the firefighter at the tip rests the ladder beam on a shoulder (Figure 5.74).

Step 2: The firefighter at the butt heels the ladder by standing on the bottom rung, crouches down to grasp a convenient rung with both hands, and leans back. The firefighter at the tip steps beneath the ladder and grasps a convenient rung with both hands (Figure 5.75).

Figure 5.75 Pivot the ladder flat, and heel it by standing on the bottom rung as the firefighter at the tip steps beneath.

Step 3: The firefighter at the tip advances hand-over-hand down the rungs toward the butt until the ladder is in a vertical position. As the ladder comes to a vertical position, the heelman grasps successively higher rungs until standing upright, or rides the bottom rung using the body as a counter balance (Figure 5.76).

Figure 5.74 At the desired location the heelman places the ladder butt on the ground.

Figure 5.76 The two firefighters bring the ladder to a vertical position.

CAUTION: The area overhead should be visually checked for obstructions before bringing the ladder to a vertical position. The terrain in front of the firefighter should also be visually checked before stepping forward.

Step 4: Both firefighters face each other and heel the ladder by placing their toe against the same beam. When raising extension ladders, pivot the ladder to position the fly away from the building. Grasp the halyard and extend the fly section with a hand-over-hand motion (Figure 5.77). When the tip is at the desired elevation, make sure the ladder locks are in place.

Step 5: To lower the ladder, the firefighter on the outside of the ladder places one foot against a butt spur or on the bottom rung and grasps the beams. Both firefighters gently lower the ladder into the building (Figure 5.78).

Figure 5.77 After pivoting the ladder to a fly out position, both firefighters face each other, heel the ladder, and extend the fly in a hand-over-hand motion.

Figure 5.78 Both firefighters lower the ladder in by the beams, and the outside one heels the ladder.

BEAM RAISE

Step 1: When the desired location for the raise has been reached, the heelman places the ladder beam on the ground. The firefighter at the tip rests the beam on one shoulder while the heelman places one foot on the lower beam at the butt spur (Figure 5.79).

The heelman then grasps the upper beam with hands well apart and the other foot extended well back to act as a counterbalance (Figure 5.80).

An alternate method of heeling the ladder is to stand parallel to the ladder at the butt. Place one foot against the butt spur and the other positioned forward toward the tip of the ladder (Figure 5.81).

Step 2: The firefighter at the tip advances hand-over-hand down the beam toward the butt until the ladder is in a vertical position.

CAUTION: The area overhead should be visually checked for obstructions before bringing the ladder to a vertical position.

Figure 5.79 At the desired location, the heelman places the ladder beam on the ground and places one foot at the butt spur.

The terrain in front of the firefighter should also be visually checked before stepping forward.

Step 3: Pivot the ladder to position the fly section away from the building and complete the extending and lowering procedures as described for the flat raise.

Figure 5.80 The heelman grasps the upper beam and extends the other foot as a counterbalance.

Figure 5.81 An alternate heeling method is to place one foot at the butt spur and the other forward while pushing down with the hand at the butt and pulling upward with the other.

Three-Firefighter Raise

As the length of the ladder increases, the weight also increases. This requires the use of more personnel for raising the larger extension ladders. The firefighter at the butt is responsible for placing the butt at the desired distance from the building and determining whether the ladder will be raised parallel with or perpendicular to the building. The procedure for raising ladders with three firefighters is as follows.

Step 1: When the desired location for the raise has been reached, the heelman places the ladder butt on the ground while the firefighters at the tip rest the ladder flat on their shoulders (Figure 5.82).

Step 2: The firefighter at the butt heels the ladder by standing on the bottom rung, crouches down to grasp a convenient rung with both hands, and leans back (Figure 5.83).

Step 3: The firefighters at the tip advance in unison, outside hands on beams, inside hands on rungs until the ladder is in a vertical position (Figure 5.84).

CAUTION: The area overhead should be visually checked for obstructions before bringing the ladder to a vertical position. The terrain in front of the firefighters should also be visually checked before stepping forward.

Figure 5.83 The heelman stands on the bottom rung, grasps a rung, and leans back to serve as a counterbalance.

Figure 5.82 With a three-firefighter carry there is a single heelman who places the ladder butt on the ground.

Figure 5.84 The tip firefighters bring the ladder vertical using one hand on the beams and the other on the rungs.

Step 4: Pivot the ladder to position the fly section away from the building, using the procedure for three firefighters.

Step 5: To extend the ladder, grasp the halyard and extend the fly section with a hand-over-hand motion (Figure 5.85). When the tip is at the desired elevation, make sure the ladder locks are in place.

Step 6: To lower the ladder, the firefighter on the outside of the ladder places one foot against a butt spur or on the bottom rung and grasps the beams or a convenient rung. All firefighters gently lower the ladder into the building (Figure 5.86).

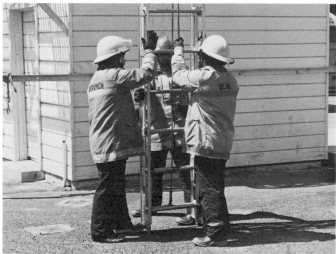
Figure 5.85 Pivot the fly out, steady the ladder with the feet, and extend the fly.

Figure 5.86 All firefighters assist in lowering the ladder with the one on the outside heeling it by the bottom rung or butt spur.

To raise a ladder using the beam method with three firefighters, follow the same procedures for the two-firefighter raise. The only difference is that the third firefighter is positioned along the beam (Figure 5.87). Once the ladder has been raised to a vertical position, follow the procedures described for the flat raise.

Figure 5.87 For a beam raise with three firefighters, the third firefighter assists along the beam.

Four-Firefighter Raise

When personnel are available, four firefighters are desirable because they can better handle the larger and heavier extension ladders. A flat raise is normally used and the procedures for raising the ladder are similar to the three-firefighter raise except for the placement of personnel. A firefighter at the butt is responsible for placing the butt at the desired distance from the building and determining whether the ladder will be raised parallel with or perpendicular to the building. The procedure for raising ladders with four firefighters is as follows.

Step 1: When the desired location for the raise has been reached, the heelmen place the ladder butt on the ground while the firefighters at the tip rest the ladder flat on their shoulders (Figure 5.88).

Step 2: The firefighters at the butt heel the ladder by placing their inside foot on the bottom rung and the outside foot on the ground outside the beam. They then grasp a convenient rung with the inside hand and the beam with the other hand and pull back (Figure 5.89).

Figure 5.88 With four firefighters, the heelmen place the ladder butt on the ground at the desired location.

Figure 5.89 The heelmen place the inside foot on the bottom rung, grasp a rung with the inside hands and the beams with the outside hands.

Figure 5.90 The tip firefighters raise the ladder with inside hands to the rungs and outside hands to the beams.

Step 3: The firefighters at the tip advance in unison, outside hands on the beams, inside hands on rungs until the ladder is in a vertical position (Figure 5.90).

CAUTION: The area overhead should be visually checked for obstructions before bringing the ladder to a vertical position. The terrain in front of the firefighters should also be visually checked before stepping forward.

Step 4: Pivot the ladder to position the fly section away from the building using the procedure for four firefighters.

Step 5: To extend the ladder, grasp the halyard and extend the fly section with a hand-over-hand motion (Figure 5.91). When the tip is at the desired elevation, make sure the ladder locks are in place.

Step 6: To lower the ladder, the firefighters on the outside of the ladder place their inside foot against the butt spur or bottom rung and grasp the beams. All firefighters gently lower the ladder into the building.

Figure 5.91 Pivot to the fly out position, steady the ladder from all sides, and extend the fly. Secure the locks in place.

Raising Pole Ladders

Pole ladders generally range from 40 to 60 feet (12 to 18 m) in length. Staypoles are used to assist in raising, lowering, and stabilizing the ladder. They may be permanently attached to the ladder or they may be removable.

Pole ladders may be raised either parallel or at right angles to a building because they can always be pivoted after reaching the vertical position. It is the responsibility of the firefighter at the butt of the ladder to select the location for the raise, and position the butt from the building.

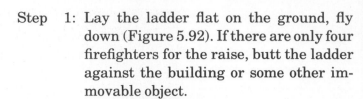

Figure 5.92 Place the ladder flat on the ground from the carrying position.

Figure 5.93 Take positions for passing and receiving the staypoles at the middle and tip of the ladder.

Step 1: Lay the ladder flat on the ground, fly down (Figure 5.92). If there are only four firefighters for the raise, butt the ladder against the building or some other immovable object.

Step 2: The firefighters take their positions for passing and receiving the staypoles. Two stand at the tip, ready to receive the poles, and two stand mid-ladder to pass the poles. The fifth stands at the butt, ready to heel the ladder (Figure 5.93). If there are six firefighters, both the fifth and sixth firefighters stand at the heel and remove the keeper pins from the staypoles (Figure 5.94).

Step 3: The firefighters at mid-ladder pass the staypoles to the firefighters at the tip who then move out from the tip of the ladder (Figure 5.95). If the staypoles are permanently attached, the poles are passed in a vertical arc. If the staypoles are not attached, they are passed butt first to the firefighters at the tip, after which the firefighters at the beams attach the toggle (swivel) ends to the bed section.

Step 4: The firefighters take their positions for the raise. The heelman crouches on the bottom rung and holds onto the beams. The firefighters at the beams kneel (inside knee on the ground) beside the ladder and grasp the beam or a rung just below the toggles. The firefighters with

Figure 5.94 The heelman releases the staypoles and initiates their passing.

Figure 5.95 The tip firefighters receive the poles and move out from the ladder.

the poles stand at the outside of the poles, one hand holding the butt end, the spur extending between the fingers, and the other hand holding the pole at a comfortable distance from the butt end. The poles should be as nearly in line with the beams as is possible (Figure 5.96).

Step 5: The firefighter at the butt gives the command to raise the ladder and leans back so the body weight helps the raise. The firefighters at the beams rise, pivoting and bringing the ladder to shoulder level (Figure 5.97).

Figure 5.96 Take positions for raising the ladder, heelman on the bottom rung, beammen kneeling just below the toggle, and the polemen with the poles in line with the ladder beams. Polemen should be on the outside with the spur between the fingers.

Figure 5.97 As the heelman counterbalances, the beammen bring the ladder to their shoulders.

Step 6: The firefighters under the ladder advance in unison, outside hands on the beams, inside hands on the rungs until the ladder is in a vertical position (Figures 5.98). When the ladder is at an angle of about 45 degrees, the pole firefighters assume the ladder's weight and push the ladder up. The firefighters at the beams help steady the ladder.

CAUTION: The area overhead should be visually checked for obstructions before bringing the ladder to a vertical position. The terrain in front of the firefighters should also be visually checked before stepping forward.

Figure 5.98 The beammen raise the ladder to a 45 degree angle before the polemen assume the weight and finish the raise.

Figure 5.99 When the ladder is vertical, the polemen take positions at right angles and watch the tip to control forward and backward movement, while communicating with each other on the position of the ladder tip.

Figure 5.100 The firefighters at the base of the ladder heel it, extend the fly, and set the locks.

Figure 5.101 All firefighters assist in lowering the ladder to the building.

Step 7: When the ladder is vertical, one of the pole firefighters walks toward the building so the pole is held in line with the lateral plane of the ladder. The other pole firefighters move to stand directly in front of the ladder. The two poles now give four-way stability to the ladder. The two pole firefighters must watch the tip of the ladder and control only their respective forward-and-back movement of the ladder (Figure 5.99).

Step 8: One of the beam firefighters raises the fly section to the desired height while the other beam firefighters help to steady the ladder (Figure 5.100).

Step 9: The ladder is lowered gently into the building (Figure 5.101).

Step 10: The pole firefighters walk their poles to the building and let them rest on the ground (Figure 5.102). The poles must not be wedged. Staypoles are not designed to carry the stresses put on the ladder. They are only for helping to raise and lower the ladder.

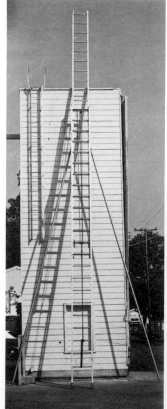

Figure 5.102 The poles are placed parallel to the building in line with the toggle at the side of the ladder. Do not wedge the poles, but properly place them to allow the ladder to have the desired natural sway.

Securing the Ladder

A ladder should be kept from slipping whenever firefighters climb it, especially if the ladder is at a lower-than-desirable angle or if there are strong winds or the ground is icy or unstable. Added stability might also be necessary when operating hoselines from the ladder or when using the ladder for rescue.

HEELING

There are several methods of properly heeling a ladder. One method, as illustrated in Figure 5.103 shows the firefighter underneath the ladder. The firefighter stands with the feet about shoulder width apart, grasps the ladder beams at about eye level, and pulls backward to press the ladder against the building. When using this method the firefighter must wear head protection and must not look up when there is someone climbing the ladder. Be sure to grasp the beams, not the rungs.

Another method of heeling a ladder is illustrated in Figure 5.104. With this method either the firefighter's toes are placed against the heel of the ladder or one foot is placed on the bottom rung. The hands grasp the beams and the ladder is pressed against the building. The firefighter heeling the ladder must stay alert for descending firefighters.

Figure 5.103 A ladder can be heeled by a firefighter underneath the ladder.

Figure 5.104 A recommended heeling method is with the foot against the butt spur or on the bottom rung and the hands pushing on the beams.

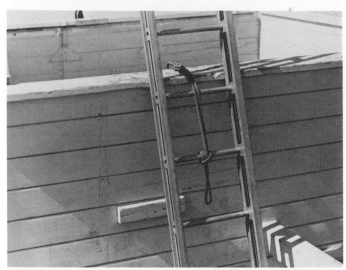

Figure 5.105 A rope hose tool or safety strap in a clove hitch can be used to secure the tip of the ladder.

Figure 5.106 The rope hose tool can also be placed over the beams and brought back between the rungs to secure the ladder. Ladders can be tied to a tool placed inside a window.

TYING IN

Whenever possible a ladder should be tied securely to a fixed object. A rope hose tool or safety strap can be used, as shown in Figure 5.105. Alternate methods of tying off a ladder are shown in Figure 5.106. Tying a ladder in is simple, can be done quickly, and is strongly recommended to prevent the ladder from slipping or pulling away from the building. Tying in also frees personnel who would otherwise be holding the ladder in place. For complete safety the ladder should be tied in at both the top and the bottom (Figure 5.107).

TYING THE HALYARD

Before an extension ladder is climbed, the excess halyard should be tied to the ladder to prevent the fly from slipping and to keep anyone from trip-

Figure 5.107 The bottom of the ladder should also be secured. A clove hitch tied to any available object will do the job.

ping over it. The same tie can be used for either a closed- or open-ended halyard. The procedure for tying the halyard is described as follows.

Step 1: Wrap the excess halyard around two convenient rungs and pull it taut (Figure 5.108).

Step 2: Hold the halyard between the thumb and forefinger with the palm down (Figure 5.109).

Step 3: Turn the hand palm up and push the halyard underneath and back over the top of the rung (Figure 5.110).

Step 4: Grasp the halyard with the thumb and fingers and pull it through the loop, making a clove hitch (Figure 5.111). Finish the tie by making the half hitch on top of the clove hitch.

Figure 5.109 Hold the halyard between the thumb and forefinger with the palm down.

Figure 5.110 Turn the palm up bringing the halyard around the rung.

Figure 5.108 Wrap the halyard around two convenient rungs and pull it tight.

Figure 5.111 Catch the halyard with the thumb and bring it through forming a clove hitch. Secure it with a half hitch.

CLIMBING LADDERS

General Procedures

Ladder climbing should be done smoothly and with rhythm. The climber should ascend the ladder so there is the least possible bounce and sway. This smoothness is accomplished if the climber's knee is bent to ease the weight onto each rung. Balance on the ladder will come naturally if the ladder is properly spaced from the building, for the body will be perpendicular to the ground.

The climb may be started after the climbing angle has been checked and the ladder is properly heeled. The eyes of the climber should be focused straight forward, with an occasional glance at the tip of the ladder. The climber's arms should be kept straight during the climb, which will keep the body away from the ladder and permit free knee movement during the climb. The hands should grasp the rungs with the palm down and the thumb beneath the rung (Figure 5.112). Some persons find it natural to grasp every rung with alternate hands while climbing; others prefer to grasp alternate rungs. A person should try both methods and select the one that is most natural.

If the feet should slip, the arms and hands are in a position to stop the fall. All of the upward prog-ress should be done by the leg muscles, not the arm muscles. The arms and hands should not reach upward during the climb, because reaching upward will bring the body too close to the ladder.

Practice climbing should be done slowly to develop form rather than speed. Speed will be developed as the proper technique is mastered. Too much speed results in lack of body control, and quick movements cause the ladder to bounce and sway.

Many times during fire fighting, a firefighter is required to carry equipment up and down a ladder. This procedure interrupts the natural climb either by the added weight on the shoulder or the use of one hand to hold the item. If the item is to be carried in one hand, it is desirable to slide the free hand under the beam while making the climb. This method permits constant contact with the ladder as shown in Figure 5.113. Whenever possible, a handline rope should be used to hoist tools and equipment.

The technique used for climbing aerial ladders is basically the same as for ground ladders. The only difference is that the firefighter has the option of grasping the handrails as well as the rungs when climbing the ladder (Figure 5.114).

Figure 5.112 Climb with the body erect, eyes forward, and arms straight grasping the rungs directly in front of the chest.

Figure 5.113 When carrying equipment, slide the free hand under the beam to maintain constant contact with the ladder.

Figure 5.114 The firefighters climbing an aerial can grasp the handrails or the rungs.

Climbing to Place a Roof Ladder

There are a number of ways to get a roof ladder in place on a sloped roof, using either one or two firefighters. A roof ladder can be carried conveniently by one firefighter using the shoulder method. The hooks should be closed while the ladder is being carried. The following procedure shows one method of placing a roof ladder in position.

Step 1: Carry the roof ladder to the desired location.

Step 2: One firefighter ascends the climbing ladder and locks in. The other firefighter heels the climbing ladder and opens the hooks on the roof ladder (Figure 5.115).

NOTE: For a one-firefighter operation, once the roof hooks are opened, the ladder can be raised to a vertical position and leaned against the beam of the climbing ladder (Figure 5.116). To place the ladder onto the roof, follow Steps 3-6.

Step 3: The firefighter on the ground passes the roof ladder to the firefighter on the climbing ladder (Figure 5.117).

Figure 5.115 As the first firefighter ascends the ladder and locks in, the second heels the ladder and opens the hooks on the roof ladder.

Figure 5.116 When there is only a single firefighter, open the hooks, lean the roof ladder against the beam of the extension ladder, and proceed up the ladder.

Figure 5.117 The firefighter on the ground raises the roof ladder to the firefighter at the roof edge.

Figure 5.118 The roof ladder is placed on a beam on the roof.

Figure 5.119 Slide the ladder on the beam up the roof until the hooks pass the ridge.

Figure 5.120 Place the ladder flat with the hooks down and pull the hooks into the roof.

Step 4: The roof ladder is then hoisted onto the roof and placed on the beam (Figure 5.118).

Step 5: Slide the beam of the ladder up the roof until the hooks pass the ridge (Figure 5.119).

Step 6: Turn the ladder flat with hooks down and pull hooks into opposite side of the ridge to secure the ladder (Figure 5.120).

An alternate method of taking a roof ladder aloft is described in the following procedure.

Step 1: Carry the roof ladder to the base of the climbing ladder, open the hooks, raise the ladder to a vertical position, and lean it against the beam of the climbing ladder.

Step 2: Climb ladder until just past the mid point of the roof ladder. Place ladder on shoulder and continue up ladder to the roof edge (Figure 5.121).

Figure 5.121 The roof ladder can be picked up while ascending an extension ladder. Pick the roof ladder up near mid point and place it on the shoulder, then proceed to the roof.

Working on a Ladder

Firefighters must sometimes work while standing on a ground ladder and must have both hands free. A safety belt or a leg lock can be used to safely secure oneself to a ladder while working from it.

The safety belt, or life belt, is strapped tightly around the waist. The hook may be moved to one side, out of the way, while the firefighter is climbing the ladder. After reaching the desired height, return the hook to the center and attach it to a rung (Figure 5.122). A rope hose tool may be used as a safety belt.

To secure oneself with a leg lock while facing the ladder, stand with both feet on one rung. Then raise the leg that is opposite the side from which

Figure 5.122 The safety belt should be hooked over the center of a ladder rung.

the work is to be done up and over two consecutive rungs. The bend of the knee should be over the second rung. Lean forward and bring this foot and leg back through the ladder just below the rung that supports the knee. This foot can be placed over the top side of the beam or rung; the other foot against the opposite beam for added stability. The completed leg lock is illustrated in Figure 5.123. An extremely long- or short-legged person may need to alter this procedure for comfort.

To secure oneself with a leg lock while facing away from the ladder, first secure the top of the ladder. Then descend to the level from which the work is to be done. While facing away from the ladder, lift one leg and insert it through the ladder, over the second rung above the one on which the other foot rests. Bring the foot back through, over the rung above the one on which the other foot rests, and hook the foot over the beam.

ASSISTING A VICTIM DOWN A LADDER

When it is known in advance the ladder will be used for a window rescue, the ladder tip is raised only to the sill. This gives the victim easier access to the ladder. All other loads and activity should be removed from the ladder, which should be securely anchored at both the top and bottom if possible.

Use three or four firefighters on a ladder rescue. For ground ladder rescues, at least two fire-

Figure 5.123 Use the leg on the side opposite the working side to form a leg lock, hooking the toes over the beam or a rung.

Figure 5.124 The rescuer should place both arms around the victim with the hands near the center of the rungs. This may not be possible on aerial ladders. In all cases, constantly reassure the victim.

fighters should be in the building and one on the ladder. If the victim is conscious, the firefighters in the building should lower the victim feet first from the building to the ladder. The rescuer on the ladder supports the victim and descends the ladder. The rescuer descends first, keeping both arms around the victim under the armpits, with hands on the rungs in front of the victim for support in case the victim slips or passes out (Figure 5.124). When descending aerial ladders it may not be possible to place the arms around the victim as the side rails may be in the way. Reassure the victim constantly while descending because the victim probably will be nervous and panicky.

An unconscious victim is held on the ladder in the same way as a conscious victim except that the body rests on the rescuer's supporting knee. Place the victim's feet outside the rails to prevent entanglement (Figure 5.125).

Another way to lower an unconscious victim involves the same hold by the rescuer, except that

Figure 5.125 Support an unconscious victim with the knee and arms. Place the victim's feet over the beams.

the victim is turned around to face the rescuer (Figure 5.126). The position lessens the chance the victim's limbs will catch between the rungs.

In another method, the victim is lifted out to the rescuer on the ladder. The victim is supported at the crotch by one of the rescuer's arms and at the chest by the other arm (Figure 5.127).

If the victim is heavy or if several rescuers must help, two ground ladders can be placed side by side. One firefighter descends first and supports the victim. A second firefighter on the other ladder descends slightly above the victim, supporting the upper torso.

Smaller sized adults and children can be brought down a ladder by cradling them across the arms (Figure 5. 128).

Figure 5.126 The unconscious victim can be brought down turned toward the rescuer. This removes the possibility of catching the victim's limbs in the rungs.

Figure 5.127 An unconscious victim can be lifted out and supported by the arms at the crotch and chest.

Figure 5.128 Small adults and children can be brought down by cradling in the arms and sliding the beam with the hands.

LADDER SAFETY

A firefighter's safety and well-being while on a ladder depend on common sense precautions. Firefighters should check important items at every opportunity. Points to insure a safe climb are:

- Check the ladder for the proper angle.

- Make sure the ladder is secure at the top or the bottom (preferably both) before climbing.

- Check the locks to be sure they are seated over the rungs (Figure 5.129).

- Check staypoles to be sure they are set properly.

- Climb smoothly and rhythmically.

- Always tie in with a leg lock or a safety belt when working from a ladder.

- Do not overload the ladder.

- Always wear protective gear, including gloves, when working with ladders.

- Choose the proper ladder for the job.

- Use the proper number of firefighters for each raise.

- Use leg muscles, not back or arm muscles, when lifting ladders.

- Make sure ladders are not raised into electrical wires.

- Inspect ladders for damage and wear after each use.

Figure 5.129 For ladder safety, always check that the ladder locks are over the rungs securely.

NFPA STANDARD 1001
FORCIBLE ENTRY
Fire Fighter I

3-2 Forcible Entry

3-2.1 The fire fighter shall identify and demonstrate the use of each type of manual forcible entry tool.

3-2.2 The fire fighter shall identify the method and procedure of properly cleaning, maintaining, and inspecting each type of forcible entry tool and equipment.*

Fire Fighter II

4-2 Forcible Entry

4-2.1 The fire fighter shall identify materials and construction features of doors, windows, roofs, floors and vertical barriers and shall define the dangers associated with each in an emergency situation.

4-2.2 The fire fighter shall identify the method and technique of forcible entry through any door, window, ceiling, roof, floor, or vertical barrier.*

*Reprinted by permission from NFPA Standard No. 1001, *Standard for Fire Fighter Professional Qualifications.* Copyright © 1981, National Fire Protection Association, Boston, MA.

IFSTA's Forcible Entry Transparencies are designed to complement this chapter.

Chapter 6
Forcible Entry

Although there are specific places through which firefighters usually force entrance into a building, a general knowledge of how buildings are constructed is also essential. It is important for firefighters to be familiar with the various trade terms in construction so that they can better judge both where and how to force entrance. Likewise, various tools and devices best suited for forcible entry must be thoroughly understood by firefighters.

BUILDING CONSTRUCTION FEATURES

Most modern buildings have continuous masonry foundations of concrete, brick, or stone upon which the building proper rests. The foundation walls which support frame construction may extend well above the ground. Exterior walls may be constructed of masonry, masonry veneer, metal, or wood frame. Masonry exterior walls are the most desirable from a fire protection point of view. They are usually 8 to 12 inches (20 to 30 cm) thick depending upon the particular material used. Masonry veneered walls are essentially frame walls in which the supporting members are wood with a veneer of one layer of brick or stone on the exterior to give the appearance of a solid brick or stone wall. The upright wood supporting members in masonry veneered walls are called "studs" which are usually 2 x 4's (5 x 10 cm) spaced at 16-, 18-, or 24-inch (41, 46, or 61 cm) intervals. The presence of these studs creates hollow spaces in the walls through which fire can spread. Fire stops should be provided in these hollow spaces. Frame walls are constructed of wood throughout, with wood or wallboard sheathing fastened to the studs. Over the sheathing is fastened the exterior siding, which may be of wood clapboards, wood "board and batten" siding, asbestos shingles, stucco, or other types of exterior finish. Frame walls also have hollow spaces in them that are created by the wood studs, which should be blocked off with fire stops.

Fire stops are usually short pieces of 2 x 4's (5 x 10 cm) that are placed in walls, partitions, ceilings, stairways, between the studs at each floor level, and at the upper end of the stud channels in the attic as shown in Figure 6.1. These wood fire stops cut off the draft within the walls and help prevent the spread of fire and smoke. When fire stops are not used or the spaces are not filled with noncombustible insulation, it is possible for a fire which originated in the basement to spread throughout the walls of the building before it is discovered. Some insulations give off poisonous gases

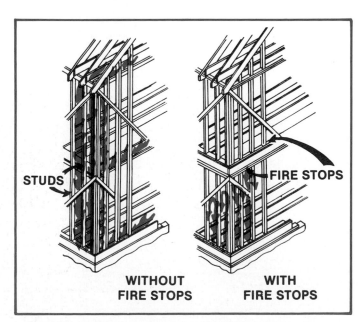

WITHOUT FIRE STOPS **WITH FIRE STOPS**

Figure 6.1 Fire stops are short pieces of wood 2" x 4" (5 x 10 cm) that cut off the draft within walls to reduce the spread of fire, heat, and smoke. Without them, even a small seemingly controlled fire might be extending unseen.

and lose their fire resistance when heated to a certain degree. Modern type construction uses top and bottom plates on the studs which will prevent fire from traveling from floor to floor. However, it can still travel the height of the partition or wall.

Dwelling roofs are usually built of wood supporting members called "rafters" to which roof sheathing is fastened. The roof covering material is then applied to the roof sheathing. Roof coverings may be wood shingles, composition shingles, composition roofing paper, tile, slate, or a built-up tar and gravel surface. The roof covering is the ex-

posed part of the roof, and its primary purpose is to afford protection against the weather. The selection of a proper roof covering is important from a fire protection standpoint because it may be subjected to sparks and blazing brands if a nearby building should burn. A more thorough development of roof construction and the types of roofs found on industrial and mercantile buildings will be studied in the section on opening roofs. Some terms which are used in the building trade and are very useful in this study are illustrated in Figure 6.2.

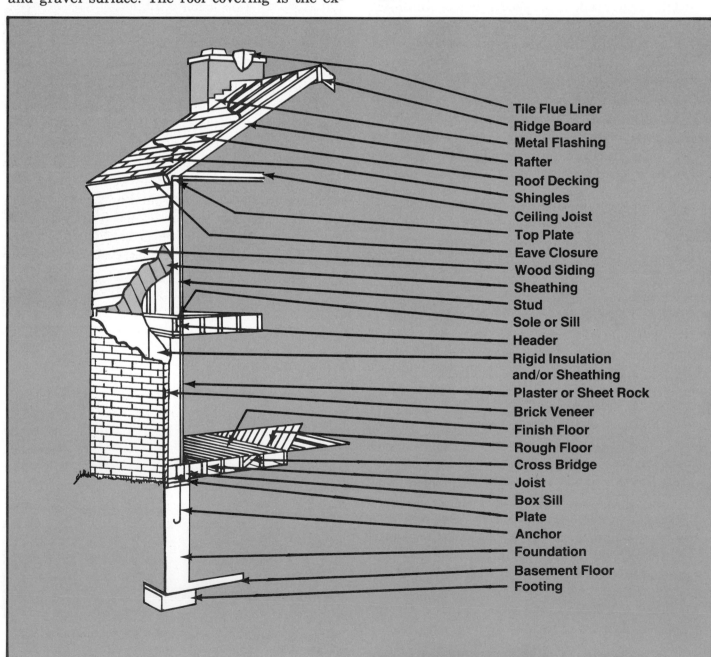

Tile Flue Liner
Ridge Board
Metal Flashing
Rafter
Roof Decking
Shingles
Ceiling Joist
Top Plate
Eave Closure
Wood Siding
Sheathing
Stud
Sole or Sill
Header
Rigid Insulation and/or Sheathing
Plaster or Sheet Rock
Brick Veneer
Finish Floor
Rough Floor
Cross Bridge
Joist
Box Sill
Plate
Anchor
Foundation
Basement Floor
Footing

Figure 6.2 Common building construction terms.

Fire Doors

Fire doors are used principally to protect openings in division walls and walls of vertical shafts, but for certain exposures they may be found on door openings. Openings for fire doors are classified as to the character and location of the wall in which they may be situated. Although classifications exist from "A" through "F," only Classes A and B will be discussed here. Class A openings are those which are located in walls separating buildings or in walls within a building that is separated into distinct fire areas. Class B openings are vertical enclosures, such as elevators, stairways, dumbwaiters, which may allow fire to spread within a building. The following types of standard fire doors are: horizontal and vertical sliding, single and double swinging, and overhead rolling. They may or may not be counterbalanced. There are two standard means by which fire doors operate: (1) Self-closing doors are those which when opened return to the closed position; and (2) Automatic closing fire doors are those which normally remain open but which close when heat actuates their closing device (Figure 6.3).

Swinging fire doors are generally used on stair enclosures and in other situations where they must be opened and closed frequently in normal service. Vertical sliding fire doors are also normally open and so arranged as to close automatically. They are employed where horizontal sliding or swinging fire doors cannot be used. Overhead rolling fire doors may be installed where space limitations prevent the installation of other types and they, like vertical sliding doors, are arranged to close automatically. They are all similar in structural design and they may be mechanically, manually, or electrically operated. The barrel, on which the door is wound, is usually turned by a set of gears that are located near the top of the door on the inside of the building. This feature makes the door exceptionally difficult to force and, therefore, entrance should be gained at some other point and the door operated from the inside. Counterbalanced fire doors are generally employed on openings to freight elevators and they are mounted on the face of the wall inside the shaft.

Fire doors, because of their protective qualities, are generally not considered as doors that

Figure 6.3 Self-closing fire doors may be single or double swinging. Automatic closing fire doors may be magnetic controlled, sliding, or the overhead rolling type.

need to be forced. Fire doors that close automatically need only to be pried open with suitable forcible entry tools. Fire doors that are used on exterior openings may be locked and, therefore, the lock must be forced. A precautionary measure that firefighters should take, when passing through an opening protected by a fire door, is to block the door open to prevent its closing and trapping them. Fire doors have also been known to close behind firefighters and cut off the water supply in a hoseline.

FORCIBLE ENTRY TOOLS

Many tools on the market have been devised to make forcible entry possible. Such tools are often designed primarily for the fire service. Many of the firefighter's best tools have come from the develop-

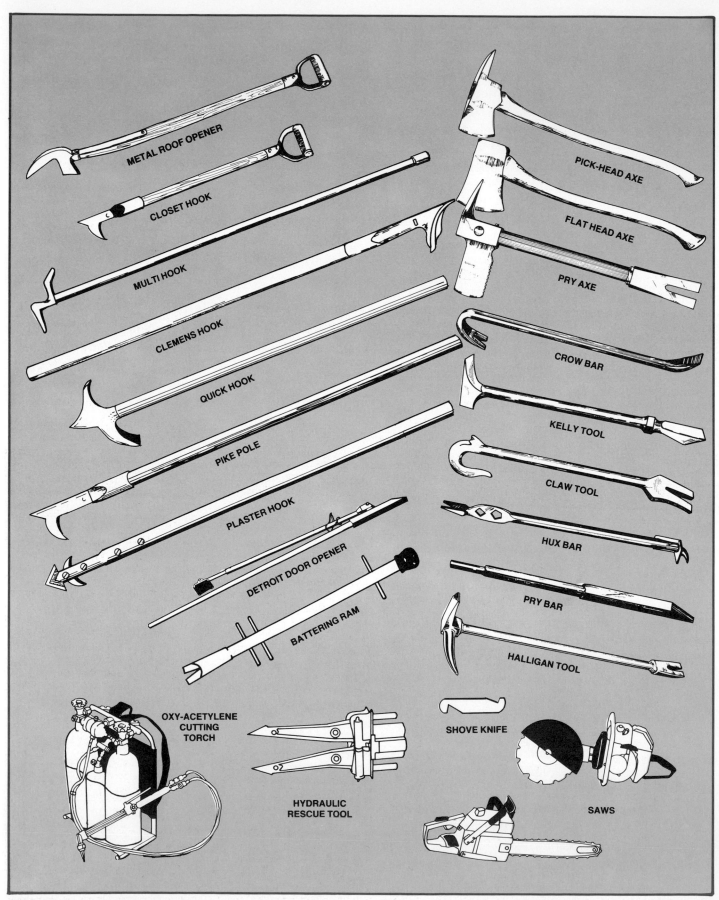

Figure 6.4 Sample forcible entry tools.

ment of and experimenting with individual ideas. Every fire department should set up its own rules concerning the care of forcible entry tools. Developing an appreciation for the tools a firefighter works with will instill a desire to keep them properly maintained.

The factor of safety, more than any other factor, should determine the method by which forcible entry tools are carried. When these tools are carried in the hands, precautions should be taken to protect the carrier, other firefighters, and bystanders. In atmospheres which could be explosive, extreme caution should be taken in the use of power and hand tools that may cause arcs or sparks. When tools are not in use, they should be kept in their properly designated place on the apparatus. Some forcible entry tools are shown in Figure 6.4.

Cutting Tools

AXES AND HATCHETS

Firefighters use the flat-head and pick-head axe for many purposes. The flat-head axe is more adaptable to striking, but the pick-head axe is more versatile for fire fighting. Firefighters sometimes carry special hatchets which are equipped with a pick head. Although they are not designed for heavy work, these hatchets are used in operations similar to those in which a pick-head axe is used.

A fire axe should be carried so that there will be little danger of an injury. An axe or any other edged or pointed tools should never be carried on the shoulder. A firefighter should carry the axe head close to the body. Three methods of carrying an axe are shown in Figure 6.5. In these carries, the point of the pick is shielded and the sharp blade protected or pointed away from the body.

If the blade is extremely sharp and the body of the blade is ground too thin, pieces of the blade may be broken out when cutting gravel roofs or striking nails and other materials in flooring. If the body is too thick, regardless of the sharpness of the blade, it is difficult to drive the axe through ordinary ob-

Figure 6.5 Three methods of safely carrying an axe.

jects. Medium thickness of the body of the axe is necessary. The body thickness should be about ¼-inch (6 mm) at ¾-inch (19 mm) from the edge, ⅜-inch (10 mm) at 1-¼ inches (32 mm), and ½-inch (12.5 mm) at 2 inches (51 mm) from the edge. The measurements are to be taken at the center of the blade. The temper should be such that the blade will not bend easily or break off when nails, gravel, or other foreign matter are struck. Care must be used when grinding to prevent heating and the consequent softening of the steel. The axe should be ground to preserve the body thickness and not merely to sharpen the blade. After sharpening the blade, the firefighter should rub it lightly over a stone to take off the keen edge and to lessen the possibility of cutting firefighters who may rub against the blade at fires. A keen edge is not necessary since it is soon lost during use. The effectiveness of an axe depends more on its proper usage than on the sharpness of the blade.

The shoulder or the handle should be rather thick to prevent breaking. The grip should be thin to provide elasticity and it should be smooth to prevent splinters. Wooden handles of axes and other similar tools should be sanded to minimize hand injuries. After usage, they should be cleaned with soapy water, rinsed, and dried. To prevent roughness and warping, a coat of boiled linseed oil is recommended for wood handles. Paint or varnish on the handles tends to blister when exposed to heat. Axes and hatchets equipped with fiber glass handles require a minimum of care. Under normal usage they will not crack or splinter. These handles will not absorb moisture, dry rot, or warp. After usage, they should be dried with a clean, soft cloth. Rust spots on the metal parts of such tools should be removed with emery paper and the spot covered with a light film of oil.

HAND AND POWER SAWS

Wood and metal hand saws are useful where entry must be made through wood or light metal. Power saws can be equipped with carbide tip blades to cut metal or concrete. Chain saws and saber saws are also useful. These saws may be run by electrical or gasoline power. Also some saws will operate from compressed gas bottles.

The ring saw is a heavy-duty versatile saw with an open center circular blade for cutting a variety of substances. The K-12 saw uses a solid circular blade of different hardnesses to cut many substances seen in forcible entry situations. Both use gasoline motors.

METAL CUTTERS AND CUTTING TORCHES

Bolt cutters are used as a forcible entry tool to cut iron bars, bolts, cables, padlock hasps, and other objects. Wire cutters are used to cut fences and wire other than electrical.

An acetylene cutting torch is classified as a burning tool and is very useful to enter an area blocked by heavy iron bars or iron plates. These cutting units are available as a back pack, and they can be carried on fire apparatus as a special forcible entry or rescue tool. **CAUTION:** Safeguards should be taken against the hazard of molten metal.

PRYING TOOLS

There are many different designs of prying tools used in the fire service. These tools are very effective in breaking locks, opening doors, forcing windows, and prying up objects. Prying tools should be used in a manner that will protect the firefighter and others from injury. Prying tools require little maintenance. The bit part of a prying tool should have a long narrow taper to permit entrance into narrow spaces, such as doors and windows.

HYDRAULIC SPREADERS

This rescue tool has proved very effective in extrication rescues, and the powerful jaws capable of exerting 10,000 psi (70,000 kPa) of pressure, are also useful in forcible entry situations. The tool is operable by one person. Its ability to exert pressure in both spreading and pulling modes makes it a versatile device.

Pushing/Pulling Tools
PIKE POLES AND PLASTER HOOKS

Pike poles are useful as striking and pulling tools. They can be used to open windows, ceilings, and partitions. The plaster hook has two knifelike wings which depress as the head is driven through an obstruction and reopen or spread outward under the pressure of self-contained springs. Pike poles and plaster hooks should be carried with the sharp ends toward the front and lowered. The wooden handles of pike poles and plaster hooks should be cared for in the same manner as previously described for axe handles.

BATTERING RAM

This forcible entry tool is used to open heavy doors and to breach walls. The battering ram may be safely carried by two persons. It is designed to be used by two or four persons. There is little maintenance required for this tool. Metal fragments may protrude from the head of the tool after it has been used on hard objects like concrete. These fragments should be filed off, the pick end examined and renewed, and the entire tool painted if necessary.

Striking Tools

HAMMER-HEAD PICK AND SLEDGE HAMMER

The hammer-head pick serves as a driving device and its long, sharp-pointed pick may be used for digging dirt, concrete, or debris. The hammer head can be used as a sledge although, like the flat-head axe, it does not have as much striking power as a sledge or mall. The sledge hammer may be effectively used to break out tile, dead lights, walls, concrete, terrazzo, or to free iron bars set in masonry.

CARE AND MAINTENANCE OF FORCIBLE ENTRY TOOLS

Proper cleaning, drying, and repairing of forcible entry tools is very important. The proper care of forcible entry tools will increase their span of service depending upon the intensity of their use. Some procedures for forcible entry tool care are listed below.

Wooden Handles

- Check for cracks in the wood.
- Check to see that the head is on tight.
- Check for splinters.
- Sand down if burned.
- Do not completely paint; it hides cracks. A

½-inch (12.5 mm) stripe painted around the handle for identification purposes is suggested.

Cutting Edges

- Check to see it is free of nicks or torn edges.
- Replace cutting edge of bolt cutters.
- Provide a dull edge instead of a sharp edge.
- File the edges. Grinding takes the temper out of the metal.

Painted and Plated Surfaces

- Keep painted.
- Inspect for damage.
- Plated surfaces should be wiped clean or washed with soap and water.

Unprotected Metal Surfaces

- Keep clean of rust.
- Keep oiled when not used.
- Do not completely paint; it hides cracks.
- Should be free of burred or sharp edges. File off when found.

Power Equipment

- Check to see if they start manually.
- Check blades and equipment.
- Check electric tool cords.
- Check electric tool cutting blades.

OPENING DOORS

Doors can also be considered as obstacles to firefighters when they endeavor to reach various areas of a building. This study of opening locked doors discusses each type of door on the basis of its construction features and the opening techniques that have been proved through practice. No attempt will be made to designate one product better than another, and the illustrations used herein have been selected for clarity of pattern rather than comparison of design.

The construction features of some doors render them practically impossible to force, and entry may be more easily made by some other means.

From a firefighter's standpoint of forcible entry, doors may be classified as swinging, revolving, sliding, and overhead doors. Regardless of the class of door, firefighters should try the door to make sure it is locked before force is used. A good rule is try before you pry. If the door is locked, examine it to determine which way it swings and which method of forcible entry will prove most effective.

Swinging Doors

LOCKS AND FASTENERS

Door locks and fasteners for swinging doors consist of a bolt or bar that protrudes from the door into a metal keeper which is mortised into a door jamb. This bolt or bar may be part of the lock assembly or it may be entirely separate, but in either case the jamb must be sprung enough to permit the bolt to pass the keeper during forcible entry. Some special installations place two bars, one at the top of the door and one at the bottom, and such door locks are exceedingly difficult to force. A record of the type of door and how it is locked can be valuable to firefighters if such information is collected during inspection surveys. Some of the more common types of locks for swinging doors are shown in Figure 6.6.

Figure 6.6 Due to the many door locks in service prior knowledge of proper entry techniques is often required.

WOOD DOOR CONSTRUCTION

Three general kinds of wood swinging doors are panel, slab, and ledge. Front doors in residences may be either panel or slab with glass panels in the upper area. Some panels of glass are held in place with moldings, which may be pried off to remove the panel. Residence doors generally open inward; in this they differ from outside doors in public buildings, where they should open outward.

Slab doors are very popular and they may be constructed as either hollow core or solid core doors. The term "hollow core" implies that the entire core of the door is hollow, without fillers, which is not correct. Instead, the core is made up of an assembly of wood strips formed into a grid or mesh. These strips are glued within the frame, forming a rigid and strong core. Over this framework and grid are glued several layers of plywood veneer paneling. The purpose of the hollow core grid is to decrease the weight and cost of the door. Most exterior slab doors that are found on newly constructed residences will be hollow core slab doors, but the exterior slab doors on older homes may be solid core.

The term "solid core" means that the entire core of the door is constructed of solid material. Some solid core slab doors have a core built up of tongue and groove blocks or boards, which are glued within the frame. Other solid core doors may be filled with a compressed mineral substance that is fire resistant. In either case, the door is solid with a plywood veneer covering, a construction which adds considerably to the weight of the door. Some exterior slab doors on industrial and mercantile buildings are solid core slab doors. Examples of hollow core and solid core construction are shown in Figure 6.7.

Ledge doors, sometimes called batten doors, are found on warehouses, storerooms, barns, and the like. They are made of built up material and are locked with either surface locks, hasps and padlocks, bolts, or bars. Hinges on ledge doors, generally of the surface type with stationary pins, are fastened to the door and facing with screws or bolts.

Door jambs are the sides of the doorway's openings. Door jambs for wood swinging doors may be

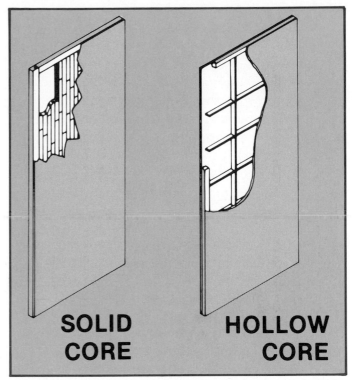

Figure 6.7 Hollow core doors are of lighter construction than a solid core door.

rabbeted jambs or stopped jambs. The rabbeted jamb is one into which a shoulder has been milled to permit the door to close against the provided shoulder. Stopped door jambs are provided with a wooden strip or door stop that is nailed inside the jamb against which the door closes. This stop may be easily removed with most prying tools. Construction of both rabbeted and stopped door jambs are shown in Figure 6.8.

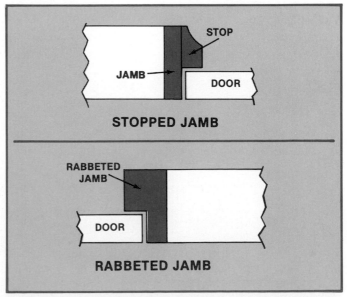

Figure 6.8 Doors with rabbeted jambs (milled) are more difficult to force open than stopped jambs (attached) that can be easily removed.

METAL DOOR CONSTRUCTION

Metal swinging doors may be classified as hollow metal, metal covered, and tubular. Metal swinging doors are generally more difficult to force because of the manner in which the door and door jamb are constructed. This difficulty is also more acute when the jamb is set in masonry. It is generally considered impractical to force metal doors. The framework of hollow metal doors is constructed entirely of metal. The jambs are hollow and are fastened to the walls by specially designed metal anchors.

Metal covered doors are constructed essentially as hollow metal doors except that there may be a wooden core or metal ribs over which the metal covering is placed. The paneling sometimes consists of metal-covered asbestos.

The structural design of tubular metal is of seamless rectangular tube sections. A groove is provided in the rectangular tube for glass or metal panels. The tube sections form a door with unbroken lines all in one piece and are sometimes found on exterior openings of the more recently constructed buildings. The tubular doors are hung with conventional hardware except that the balance principle of hanging is sometimes used. The operating hardware consists of an upper and lower arm connected by a concealed pivot. The arms and pivots are visible from the exterior side only, and from the interior side the balanced door resembles any other door.

Tubular aluminum doors with narrow stiles are also quite commonly used. The panels of these doors are generally glass but some metal panels are used. Tubular aluminum doors are comparatively light in weight, strong, and are not subject to much spring within the aluminum frame. Some construction features of the tubular aluminum door are shown in Figure 6.9.

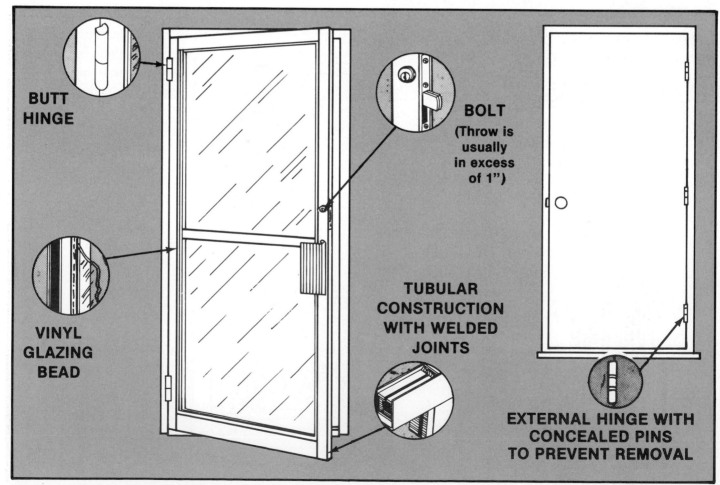

BUTT HINGE

VINYL GLAZING BEAD

BOLT (Throw is usually in excess of 1")

TUBULAR CONSTRUCTION WITH WELDED JOINTS

EXTERNAL HINGE WITH CONCEALED PINS TO PREVENT REMOVAL

Figure 6.9 Since the butt hinges are difficult to force and the bolt has a throw in excess of 1" (2.5 cm), it may be easier, faster, and cheaper to break out the glass than damage the door. The vinyl glazing may be taken out and the glass removed if time is not critical.

OPENING TECHNIQUES

The method used to force a swinging door is determined, first, by how the door is hung, and secondly, by how it is locked. Before attempting to force any door, check to see if the door is locked, and whether or not the hinge pins can be removed. Also, the conditions of the building should be observed, and hoselines should be made available for use. Firefighters should then feel the door for heat by using the back of the hand, which is more sensitive to heat. The temperature of the door will indicate whether a backdraft (explosion) is likely when the door is forced or opened.

Breaking Glass

In some cases, less damage may be done by breaking a small glass near the lock, through which the door can be opened from the inside. The act of breaking glass must be done in a certain manner to assure safety to the firefighter, because glass will shatter into fragments of keen cutting edges. Some of the principal safety features for breaking glass are:

- Stand to the windward side of the glass pane to be broken if possible.
- Strike the tool at the top of the pane.
- Keep the hands above the point of impact.

This procedure permits the broken particles of glass to fall downward away from the hands and to the side of where the firefighter stands. The glass may be broken with an axe or other tools as shown in Figure 6.10. **CAUTION:** never break glass with the hands. Full protective clothing should always be worn when breaking glass. After the glass is broken, all jagged pieces should be removed from the sash. Glass removal may be done with the same tool that was used to break the glass. Removing all pieces of glass will avoid cutting anyone who goes through or reaches through and will prevent injury to hose, ropes, or other material that may be passed through the opening.

Figure 6.10 Follow all safety precautions including the removing of all pieces of glass from the sash when breaking windows.

Special precautions should be taken when breaking windows above the ground floor to prevent a "flying-guillotine" hazard to civilians and firefighters below. Winds may cause heavy shards of glass to travel great distances. This flying-guillotine hazard must not be disregarded and close coordination between ground forces is required to prevent serious injury to persons below.

Breaking the Lock

A lock may sometimes be removed or destroyed with less damage and expense than might occur if the door is forced. A lock puller or K-Tool (Figure 6.11) may be used to pull or force a cylinder lock. These devices require little force and accomplish a fast, neat job of forcible entry. It is used in conjunction with a flat-head axe or another suitable driving tool. A "key tool," which has a straight and bent end, is used to move the bolt or plunger from the keeper.

One way to break a lock is to physically pull the cylinder out of the door. The new K-Tool invented by a firefighter, is especially useful in pulling all types of lock cylinders, either rim, mortise or tubular. Used with a Halligan or other prying tool, the K-Tool is forced behind the ring and face of the cylinder until the wedging blades take a bite into the cylinder. Light blows of a hammer help this forcing. The K-Tool is positioned correctly when the blades are engaged into the body of the cylinder, thus eliminating the likelihood of breaking off the ring and face of a cylinder made of white metal. The front metal loop of the tool acts as a fulcrum for leverage and holds the fork end of the prying tool.

When a cylinder is found close to the threshold or jamb, the narrow blade side of the tool will usually still fit behind the ring. A close-clearance situation will often be found on a glass sliding door, but only a half-inch clearance is needed.

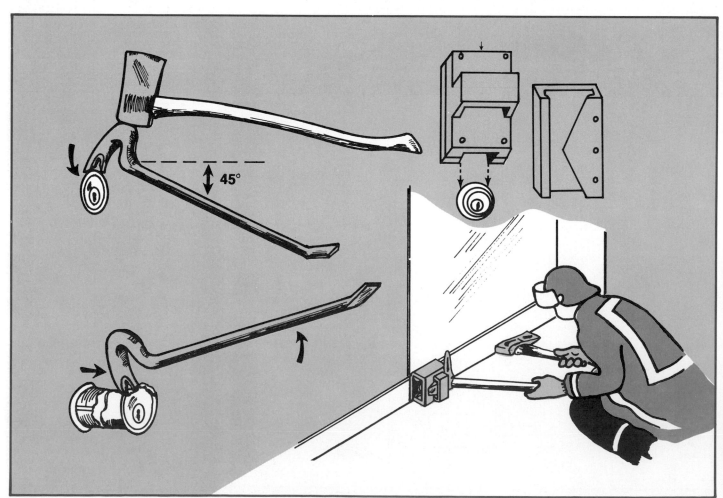

Figure 6.11 A lock puller or K-Tool can be used for removing lock cylinders.

When the cylinder is removed, use a key tool in the hole to move the locking bolt to the open position. Key tool operations are discussed later in this section.

The lock puller is a different tool which accomplishes the same job as the K-Tool. Its jaws are also forced around and behind the protruding rim of a cylinder. Then the curved head and long handle are used to provide the leverage for pulling the cylinder. On the other end of the tool, a chisel head is used when necessary to gouge out the wood around the cylinder for a better bite of the working head. The lock puller is slightly more damaging to the door than a K-Tool but it will rapidly pull the cylinder, and sometimes the fire situation will necessitate quick entry at whatever the damage to a door. Insert either the straight or bent end of the key tool into the hub of the lock and turn (Figure 6.12). If for any reason this fails, insert the straight end of the lock puller through the hole, and drive the lock off the door with an axe.

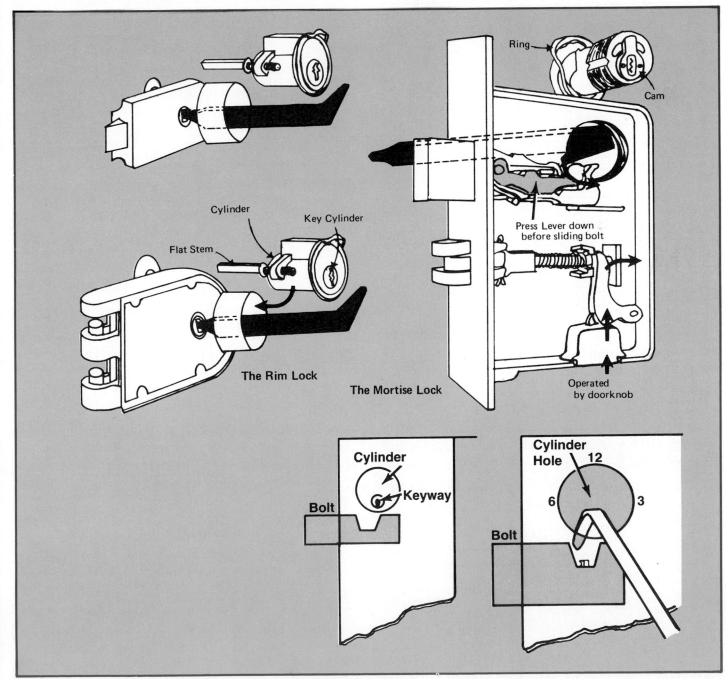

Figure 6.12 With the cylinder removed the key tool is used to trip the bolt mechanism.

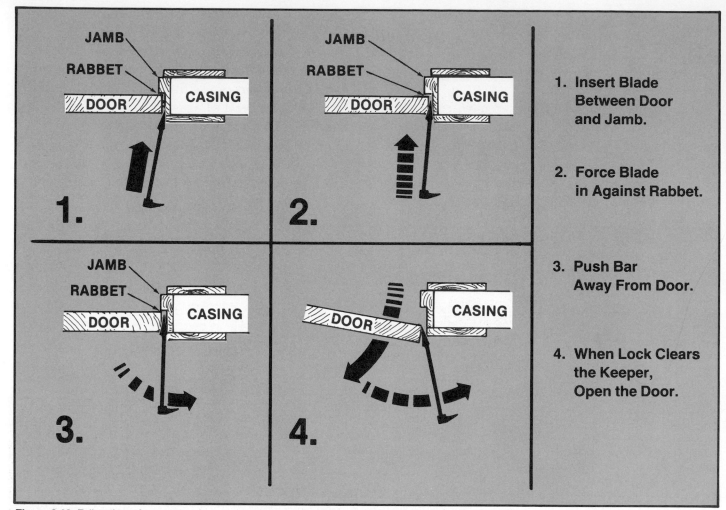

Figure 6.13 Follow these four steps to force a door that opens toward you.

Forcing Doors that Open Toward the Firefighter

The door hinges and jamb may be checked to determine the direction the door swings. If the door opens toward the firefighter, it must be forced in a certain manner. Several forcible entry tools may be used for this operation. The techniques of their use are similar and the steps are as follows and shown in Figure 6.13.

Step 1: Insert the blade of the tool between the door and jamb near the lock.

Step 2: Force the blade in and against the rabbet or stop by working and pushing on the tool (the tool may be hammered with another object).

Step 3: Pry on the tool bar away from the door to move the door and jamb apart.

Step 4: Pull the door open or pry open with another tool when the lock has cleared the keeper.

The use of different tools for this operation is shown in Figure 6.14. These methods may work very well on ordinary and inexpensive doors, but variations may need to be made to force more sturdy doors. Two tools may be used together to open a door. If two tools are used, insert both tools between the door and jamb, one just above the lock and the other just below the lock. By alternately prying with one and catching bite with the other, more force can be applied.

Forcing Doors that Open Away from the Firefighter

Swinging doors that open away from an operator present greater difficulties. If the door is in a stopped frame, the blade of the tool may be inserted between the stop and the jamb, the stop lifted, and the tool inserted between the door and jamb near the lock. By separating the door away from the jamb, the operator may spring it sufficiently to permit the bolt to pass the keeper.

Figure 6.14 Different tools may be used to force open or remove a door. Removing the hinges and then the door cause the least damage.

Under certain conditions, it may be better to remove the stop completely. The steps in forcing a door that opens away from an operator when stops are used on the jamb are as follows and shown in Figure 6.15. (As shown in the illustrations, the stop has not been completely removed from the jamb.)

Step 1: Bump the cutting edge of the tool against the stop to break the paint or varnish so the blade can be inserted.

Step 2: Loosen the stop at the lock or remove the stop completely.

Step 3: Start the blade between the door and the jamb.

Step 4: Make the initial pry only after the blade is halfway in, to permit the blade to be worked and pushed.

Step 5: With a full bite behind the door, pry the door away from the jamb until the bolt passes the keeper.

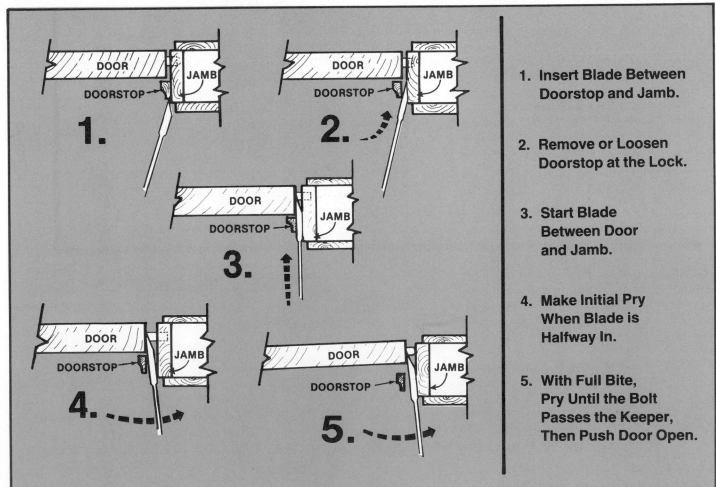

Figure 6.15 Follow these five steps to force a stopped jamb door that opens away from you.

If the door opened away from the operator is in a rabbeted jamb, it can more easily be forced by using two tools. Prying against the door with one tool should open a crack between the door and the rabbet, into which the blade of the second tool can be inserted. After the blade of the second tool has been forced well into this opening between the door and jamb, the door may be pried sufficiently to permit the bolt to pass the keeper. Even with two tools, forcing this type of door construction may be quite difficult. The steps in working the blades of the tools into the crack between the door and the rabbeted jamb are as follows and illustrated in Figure 6.16.

Step 1: Lay the blade of the tool flat against the door and insert the blade between the rabbet and the door.

Step 2: Make short pries with the first tool to spread the jamb.

Step 3: Work the blade of the second tool between the door and the jamb hammering the blade well into the opening.

Step 4: With a full bite behind the door, pry the jamb away from the door until the bolt passes the keeper.

Double Swinging Doors

Double swinging doors may be forced with most pry tools by prying the two doors sufficiently apart at the lock to permit the lock bolt to pass the keeper. Sometimes a wood molding is fastened to one or two wooden doors where they come together at the center. The purpose of this molding is to cover the crack between the doors when they are closed. In this event, this molding must be removed before the blade of the tool can be inserted. Swinging double doors are sometimes secured with a bar on the inside wall. If the opening between the doors

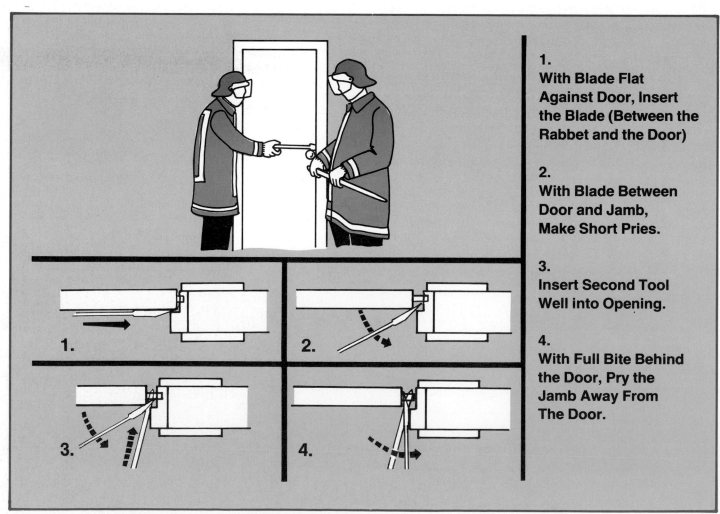

1.
With Blade Flat Against Door, Insert the Blade (Between the Rabbet and the Door)

2.
With Blade Between Door and Jamb, Make Short Pries.

3.
Insert Second Tool Well into Opening.

4.
With Full Bite Behind the Door, Pry the Jamb Away From The Door.

Figure 6.16 Follow these four steps to force a rabbeted jamb door that opens away from you using two tools.

is sufficient to permit the insertion of a flat tool or object, a bar can sometimes be lifted or bumped from the stirrups. When this process is not possible, doors secured with a bar may need to be battered in if forcing is necessary. It may be possible, however, to cut a hole in a door panel or break glass and remove the bar.

Tempered Plate Glass Doors

Recent trends in building construction and the modernizing of older structures have increased the use of tempered plate glass doors. Such installations are frequently encountered in fire fighting operations. The breakage characteristics of tempered plate glass are quite different from those of ordinary plate glass. This difference is due to the heat treatment that is given to the glass during tempering. The results of heat tempering plate glass produce high tension stresses in the center of the glass and high compression stresses in the exterior surfaces. These tension and compression stresses balance each other and may be visualized when the glass is subjected to polarized light. Under certain conditions, sunlight may produce polarized light and the characteristic multicolored lines or bands may be seen.

The heat treatment given to tempered plate glass increases its strength and flexibility. Its resistance to shock, pressure, impacts, and temperatures is also increased. Tempered plate glass is said to be approximately four times as strong after tempering as before. It is several times more resistant to impact and also will withstand without breaking, a temperature of 650°F (343°C) on one side while the other side is exposed to ordinary atmospheric temperature. When broken, the sheet of glass suddenly disintegrates in relatively small pieces. The glass should be shattered about 14 inches (36 cm) or knee high from the bottom and then cleaned out of the frame. To break the glass, use the pick head of a pick-head axe.

Although tempered plate glass doors may be locked at the center, top, or bottom of the door, its resistance to shock and its rigid characteristic make it almost impossible to spring with forcible entry tools. Tests that have been conducted warrant the basic conclusion that firefighters should use every other available means of forcible entry

before deciding to gain entrance through an opening that is blocked by a tempered plate glass door. Tempered plate glass door panels are considerably more expensive than any other glass-paneled doors of similar size. Each door is, in a sense, custombuilt and the cost of installation varies. The time necessary to prepare a replacement and install it may be considerably longer than for other type doors.

Plate glass doors frequently have narrower tempered plate glass wing panels installed in the doorway opening. These panels should be regarded with the same precaution as the door. Whenever it may be necessary to break tempered plate glass door panels, such breakage can be effected more easily by the pick point of a standard fire axe. A firefighter should wear a suitable faceshield to protect against eye injury, or turn away from the door as the glass is being broken. Some departments place a shield made of a salvage cover and two pike poles as close to the glass as possible, and the blow is struck through the cover.

Fragmentation of the tempered plate glass is in small granules with relatively blunt points and dull edges, while plate glass breaks into much larger sharp and pointed pieces. Plate glass fragments have sufficient weight and force to cause serious cuts or stabbing injuries. It is generally agreed that there is considerably less hazard involved in breaking a closed and locked tempered plate glass door panel than there is in breaking ordinary plate glass panels of the same size.

Nondestructive Rapid Entry Method

Doors in most occupancies are locked after normal business hours. While this security precaution is necessary, it often causes complications for emergency services. Even when the front door of a building can be opened with a minimum amount of damage or time delay, firefighters may find numerous interior doors locked. Lack of immediate access into a building will prevent firefighters from reaching the seat of the fire. This may lead to an increase in fire spread and the resulting property damage from extended fire fighting activities.

The problem of gaining rapid entry without destruction has confronted fire departments for as long as locks have existed. In trying to find a solution, many departments have attempted to keep an inventory of keys to all the buildings in their area. While this procedure does reduce damage from forcible entry, it also presents a problem of maintaining an inventory of keys and gaining quick access to the right key at the right time. Fortunately, the problems presented by locked doors can be eliminated through the use of a rapid entry key box system (Figure 6.17). This system provides the security that the building owner needs and is available at no cost to the department because the building owner purchases the equipment. Rapid entry key box systems are easily installed on any building. All necessary keys to the building, storage areas, gates and elevators are kept in a key box mounted at a high visibility location on the building's exterior. Only the fire department carries the master key that will open all the boxes in their jurisdiction.

Figure 6.17 Rapid entry can be gained without destruction by using a security box to hold needed keys. Only the fire department maintains a master key. *Courtesy of the Knox Company.*

Proper mounting is the responsibility of the property owner. The fire department should indicate the desired location for mounting, inspect the completed installation, put the building keys inside, then lock the box with the department's master key. Unauthorized duplication of the master key is prevented because key blanks are *not* available to locksmiths and cannot be duplicated with conventional equipment. They are provided by the factory when the system is first put into use and are controlled strictly through an authorized signature release method. This approach provides a high degree of security and eliminates the need to carry individual building keys on the apparatus, and perform destructive forcible entry.

Revolving Doors and Forcing Techniques

Revolving doors consist of quadrants that revolve around a center shaft. The revolving wings turn within a metal or glass housing which is open on each side and through which pedestrians may travel as the door is turned. The mechanism of the revolving door is usually collapsible and panic-proof, and each of the four revolving wings is held in position when the hangers are collapsed. Some revolving doors will collapse automatically when forces are exerted in opposite directions on any two wings. All revolving doors do not collapse in the same manner, and it is a good policy to collect such information when fire department inspection surveys are made. Three methods of collapsing revolving doors are as follows.

PANIC-PROOF TYPE

The panic-proof collapsible mechanism has a ¼-inch (6 mm) cable holding the wings apart. To collapse the mechanism, push or press the doors or wings in opposite directions.

DROP-ARM TYPE

The drop-arm mechanism has a solid arm passing through one of the doors and a pawl will be found on the door through which the arm passes. To collapse the mechanism, press this pawl to disengage it from the arm and then push the wing to one side.

METAL-BRACED TYPE

The metal-braced mechanism is held in position by arms that resemble a gate hook with an eye. To collapse the mechanism, it is only necessary to lift a hook and fasten it back against the fixed door or wing. The hooks are on both sides of these doors. The pivots are, in most cases, cast iron and can be broken by forcing the door with a bar at the pivots.

Revolving doors may be locked in various ways and, in general, they are considered difficult to force when they are locked. Swinging doors are usually found on either side of a revolving door and large transom lights may be over the doors.

Sliding Doors and Opening Techniques

Sliding doors are generally considered to be those which can travel either to the right or left of their opening and in the same plane. Sliding doors are usually supported upon a metal track and their side movement is made easier by smaller rollers or guide wheels. The ordinary sliding door travels into a partition or wall when it is pushed open. This type of installation is more common for interior openings. These doors may be forced similarly to the swinging door except that they must be pried straight backward from the lock. From a forcible entry point of view, the sliding patio door presents the greatest problem. These units consist of heavy duty, full panel glass that is set into a metal or wood frame. These glass panels are sometimes doubled, "thermopane," or tempered which adds to their value. Patio sliding doors usually slide past stationary glass panels instead of disappearing into a wall. Breaking these glass panels to gain entrance is not recommended. Sliding patio doors can, however, be forced open by inserting a wedge tool between the jamb and door near the lock and prying the door away from the frame. Patio sliding doors may sometimes be barred or blocked by a metal rod or a special device. This feature can easily be seen from the outside and it practically eliminates any possibility of forcing.

Overhead Doors and Opening Techniques

Doors that open by moving upward may be classified into upward action sectional, rolling steel, and slab doors (Figure 6.18). Each one will be considered separately for clarity of forcible entry practices. Overhead doors may be constructed of metal, fiber glass, or wood framework with wood or glass panels. From a forcible entry point of view, the sectional door does not present a serious problem unless it is motor driven or remote controlled.

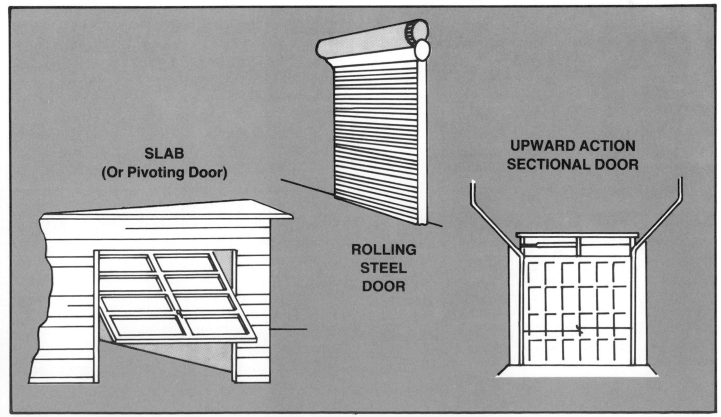

SLAB
(Or Pivoting Door)

ROLLING STEEL DOOR

UPWARD ACTION SECTIONAL DOOR

Figure 6.18 Overhead doors constructed of various materials roll, fold, or raise as a single slab panel.

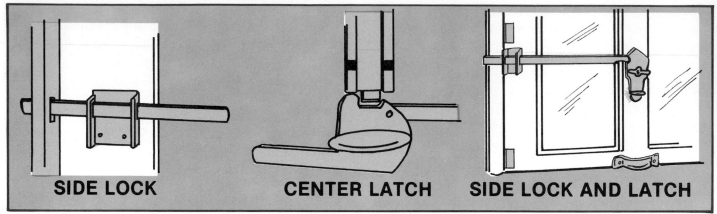

SIDE LOCK   **CENTER LATCH**   **SIDE LOCK AND LATCH**

Figure 6.19 Overhead door locks and latches may be in different locations.

The latch is usually in the center of the door and it controls the locks which are on each side of the door. The lock and latch may also be located on only one side. These latches and locks are illustrated in Figure 6.19. Overhead doors may be forced by prying upward at the bottom of the door with a good prying tool, but less damage will be done and time will be saved if a panel is knocked out and the latch is turned from the inside (Figure 6.20). Some overhead doors may be locked with a padlock through a hole at either end of the bar, or the padlock may even be in the track. These systems of locking may make it necessary to batter the door down with a battering ram or other ramming device if access cannot be gained through a panel to remove the padlock.

Pivoting or Overhead slab doors are locked similarly to the previously described overhead folding door. Slab doors may be either metal, fiber glass, or wood, and unless glass windows are present, it is practically impossible to reach the latch on the inside. Sometimes it is possible to pry outward with a bar at each side near the bottom. This action will tend to bend the lock bar enough to pass the keeper. **CAUTION:** all overhead doors

Figure 6.20 Quick entry is possible by knocking out a panel and turning the latch.

should be blocked open (up position) to prevent injury to firefighters should the control device fail.

FENCES

Fences of wood, metal, masonry, and woven wire often pose an entrance problem. The gates on these fences are usually locked with padlock and hasps. These locks may be pulled apart by a claw tool or cut by some available cutting tool. The staple of the lock can sometimes be pried or twisted off by using a pry tool or the point of the claw tool. Another quick method, which applies to inexpensive locks, is to brace the bow of the lock against the hasp or staple and then strike the bottom of the lock with the back of the axe. The lock will usually either spring open or the bow will snap. Where chains and locks are used for gates, the best method of entry would be cutting the chains.

A wooden fence may require that several boards be pried off or cut for safe entrance to the area. Wovern wire fences may sometimes be cut, but woven wire of the cyclone fence type should generally be crossed by some provision as shown in Figure 6.21. Wire fences should be cut near posts to provide adequate space fore fire apparatus and to lessen the danager of injury from the whip coil of loosened wires.

It is extremely difficult, costly, and time-consuming to breach masonry fences, and it should be done only as a last resort. If breaching a masonry fence is required, it should be done as described in a later item on Breaching Masonry Walls. **CAUTION:** The firefighter should be alert to animals confined inside enclosures.

OPENING WINDOWS

There are many different types and designs of windows through which firefighters must force entrance to perform their duties of rescue and fire extinguishment. Each type presents a different technique if effective forcible entry is to be accomplished through them. It is often easier to force a window than to force a door, and entrance through a window may permit a door to be opened from the inside. The types of windows which will be studied

Figure 6.21 Wire fences may be cut, or the fence crossed with ladders.

here are double hung or checkrail, hinged or casement, factory or projected, and awning or jalousie windows.

Checkrail Window and Opening Techniques

Checkrail or double hung windows may be made of either wood or metal, but the construction design is quite similar. They usually consist of two sashes that meet in the center of the window, known as the upper and lower sashes. These two sashes may be locked together by a latch or bolt on the inside (Figure 6.22). Wood checkrail windows

CHECKRAIL (DOUBLE HUNG) WINDOW

Figure 6.22 Wood double hung windows are easy to pry since there are only two small screws to force.

Figure 6.23 Break the lower glass pane, unlock the latch, and operate the mechanism to open the window.

are not difficult to pry open if the latch is on the checkrail, for the screw of the lock will pull out and the sashes will separate. Practically any prying tool may be used, such as an axe or spanner wrench. The pry should be made at the center of the lower sash if the sashes are locked at the center of the checkrail.

Forcing sash bolt locks in metal windows presents a different problem. The latch or lock is not likely to give under pry, and excessive damage may be done and more time expended than if the glass were broken near the lock and the window unlocked from the inside. Wire glass often requires more force to break, and the use of the blade or pick of the axe instead of the flat part may sometimes prove advantageous.

Casement Windows and Opening Techniques

Casement or hinged windows are usually made of metal, but wood casement windows are used. They consist of one or two sashes which are hinged on the side and they swing outward from the opening. If screens are employed, they are located on the inside opposite the direction which the windows swing. Various kinds of casement windows, locks, and operating devices are available. Figure 6.23 further points out that casement win-

dows are not only latched, but that the operating mechanism must be reached to open the window. To reach the latch involves cutting the screen and it is quite obvious that the screen must be removed if entrance is to be made at this point. Because of these conditions, the most practical way to force entrance through casement windows is as follows:

- Break the lowest pane of glass and clean out sharp edges.
- Force or cut the screen in the same area.
- Reach in and upward to unlock the latch and then operate the cranks and levers at the bottom.
- Remove the screen completely and enter.

Projected Windows and Opening Techniques

Projected or factory windows are ordinarily made of metal and they may project in or out from an opening (Figure 6.24). They may be pivoted in the center or they may pivot at the top or bottom. "Projected-Out" factory windows swing outward at the bottom and slide down from the top in a groove which is provided for that purpose. "Projected-In" factory windows swing inward at the top and they are usually hinged at the bottom. Pivoted projected windows are usually operated by a push bar that is

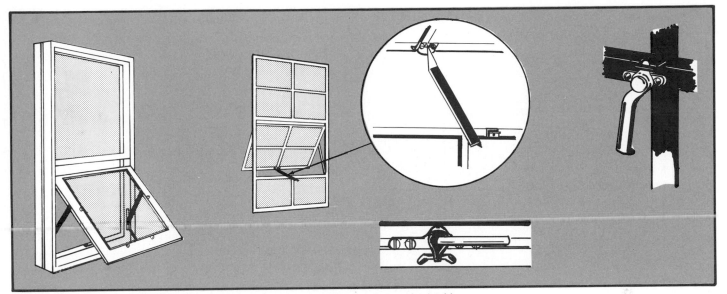

Figure 6.24 Projected or factory windows will generally require breaking the glass and unlatching.

notched to hold the window in place. Screens are seldom used with this type of window, but when present, they are on the side opposite the direction of the projection. The most practical method of forcing factory-type windows is the same as has been previously described for casement windows, except that the crank-operated utility windows are often of the projected window type and they are locked similarly to all other projected windows. Entrance may be made by applying similar techniques.

Awning or Jalousie Window and Opening Techniques

Although awning and jalousie windows are often considered to be the same type, there are two main differences that should be considered in a study of forcible entry. Both types are sometimes referred to as louver windows, because of their methods of operation (Figure 6.25).

Awning windows consist of large sections of glass about one foot (10 cm) wide and as long as the window width. Jalousie windows consist of small sections about four inches wide and as long as the window width.

Awning window sections are constructed with a metal or wood frame around the glass panels, which are usually double strength glass. Jalousie window sections are usually without frames, and the glass is heavy plate that has been ground to overlap when closed.

The glass sections of both awning and jalousie windows are supported on each end by a metal operating mechanism. This mechanical device may be exposed or concealed along the sides of the window, and each glass panel opens the same distance outward when the crank is turned. The operating crank and gear housing are located at the bottom of the window. Awning or jalousie windows are the most difficult of all types to force. Even with the louvers open, it is obvious that there is not enough room between the louvers to permit a

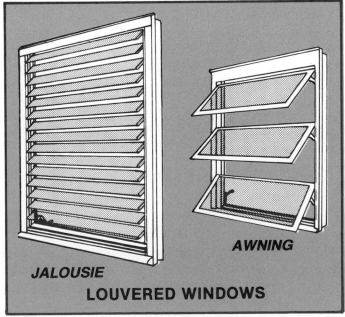

Figure 6.25 Louvered windows of the awning or jalousie type are opened by breaking or removing a panel and cranking open. They can be used during fire fighting only for ventilation purposes.

person to enter. Entrance through these windows requires several panels to be broken out. Because of the cost of jalousie windows, these openings should be avoided.

Lexan Windows and Opening Techniques

A current trend in the construction industry is the use of Plexiglas and other thermoplastics in place of glass windows. "Lexan" is an example of one such polycarbonate which has seen wide application as a glass substitute because of its ability to withstand abuse from vandalism or weather. Lexan is 250 times stronger than safety glass, 30 times stronger than acrylics, and is classified as self-extinguishing. It is 50 percent lighter than glass and 43 percent lighter than aluminum. Lexan is available in thicknesses ranging from ⅛ to ½-inch (3 mm to 13 mm). Various tests have been conducted by fire departments across the country to gain information on the best way of forcing entry into Lexan windows. The tests have utilized standard forcible entry equipment including pick head axes, oxyacetylene cutting torches, reciprocating type saws, and circular saws. These tests along with actual field experiences with Lexan during fire situations indicate that a circular saw with a carbide tipped blade is most effective when entry must be made through Lexan. Caution must be taken in blade selection, a blade with teeth too small will melt the Lexan and cause the blade to bind; conversely, a blade which is too coarse will cause the blade to dangerously slide over the cutting surface. A medium tooth blade (approximately 40 teeth) has been found to yield the best results.

Screened or Barred Windows and Other Openings

Heavy wire mesh guards over windows and doors present a serious problem to forcible entry, ventilation, and fire fighting practices. These mesh guards may be permanently installed, hinged at the top or side or fitted into brackets and locked securely. In either case, forcing wire mesh guards involves considerable time and should be avoided.

To free bars in masonry, a firefighter should strike the bar with a sledge about ten inches (25 cm) above the sill. As the bar bends, the end will sometimes pull free of the sill. Another method is to strike the sill with a sledge opposite the end of the bar. A blow at this point will sometimes crack masonry sufficiently to release the end of the bar. Still another method is to start a hammer-head pick in the masonry sill at the edge of the bar. Strike the head of the pick with a sledge to crack the masonry sufficiently to release the end of the bar.

Iron gratings may be found in sidewalks above basement windows, in floors, or in walls. They may be merely held in position by the friction of the grating against the sill, they may be pivoted at the rear, or they may be set in masonry and locked in position with hasp and padlock. They can sometimes be opened with the pick end of the axe by forcing it between the sill and the grating and then prying up. A tow chain connected to a motor vehicle or a winch line may be used to remove iron bars and wire mesh guards from windows. The oxyacetylene cutting torch and power saws are recommended forcible entry tools for steel construction and they will work when other tools fail. Steel doors, gratings, bars, and other steel obstructions may be cut with a torch.

OPENING FLOORS

There are almost as many kinds of floors as there are buildings. The type of floor construction is, however, limited to the two basics — wood and concrete. Either of the two may be finished with a variety of floor covering materials. With the advent of vast housing developments, a departure from the conventional subfloor and wood joist construction over foundation and grade beam footings is presently common practice for single and multiple family dwellings. Concrete slab floors over tested and rustproof plumbing are quite common. Generally speaking, the floors of upper stories of family dwellings are still wood joist with subfloor and finish construction. It is not uncommon for a floor to be classified according to its covering instead of the material from which it is constructed. The feasibility of opening a floor during a fire fighting operation obviously depends upon how and from what material it is constructed. A wood floor does not in itself assure that it can be pene-

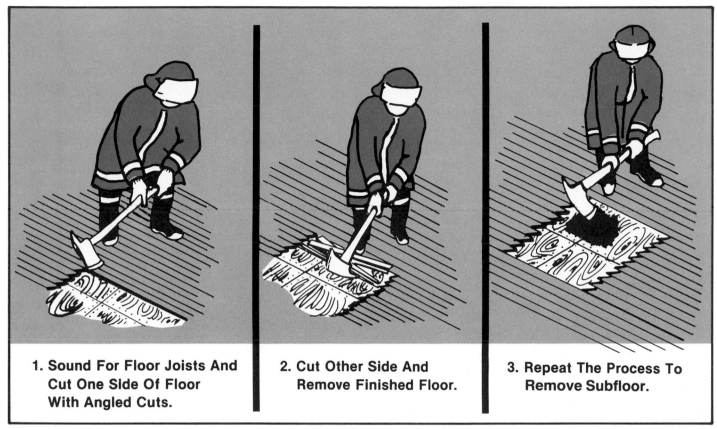

1. **Sound For Floor Joists And Cut One Side Of Floor With Angled Cuts.**

2. **Cut Other Side And Remove Finished Floor.**

3. **Repeat The Process To Remove Subfloor.**

Figure 6.26 Follow these three steps to open a wood floor.

trated easily. Many wooden floors are laid over a concrete slab. The type of floor construction can be determined by pre-fire inspection surveys of business and industrial structures, but similar information for residential structures is not easily obtained. Some accepted and recommended techniques for opening wood and concrete floors are offered. The various techniques will be discussed separately because each type of floor presents a different situation.

Wood Floors and Opening Techniques

The wooden joists of wood floor construction are usually spaced a maximum of 16 inches (40 cm) apart. A subfloor, consisting of either 1-inch (25 mm) boards or 4-foot by 8-foot (1 m by 2 m) plywood, is first laid over the joists. The finish flooring, which may be linoleum, tile, hardwood flooring, or carpeting, is laid last. The subfloor is ordinarily laid diagonally to the joists and the finished floor at right angles to the joists.

Wood floors may be opened as illustrated in Figure 6.26.

Step 1: Determine the location for the hole, sound for floor joists, and cut one side of the finished floor by using angle cuts.

Step 2: Cut the other side of the finished floor in a like manner and remove the flooring or floor covering with the pick of the axe.

Step 3: Cut the subfloor using the same technique and angle cuts. It is usually advisable to cut all sides of the subfloor before removing the boards. If just a few boards are removed before the others are cut, the heat and smoke conditions may prohibit completion of the job.

Neat cuts in wood floors can be made with power saws in a similar manner. A metal cutting blade can be provided for the circular saw, or either the saber or chain saw may be used. It is best to supply power to such saws from a portable generator, which is carried on the fire apparatus, rather than depend upon domestic power during a fire. Floors which are covered with tile, linoleum, or other such materials, should first have these materials removed before the floor is cut. Carpets and rugs should also be removed or rolled to one side before a floor is cut.

Concrete Floors and Opening Techniques

The general construction of reinforced concrete floors makes them extremely difficult to force, and opening them should be bypassed if possible. If concrete floors must be opened, the most feasible means is to use a compressed air jackhammer. Unless a jackhammer is readily available, this process is extremely slow and may not prove beneficial for fire extinguishment, but it might be the best means for rescue operations. Concrete cutting blades are available for most portable power saws. There are also special-purpose nozzles which are designed to penetrate masonry and some concrete. Although these devices are primarily nozzles, they also qualify as forcible entry tools. They are sometimes called puncture or penetrating nozzles, because of their ability to be driven into hard objects. The point of the tool is hardened steel and a place is provided, called the head, where the tool can be hammered with a mall or sledge hammer. The barrel of the tool is hollow and small holes bored through the barrel permit water to be expelled back to the point. Provision is also made for a 1½-inch (38 mm) hoseline to be attached to the nozzle which supplies the water under pressure. It is best to first strike the masonry or concrete with a sledge hammer to shatter the concrete topping and provide a center for the tool. When wood or other materials are used as a finish over concrete, it often gives the appearance of a floor other than concrete.

OPENING WALLS

Opening Masonry and Veneered Walls

The opening of masonry walls is often referred to as "breaching." One appliance which may be used is known as a battering ram. The battering ram is made of iron with handles and hand guards. One end is jagged for breaking brick and stone and the other end is rounded and smooth for battering walls and doors. The ram requires two to four firefighters for use.

Power tools prove to be the best method for breaching masonry and concrete walls. They are faster and usually require only one person to operate. Continue using the power tool until a hole of desired size is formed. Be sure that charged lines are in position before breaching a wall at a fire.

Opening Metal Walls

Metal and prefabricated metal walls are becoming more and more popular for exterior wall construction. There are several reasons why its use has recently become extensive, but one which affects forcible entry is the opportunity to add insulation between the exterior and interior surfaces. Construction of this type can be found in storage buildings, service stations, store fronts, and in other commercial structures. The metal for these walls is usually in the form of sheets, sections, or panels. These metal sheets are fastened to wood or metal studs by bolts, screws, rivets, or by welding. The metal may have a painted or a porcelain coated surface, and a damaged panel may be difficult to replace. Entrance through a door or window is usually preferred to opening a metal wall and its breaching is usually considered as a "last resort." If opening a metal wall cannot be avoided, a metal cutting power saw is normally the best way to open the wall. The panels should be examined to determine if there are studs or other supports or if the walls support the entire structure. It is extremely important to be sure that the material to be cut will not weaken the structure or that electrical wiring or plumbing is not cut. This danger can be anticipated and handled if a well-organized inspection program is maintained. A neat opening will be easier to repair and, when possible, the entire sheet or panel should be removed. The metal should be cut along the studding to provide stability for the saw and ease of repair.

After the metal is cut, it should be removed and placed in a location where it will not endanger the firefighter. Insulation or other material may be in the wall and it can usually be cut with an axe for removal. In some cases, an axe or can opener type of forcible entry tool can be used to cut this thinner metal. **CAUTION:** Proper precaution should be taken when opening metal walls covered with porcelain, because there may be flying chips of porcelain.

Opening Wood Frame Walls

Wood frame walls are constructed with wood or fiberboard sheathing nailed over the studs. The exterior siding, which may be wood clapboard, board and batten siding, asbestos shingles, stucco, or other exterior finish, is fastened over the sheathing. When opening a wood frame wall it is extremely important to watch for electrical wiring and pipes. The need for opening a wooden wall is often to gain access to the area involved by fire between studs. The procedure for opening a wood frame wall is the same as for roofs and floors except that the opening will be at a vertical instead of horizontal plane. Remove the siding, sound the wall for stud supports, and cut along a stud.

Opening Partitions

For the purpose of this study, three general types of partition construction (hollow clay tile, covered wood or metal studding, and solid concrete blocks as shown in Figure 6.27) will be considered. Solid masonry partitions should be opened in the same manner as has been described for the exterior masonry wall, and the same precautions should be taken. Unless previous inspections have been con-

ducted on the building involved, it is very difficult to determine the construction of a partition during fire fighting procedures. If an opening needs to be made in a partition, the following procedures are offered:

Step 1: Select the location of the opening and, before attempting to open the partition, check for electric wall plugs and switches.

Step 2: Have sufficient tools, such as power tools, picks, fire axes, sledges, and pry bars, available.

Step 3: If the studding in the partition is wood or metal, locate the studs by sounding.

Step 4: Cut along the studs with the fire axe.

Step 5: If the partition is hollow clay tile or concrete block, crush one or two blocks with a sledge or a jackhammer, and the other blocks or tile will remove more easily.

Partitions should not be opened unless there is an indication of fire in the wall. Three ways to determine if the partitions contain fire are feeling on wall for hot spots, looking for discolored wallpaper or blistered paint, and listening for the sound of

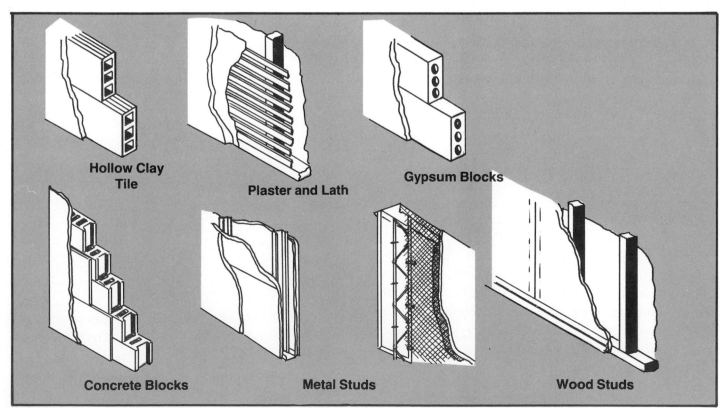

Figure 6.27 General types of partition construction. Masonry partitions should be opened in the same way as exterior walls.

burning. The partition should be opened on the side where opening will cause the least damage. If possible, place a salvage cover on the floor before opening the partition. An opening in a partition should be made only in the area where fire is indicated, and the wall covering should be cut along the studs and removed instead of ripping it off in a haphazard manner (Figure 6.28).

Figure 6.28 Lath and plaster can be raked from partitions with a pick or hook, while dry wall can be easily cut along the studs.

SPECIAL FORCIBLE ENTRY OPERATIONS

Forcible entry operations include more than opening doors, windows, floors, walls, and property barriers, for there are basements and subbasements, vault-type doors, and government lock-outs which require special operations. Basements and subbasements often have outside entrances. These openings may have stairways, elevators, or chutes leading to the ground level with iron gratings, iron shutters, or wooden doors that secure the entranceway. Vault-type doors may be found on fallout shelters, and on fur and bank vaults. An inspection and preplan of operations is suggested for such locations.

SAFETY PRECAUTIONS

Some of the many safety precautions and procedures that apply to all forcible entry operations are listed as follows:

- Always try a door or window before you pry.
- Carry tools with the safety of yourself and others in mind.
- Watch for explosive atmospheres.
- When it is possible, block a door or window open after entrance to make a safe exitway.
- Do not place tools where they will create a tripping hazard.
- Stand to the side when breaking glass and remove all jagged pieces.
- Block all overhead doors open (up position) after entrance.
- Watch for overhead obstructions and bystanders when swinging an axe and keep area clear whenever possible.
- When opening walls and ceilings, watch for electrical wiring and pipes.
- One large opening is usually better than several small openings.

NFPA STANDARD 1001
RESCUE
Fire Fighter I

3-12 Rescue
3-12.1 The fire fighter shall demonstrate the removal of injured persons from the immediate hazard by the use of carries, drags, and stretchers.

3-12.2 The fire fighter shall demonstrate searching for victims in burning, smoke-filled buildings, or other hostile environments.

3-12.3 The fire fighter shall define the uses of a life belt.*

Fire Fighter II

4-12 Rescue
4-12.1 The fire fighter shall demonstrate the techniques of removing debris, rubble and other materials found at a cave-in.

4-12.2 The fire fighter shall demonstrate the use of the following rescue tools:
- (a) Shoring blocks
- (b) Trench jacks
- (c) Block and tackle
- (d) Hydraulic jacks
- (e) Screw jacks

4-12.3 The fire fighter shall demonstrate the techniques of preparing a victim for emergency transportation by using standard available equipment, or by improvising a method.

4-12.4 The fire fighter shall identify dangers of search and rescue missions in tunnels, caves, construction sites, and other hazardous areas.

4-12.5 The fire fighter, operating as a member of a team, shall demonstrate the extrication of a victim from a vehicle accident.

4-12.6 The fire fighter shall tie a knot and lower a person from a third floor level.*

*Reprinted by permission from NFPA Standard No. 1001, *Standard for Fire Fighter Professional Qualifications*. Copyright © 1981, National Fire Protection Association, Boston, MA.

IFSTA's Rescue Transparencies are designed to complement this chapter.

Chapter 7
Rescue

Because the dangers of extreme weather conditions alone can be so widespread, with respect to tornadoes and floods for example, no community can be considered safe from incidents which may cause the need for rescue operations. Add to that the carelessness of people with fire or automobiles, and rescue situations increase. An organized local emergency plan should be made in advance that will utilize all possible local facilities and mutual aid in the event of a disaster. Planning should include the wide range of tools and methods needed by the fire department to rescue a victim who may be underground, underwater, in a wrecked automobile, or in a burning building.

Personal protection for firefighters during a rescue operation is absolutely necessary. A firefighter needs protection against the weather, excessive heat, and fire gases, and should not take unnecessary chances during rescue operations. Standard fire fighting clothing, such as helmet, coat, pants, gloves, and rubber boots will provide fair body protection, and boots should be equipped with steel insoles to protect the feet from puncture wounds.

The lungs and respiratory tract are probably more vulnerable to injury than any other body area. For this reason, special attention should be given to protective breathing equipment. Face and eye shields are sometimes necessary during some rescue operations, for these areas are also quite susceptible to injury.

Size up is an estimation or an evaluation of a condition from which an opinion or judgment can be formed. Sizing up a rescue situation is a continuous process of evaluating existing conditions. It is more than just a single step in an operation, because size up is never completed until the situation is under control. When making a size up, consider the following:

- Learn the facts of the situation.
- Understand the probabilities.
- Know your own situation (manpower, equipment).
- Determine plan of action.

RESCUE FROM BURNING BUILDINGS

When firefighters enter a burning building to perform rescue work, they must first consider their own protection. In order to protect the body surface from heat and flame, proper clothing should be worn. Standard fire fighting clothing, as mentioned, is usually adequate. The use of self-contained protective breathing equipment should be required. Rope guide lines tied to a rescuer's body are considered to be good practice when a rescue needs to be performed in the dark or under extremely hazardous situations.

A forcible entry tool should be carried by the rescuer to aid entry and most important to allow egress if a means of retreat is cut off. In addition, a helmet or belt light should be carried.

Portable radios are valuable in rescue work. Fire streams are often used in connection with protection for firefighters and victims. Victims may be trapped in a burning building and their normal means of exit cut off by fire. In order to reach such victims, fire department ladders may be needed.

Fires occuring in occupied public gathering places may present additional problems. If the normal exits are blocked, occupants must leave through exits with which they are not familiar. The very fact that fire exists in a building usually unnerves a majority of the people to an extent that panic may result and complications develop. A public gathering place should be emptied in as orderly a manner as possible. Many modern buildings are constructed with fire resistant enclosed stairways that are separated from floors by fire resistant doors. Buildings that are not constructed in this manner should be provided with fire escapes and multiple exits. Institutions such as hospitals, rest homes, sanitariums, and asylums present a

condition in which the occupants may be unable to walk. Those who perform rescue work must be prepared to move the occupants to safety without further aggravating their condition.

The most important and hazardous duty of a firefighter is the rescue of occupants from buildings involved in fire. Certain search and rescue procedures must be adhered to in order to successfully and safely find and remove fire victims. The most essential element in rescue work is time. The following procedure will help assure a swift and safe rescue operation once the firefighter enters the involved structure (Figure 7.1).

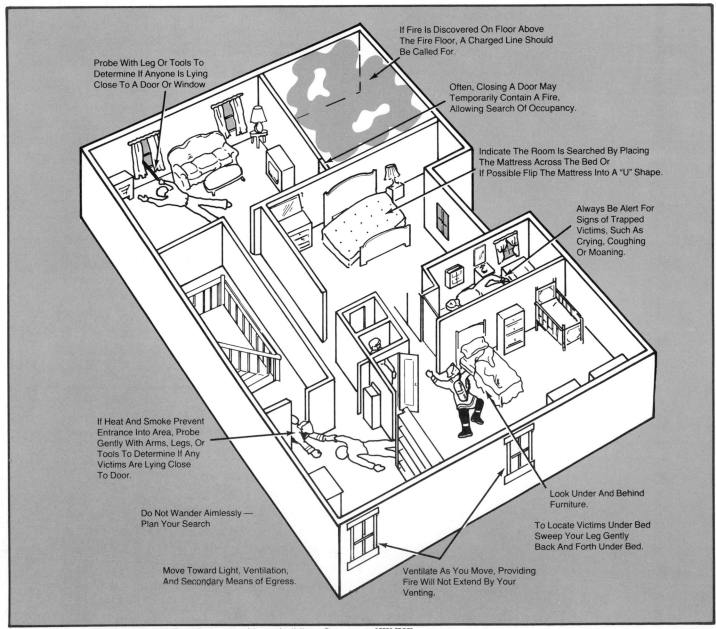

Figure 7.1 Following these hints will assist in searching a building. *Courtesy of WNYF.*

- Always wear protective breathing apparatus when performing search and rescue in fire buildings.

- If at all possible work in pairs.

- By observing exterior of building before entering, locate more than one means of egress.

- Be sure others are fighting the fire before entering building.

- Once you enter the building, visibility will be poor at best. If you cannot see your feet, do not remain standing. Search on hands and knees (Figure 7.2).

- Completely search one room before moving to the next room.

- Start your search on an outside wall. This will allow you to ventilate by opening windows as soon as possible. Ventilate only if

Figure 7.2 Search on hands and knees when visibility is obscured.

ventilation will not cause the spread of fire (Figure 7.3).

- Move all furniture, searching behind and under all furniture.

- Search all closets and cupboards including shower stalls.

- Occasionally, pause during search and listen for cries for help or other such signs or signals.

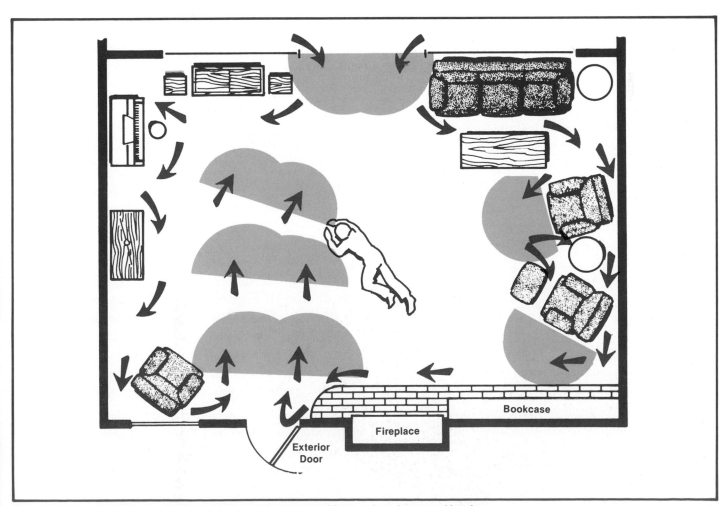

Figure 7.3 A proper team search will provide thorough coverage of the room's perimeter and interior.

- Move up and down stairs on your hands and knees, keeping head up whether ascending or descending.

- After searching a room, leave a sign or signs indicating that room has been searched. Chairs turned over, latch straps, mattress folded sideways on bed, or closet doors open. But close entry door to room to prevent the spread of fire.

- Always look for extension of fire and report any extension to commanding officer.

- If rooms or buildings are too hot to enter, reach in doorway or window with handle of a tool. Many victims will be found just inside the door or windows.

- Once you have successfully rescued a victim, place them in someone's custody so they will not try to reenter the building for any reason.

Should you become trapped in a hallway or stairs try to retreat down. Conditions will be worse on upper floors. If unable to go down and out, go to a room off the hallway. Close the door behind you. Open a window and straddle the windowsill. Call for help. Drop available articles out window to attract attention. If a window cannot be found, consciously stop and relax while considering actions for escape. Shallow breathing will extend the remaining air in the breathing apparatus.

When the tank is completely empty, disconnect the mask supply hose from the regulator and place the hose inside clothing for some filtering effect. However, this last-resort action will not improve the dangerous effects of toxic gases or insufficient oxygen. Air is usually 21% oxygen, but if the oxygen is consumed in a fire to reduce the content below 10%, collapse is probable. Increased demand from the exertions of fire fighting may cause the symptoms of disorientation and collapse even when oxygen is present at higher concentrations. Furthermore, many toxic gases present from the combustion of ordinary household materials are dangerous in concentrations of less than 1%.

RESCUE FROM COLLAPSED BUILDINGS

The difficulty that may be encountered in reaching a victim in a collapsed building depends upon conditions that are found. Immediate rescue of surface and lightly-trapped victims should be accomplished first. This immediate rescue will take care of injured persons who are not trapped and those persons who are trapped to the extent that their rescue can be quickly accomplished. Rescue of a heavily-trapped victim is a more complicated endeavor and requires more time. This type of rescue depends upon the services of trained rescue workers who have a knowledge of building construction and collapse and who are proficient in the use of rescue tools and equipment.

When floor supports fail in any type of building, the floors and roof may drop in large sections and form voids. If these sections remain in one piece, support on one side, but collapse on the other, they form a lean-to collapse as illustrated in Figure 7.4. Weakening or destruction of bearing walls may cause the floors or the roof to collapse.

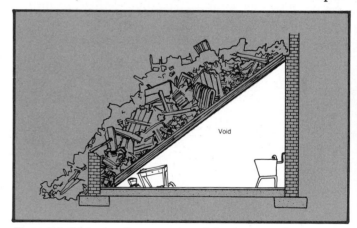

Figure 7.4 A lean-to collapse may trap victims in a precarious position.

This collapse may cause the debris to fall as far as the lower floor or basement. A collapse of this nature is referred to as a "pancake" collapse and is illustrated in Figure 7.5. Voids may be formed between the floors. When heavy loads, such as furniture and equipment, are concentrated near the center of a floor, the excess weight may cause the floor to give way. This form of collapse is known as a "V-type" collapse and is illustrated in Figure 7.6.

In order to reach a victim that is buried or trapped beneath debris, it is sometimes necessary to dig a tunnel as a means of escape. Tunneling is a process of digging one's way through debris during rescue operations. Tunneling is a slow, dangerous process and should be undertaken only after all

Figure 7.5 A pancake collapse may cause voids between fallen floors.

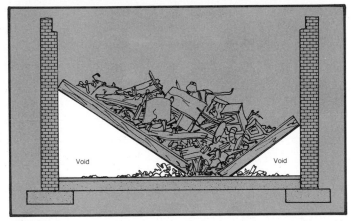

Figure 7.6 Heavy loads may cause the floor to give way creating a V-type collapse.

other methods are found impractical. Tunneling should be carried out from the lowest possible level and it should not be used for general search. Tunneling may, however, be used to reach a void under a floor where further search is to be continued. A tunnel must be of sufficient size to permit rescuers to bring victims out and it should not be constructed with abrupt turns. Tunnels as small as 30 inches (76 cm) wide and 3 feet high (1 m) have proved satisfactory for most rescue work. Tunnels should be driven along a wall whenever it is possible. This procedure simplifies the framing that is required to prevent cave-in.

Shoring is a process of erecting a series of timbers or jacks to strengthen a wall or to prevent further collapse of a building or earth opening, not to restore walls or floors to their original position. Any attempt to force beams, sections of floors, or walls back into place may cause further collapse and damage. Temporary shoring is usually all that is done by rescue squads. Since bracing, shoring, or

supporting walls and floors are basically engineering problems, fire officers should call upon the sources available for assistance when planning and conducting these operations. Shoring is difficult work and requires training and practice to be proficient. Cribbing is a process of arranging planks into a crate-like construction and this arrangement usually has separated joints. These planks may sometimes be crossed with the joints staggered. Cribbing, as used in rescue work, is usually adapted to roof and ceiling supports but it can be used on walls when necessary. Tunnel openings and passages are made secure by using the shoring and cribbing process.

RESCUE FROM CAVE-INS

A growing number of serious injuries and fatalities at construction sites are resulting from excavation and trench cave-ins. Although a buried victim must immediately receive air, continued rescue operations depend on making a site as safe as possible by shoring or cribbing to hold back other weakened earth formations. An air hose or a partly opened cylinder can be inserted in a hole dug to the victim's face. Further digging must also remove material around the victim's chest to allow for chest expansion in breathing. In an emergency, air may be directed to a victim through a garden hose. Meanwhile rescue apparatus, heavy equipment, and spectators should be moved back to avoid causing other slides or cave-ins.

While the remaining trench walls are held back, the debris must be removed from around the victim. Heavy equipment should not be used until the exact location and number of victims is known. Likewise, picks should not be used in a search for victims, or they should be used with great care. Remove debris with baskets, buckets, and wheelbarrows to a clear area. Take time for this so the material is not close enough to fall back into the cave-in site. Loose debris is unstable.

When removing debris, watch for key timbers or rocks which hold up other heavy portions of earth or other debris. Improper moving of these key pieces could cause a dangerous collapse or slide. If the key pieces must be left in place and debris tunneling used, this is one of the hardest jobs in rescue work and should be undertaken only when other means of gaining access are impractical.

During rescue operations, shoring blocks and cribbing may be used to brace the unstable walls. Spacing between the walls can be maintained by a variety of tools carried on the apparatus or available at many construction sites. Sheets of wood or metal slid into a trench to be used with jacks can help spread the force of the jack over a wider area of the trench wall.

Using Jacks for Rescue

Screw Jack. Similar to some automobile jacks, screw jacks are operated by a lever which rotates a screw to raise the load. The screw jack's main safety feature is that danger of slip back is eliminated. Many pumpers routinely carry screw jacks.

Trench Jack. A trench jack is a screw jack used between two uprights to hold trench walls apart. The screw with 2 handles (between two end plates) can be turned to adjust the opening of the jack for the width of the trench. Trench jacks can be used high and low to hold the walls and allow movement of personnel in the trench (Figure 7.7).

Hydraulic Jack. This jack is operated by means of the resistance offered when a liquid, usually oil, is forced through a small opening to move a plunger and raise a load. It requires a minimum of handle pressure to lift heavy weights, but a load should not rest for long on a hydraulic jack because of the danger of leakage of the hydraulic fluid.

JACK SAFETY MEASURES

- Work from a good foundation.
- Pack as you jack, using crosstie crib.
- Never leave a jack under load.
- Raise or lower in unison slowly when using several jacks on one load.
- Avoid overextending jack.
- Remove jack handle when not in use.

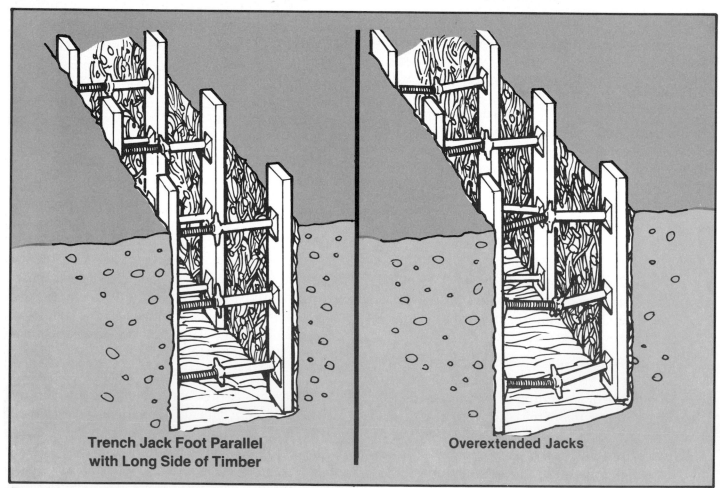

Trench Jack Foot Parallel with Long Side of Timber

Overextended Jacks

Figure 7.7 A trench jack spreads to brace trench walls against collapse. They must not be overextended.

Block and Tackle

Another lifting device used to raise or move heavy weights is the block and tackle. The block is a grooved pulley or pulleys with a hook for attachment. The tackle is an assemblage of rope and other pulleys arranged with the block to give a mechanical advantage depending upon the number of pulleys. Various riggings may be used to place the block and tackle over a cave-in site or over any object to be lifted (Figure 7.8).

Safety Precautions

There are several safety precautions firefighters and officers must remember when they are involved in cave-ins and excavation rescues:

- Fire department personnel should not enter the trench without helmets and turnout coats. Although it may seem awkward to work in a confined area while wearing these items, they may save someone from injury or suffocation if another cave-in should occur.

- Ladders should be placed in the trench on both sides of the cave-in area for a "quick exit." Ladders should extend at least three feet (1 m) above the top of the trench and be secured in place if possible.

- Firefighters should be careful with the tools they are using in the trench to avoid injuring either one another or the victim.

- Unnecessary fire department personnel should be kept out of the trench and away from its edge.

- Bystanders should be kept away from the trench. If anyone refuses to move, summon the police.

- Mechanical digging equipment that is used at the scene *must* have a guide assigned to it. This person will act as the operator's "eyes" in the trench to prevent further injuries.

- Rescuers should be aware of any other hazards that might exist at the scene, such as underground electrical wiring, water lines, explosives, or toxic or flammable gases.

- Civilian workers at the scene may be ready and willing to help, but they should be allowed to do so *only* on a volunteer basis and *only* with the permission of their supervisor. In no situation should civilians be ordered into a hazardous area to help the firefighters and rescuers.

(Courtesy of Fire Command Magazine).

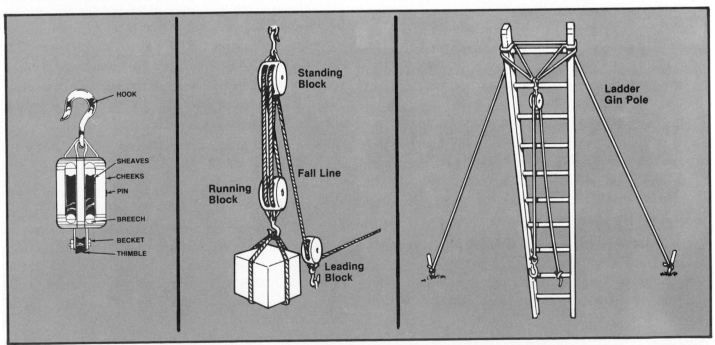

Figure 7.8 A block and tackle increases lifting power depending upon the number of sheaves. Numerous arrangements are used to suspend the block and tackle over an object.

RESCUE FROM HAZARDOUS ATMOSPHERES

Reaching a victim in a gaseous area presents a danger to respiratory organs and a danger of explosion or fire. Proper protective breathing equipment should always be worn in gaseous areas. Gases that are likely to cause a need for rescue work are (1) industrial gases which include refrigerants, fumigants, anesthetics, fuel gases, and processing gases; (2) fire gases, including smoke, carbon dioxide, carbon monoxide, and others; and (3) gases from decomposition which include ammonia, carbon dioxide, carbon monoxide, hydrogen, hydrogen sulfide, nitrogen, and sulfur dioxide.

There are two basic rules for firefighters who must work in or around hazardous atmospheres:

- Get as much information about the situation as possible, and react accordingly.

- When sufficient information has not been obtained, assume the situation to be the worst possible.

Victims trapped in a hazardous atmosphere are no better off if firefighters rush hastily to their aid and are overcome by the same hazard. The would-be rescuers must take time to size up the hazard and then take the necessary precautions before entering the area. Entry should not be attempted if the victim is obviously dead unless no risk to the firefighter is involved.

RESCUE FROM ELECTRICAL CONTACT

In order to reach a victim in contact with electrical energy, it is necessary for firefighters to contend with energized (live) wires or equipment. Firefighters should recognize the fact that an operation of this nature is extremely dangerous and that every precaution must be taken to protect themselves and others. All electrical wires and conductors should be considered as energized.

If a wire must be cut, there is always the possibility of its curling and rolling along the ground. Always secure a wire in place before cutting it so that the loose ends will be kept under control. One suggested method is to throw two objects across the wire to pin the wire to the ground until power company representatives arrive. Cut the wire with a suitable approved hot line wire cutter. Hot line cutters are not bolt cutters with insulation on their handles, but are a type of cutter that is mounted on insulated sticks and all have their limitations for safety. The wire must be cut on both sides of a victim and it must then be separated from the ends which are secured by the two objects.

If the wire is not entangled with the body of a victim or if it is not in a position that will endanger the rescuer, the quickest method is to use what is known as a "hot stick." A hot stick should be coated with a nonconductive material to prevent the handle from becoming a conductor. The protective coating should be unmarred and in good condition. When using this equipment, hook it into the victim's clothing and drag the victim clear of the wire. Do not, under any circumstances, permit yourself or your clothing to touch the wire or the victim until the victim is clear. It may sometimes be more practical to remove the wire from a victim. If such is the condition, hook the hot stick over the live wire and drag it toward you by backing away. Pulling the wire toward you rather than pushing it away permits you to back away from the live wire and thereby causes less danger to yourself in case the wire should unhook from the "hot stick." During the backing process move in a direction so that you will not be in its path should the charged wire come loose and coil toward the power line pole.

If a rope can be safely placed or thrown underneath an energized wire, this end can then be pulled well to one side. An object tied to the opposite end of the rope permits this end to be thrown over the wire. A rescuer can then secure both ends at a safe distance, and pull the wire from a victim (Figure 7.9).

TRIAGE

The word *triage* dates back to the French language of the 14th century. "Sorting according to quality" was its original meaning; however, its meaning to the firefighter and for first aid practices is that of initial examination and selection of patients and the determination of how each will be handled in order of life-saving emergencies. Obviously, a person with a severed leg will need more immediate attention than a person with a broken arm.

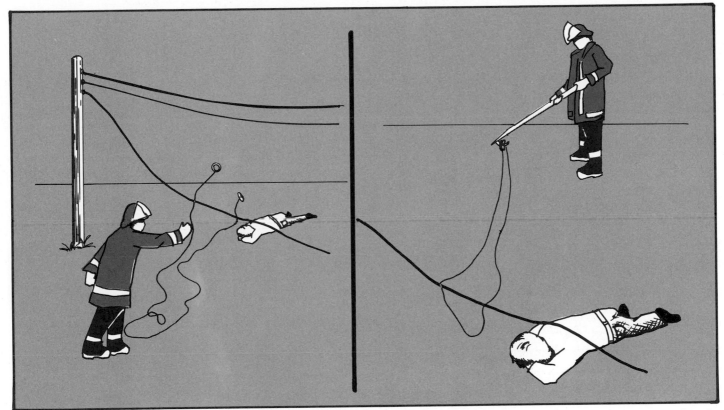

Figure 7.9 Throw one end of the rope under the wire and the other end over the wire. Safely move to the opposite side and secure both ends to a pike pole. Carefully pull the wire away.

The process of triage will be most effective at multiple-victim accidents. Large disasters, such as tornadoes or earthquakes, will make triage a necessity. Caring for and transporting those persons most seriously injured will be the responsibility of the firefighter when at the emergency scene. To make the proper decisions as to priority of injuries requires training, a good knowledge of first aid, and the ability to think rationally and logically during emergency situations. Fire departments should preplan their role in emergency situations with the local civil defense or other disaster-planning agency to utilize fully the personnel available.

Considerations to establish priorities of victim care might include the following:

HIGH PRIORITY INJURIES

- Breathing - stopped or labored
- Bleeding - severe or uncontrolled
- Suspected poisoning
- Head injuries - severe
- Cardiac arrest or heart attack
- Shock - severe
- Abdominal or open chest wounds

SECOND PRIORITY INJURIES

- Burns
- Multiple fractures - open or closed
- Spinal injuries

LOWER PRIORITY INJURIES

- Minor fractures
- Minor cuts and bruises
- Injuries that are so severe that death appears certain or has already occurred.

The process of triage will, of course, be dictated by each particular situation but should commence as soon as the firefighter arrives at the scene. Ambulatory victims not needing immediate attention should be taken from the immediate scene as soon as possible to prevent congestion of the area.

CARRIES AND DRAGS
Lone Rescuer Lift and Carry

Sometimes a lone rescuer will have to carry a victim but will have difficulty getting the victim up. The following method, injuries permitting, is often successful in such a case.

Step 1: Push the victim's feet close to the buttocks and hold them in place with a foot (Figure 7.10).

Step 2: Grasp the victim's hands and rock the victim up and down several times, the distance of the rocking becoming greater each time, for momentum (Figure 7.11).

Step 3: When ready, at the top of the upswing the rescuer jerks the victim up and onto a shoulder (Figure 7.12).

Figure 7.11 Brace the victim's legs with one foot and rock the victim up and down several times for momentum.

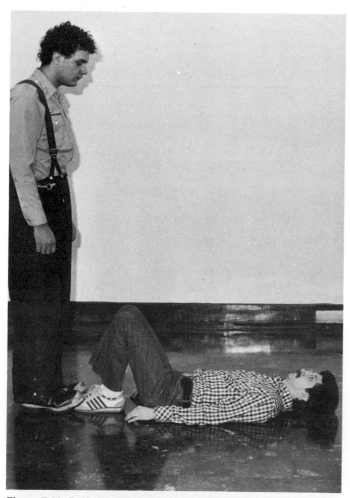

Figure 7.10 Push the victim's feet close to the buttocks.

Figure 7.12 At the height of an upswing, the rescuer jerks the victim up onto the shoulder for carrying.

Extremities Carry

The extremities carry is a two-person carry that is easy to do with conscious and unconscious victims. The steps are listed below.

Step 1: One rescuer stands at the head of the victim, and the second rescuer stands at the feet.

Step 2: The rescuer at the head kneels and slips the arms under the victim's arms and around the chest, grasping the victim's wrists (Figure 7.13).

Step 3: The rescuer at the feet kneels with feet together between the victim's legs. This rescuer grasps the victim under or just above the knees (Figure 7.14).

Step 4: The two rescuers then stand and carry the victim to a place of safety (Figure 7.15). (Remember to use your leg muscles when lifting.)

Figure 7.14 The second rescuer kneels between the victim's legs and grasps victim under the knees.

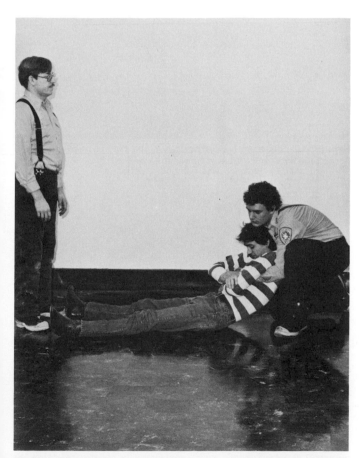

Figure 7.13 To begin the extremities carry, the first rescuer brings the victim to a sitting position and grasps the victim's wrists from behind.

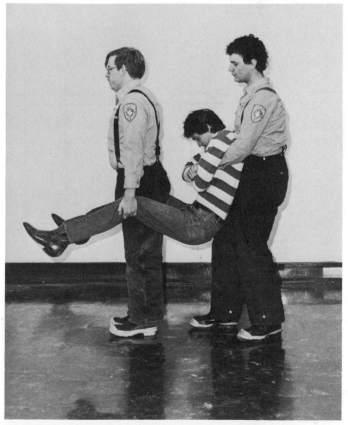

Figure 7.15 The victim is then lifted and carried to safety.

Bunker Coat or Blanket Drag

This drag can be executed by placing a blanket, bunker coat, or similar object under the victim. Follow the steps below.

Step 1: Place a bunker coat or blanket beside the face-up victim (Figure 7.16) and gather one edge close to the victim's side.

Step 2: Roll the victim toward you and, while supporting the victim, gather the coat or blanket underneath (Figure 7.17). Roll the victim onto the coat or blanket and straighten it out (Figure 7.18).

Step 3: Grasp the coat or blanket on each side of the victim's head and raise enough to clear head and shoulders off the floor. In this manner drag the victim to safety (Figures 7.19 and 7.20).

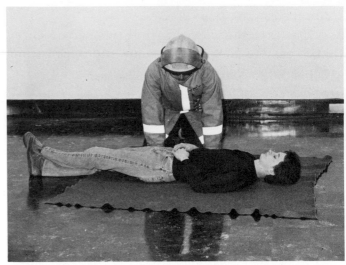

Figure 7.18 Roll the victim back onto the blanket or coat and straighten it out.

Figure 7.16 Position blanket or bunker coat next to victim.

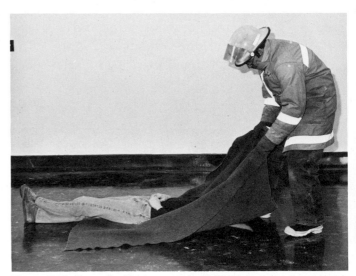

Figure 7.19 Grasp the blanket or coat, lift the victim's head and shoulders off the floor, and then pull the victim to safety.

Figure 7.17 Roll the victim toward the rescuer and gather the blanket or coat under the victim.

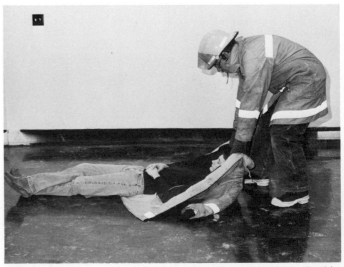

Figure 7.20 When using a bunker coat, grasp the collar yoke or shoulder area.

Placing a Victim on a Stretcher

The positions for placing a victim on a stretcher are the same as preparing to lift for a three-person carry. For placing a victim on a stretcher use the following steps.

Step 1: Each person kneels on the knee nearest the victim's feet, one person at the head, one at the hips, and one at the lower legs and another on the opposite side facing these three.

Step 2: Each person carefully works their hands under the victim until the weight is resting on the hands. The person at the feet can place their hands under the victim so that the weight is at the forearms. The person at the head has one hand under the neck and head and the other under the shoulders. The middle person has one hand under the small of the back and the other under the hips. The person at the foot has one hand under the knees and the other under the ankles. The person on the opposite side assists at the point of greatest weight or at the point of injury (Figure 7.21).

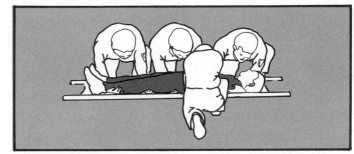

Figure 7.23 Reverse the lifting procedure and lower the victim.

Step 3: All persons slowly raise the victim, in unison, high enough to slip a stretcher in place, the fourth person then places the stretcher (Figure 7.22).

Step 4: Gently lower the victim to the stretcher by reversing the lifting procedure (Figure 7.23).

Improvised Litters

Stretchers or litters can be improvised by using materials which are available. A good litter may be made from two coats or bunker coats and two pike poles. The coats can easily be slipped onto the poles by turning them inside out over the head of a person while they hold the two poles in their hands as illustrated in Figure 7.24. The flaps

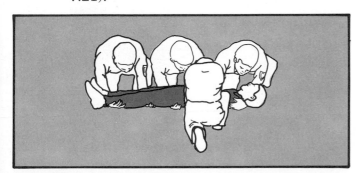

Figure 7.21 Carefully work both hands under the victim until the weight is resting on the palms.

Figure 7.22 Raise the victim slowly and rest on the knees. The fourth person then places the stretcher under the victim.

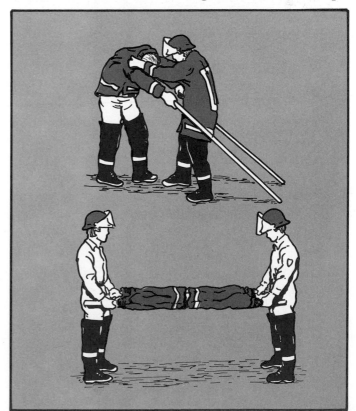

Figure 7.24 Improvising a stretcher with coats and pike poles.

should then be turned down around the poles and buttoned or snapped underneath.

Another satisfactory improvised litter can be made from a blanket and two pike poles. Place one pole on the unfolded blanket about one foot from its center and fold the short side of the blanket over the pole toward the other side. Place a second pole on the two thicknesses, about 20 inches (51 cm) from the other pole and fold the remaining side of the blanket over the second pole toward the first. When a person is placed upon the blanket, between the poles, the body weight will hold the blanket secure. The procedure is shown in Figure 7.25.

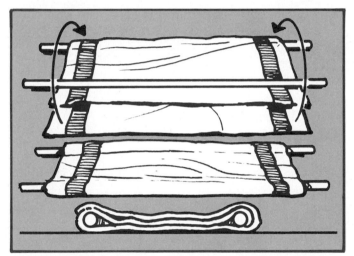

Figure 7.25 Folding a blanket or salvage cover around pike poles for an improvised stretcher.

LOWERING A VICTIM

An unconscious victim may be lowered from an upper floor by the use of a ladder and a lifeline (Figure 7.26). Although this method may appear slow compared to carrying a victim down a ladder, it may be safer under certain conditions. If several persons must be rescued from one place, this method may prove to be more rapid than expected. The rescue knot tie is the important point in this practice and the rescuer must know how to make the knot.

Rescue Knot

The bowline-on-a-bight rescue knot forms two loops in a rope which can be placed around a victim's thighs. This rescue knot must be made more secure on a victim by tying an additional loop around the victim's chest. How to tie a bowline-on-

Figure 7.26 Use a rescue knot and ladder to lower victims from upper floors when necessary.

a-bight and how to place it upon a victim is described in the following.

Step 1: Double the working end of a rope back upon itself, approximately eight feet (2 m). Stand on this double line, near the center, and with double rope in each hand, adjust it to the hips or waist of the rescuer. (If the victim or rescuer is exceptionally tall or short, this measurement must be altered) (Figure 7.27).

Step 2: Consider the loop end of the doubled rope as the working end and tie an overhand knot to form a loop approximately three feet (1 m) long, so that the loop hangs downward through the overhand knot (Figure 7.28).

Step 3: Hold the overhand knot in the left hand, reach down with the right hand and bring the three-foot (1m) loop up and over the left hand (Figure 7.29).

Step 4: Place the loop that is in the right hand under the thumb of the left hand, to be held secure (Figure 7.30).

Step 5: Grasp the two sides of the loop where they go through the overhand knot and pull upward to form a bight (Figure 7.31).

Step 6: Tighten the bight and adjust the two loops so that they are even (Figure 7.32).

Figure 7.27 Double the rope and adjust the length to the hips or waist.

Figure 7.28 Tie an overhand knot with a three-foot loop extending down through the knot.

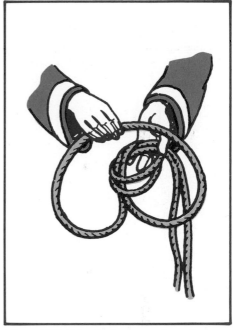

Figure 7.29 Bring the loop up and over the overhand knot.

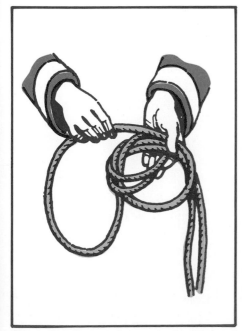

Figure 7.30 Place the loop under the right hand thumb.

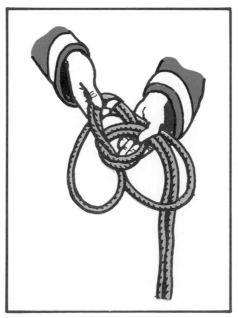

Figure 7.31 Grasp the rope where it goes through the overhand knot and pull upward to form a bight.

Figure 7.32 Tighten the bight and evenly adjust the loops.

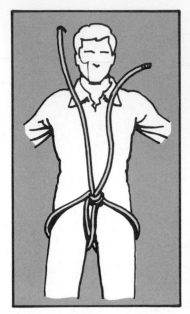

Figure 7.33 Bring the loops over the legs to the crotch.

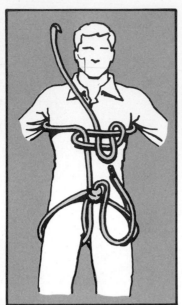

Figure 7.34 Place a half hitch around the chest and make an overhand safety with a loop to secure the hitch.

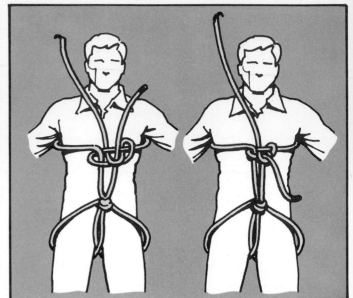

Figure 7.35 Place the loose end of the line from the bight through the loop formed at the safety, and pull on the working line until it rolls over and is secure.

Step 7: Place the two loops over the victim's legs and draw them well up in the crotch (Figure 7.33).

Step 8: Place a half hitch around the victim's chest with the working line and secure it in front with an overhand safety to form a loop (Figure 7.34).

Step 9: Place the loose end of the line through the loop that was just formed, pull on the working line to secure the tie (Figure 7.35).

VEHICLE EXTRICATION

There are millions of motor vehicle accidents in the United States every year, and the number of responses by fire departments is increasing. Vehicle extrication is a complex and demanding task requiring knowledge of motor vehicle design, hand and power tool capabilities, and patient care. In addition, the rescuer must be mentally prepared for the psychological trauma of facing victims of all ages who are burned, mortally injured, or hysterical. Serious vehicle accidents often produce mass confusion. Planning and training under simulated conditions can prepare firefighters to face these problems efficiently and professionally.

The frequent and repeated use of the tools used for extrication will increase the confidence of res-

cuers. Training sessions permit rescuers to learn how and why a tool works, when to use it, what it will do, and, most important, its limitations. Training should focus on special local hazards while refining basic skills. Cooperation with other agencies in mass casualty drills will enhance working relationships and provides an opportunity for the exchange of information.

Safety Considerations

Safety in operations is essential. Without the proper safety precautions no operation can be successful. Rescuers who disregard safety put victims in further danger and themselves in danger of injury. This applies to training or an actual emergency. Vehicle accidents offer a variety of hazards such as:

- Fire and its products
- Glass shards
- Sharp metal edges
- Flying glass and metal
- Dangerous chemicals and radiation
- Tool failure
- Unstable vehicles

Wear complete turnout gear including face-shields during any operation.

Size Up

The information received when an emergency call is received is important to the success of the operation. Information should include:

- Location of accident
- Kinds of vehicles involved
- Number of vehicles
- Condition or position of vehicle(s)
- Number of people injured and type of injuries
- Any special hazard information
- Name of person calling and call-back number

The size up or assessment of an accident theoretically begins on the training ground, but realistically it begins when the call is received. With proper information the assessment can begin while responding. Once on the scene, life-threatening hazards can be identified and correction measures can be started. If additional resources are needed, they should be ordered by the officer in charge.

Vehicle Stabilization and Access

The rescuer often finds a vehicle on its side or upside down in a gully or on a hillside. Resist the temptation to rock it or push it. Often this pressure is all that is needed to tip the vehicle over, producing disastrous results to victims and rescuers alike. A vehicle in any precarious position should first be stabilized. Stabilization may be accomplished by several means — cribbing or wedges, jacks, come-alongs, and air bags (Figure 7.36). Under no circumstances should a vehicle be tipped over while the victims are still inside. In an emergency, stabilization can be improvised with a bumper jack, ropes, or by opening the trunk lid and hood.

Choose the easiest route available to gain access to a vehicle. Try to open the doors normally, but if they are jammed, the window would be the next logical choice. If windows are broken in the accident and the frame is not bent, the rescuer has an immediate route to the victim. If not, try the rear window. The rescuer here has a large opening and

Figure 7.36 Cribbing or wedges and rope are among items that can be used to stabilize a vehicle. *Courtesy of Joel Woods.*

glass does not fall on the victim as readily as from a side window. Remember that the primary objective is to gain access and stabilize and protect the victim from further injury from sparks, glass, metal, and extrication tools.

VICTIM CARE AND REMOVAL

Once access into the vehicle has been made, the most qualified member of the rescue team must concentrate on the welfare of the victim (Figure 7.37). To stabilize the victim, rescuers first conduct

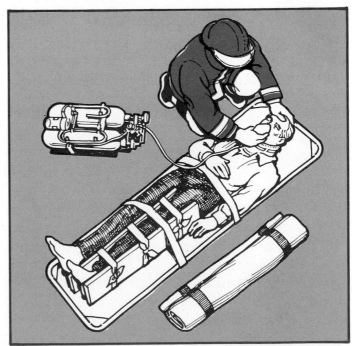

Figure 7.37 Survey the victim for injuries and stabilize before starting removal.

a primary survey to identify life-threatening injuries. Follow the steps listed below in order:

- Keep airway open
- Perform CPR if necessary
- Treat for shock
- Control bleeding
- Immobilize spinal column
- Immobilize fractures
- Position properly
- Strap in securely

In cases where the mechanisms of injury suggest back, neck, or spinal cord injury the victim should be appropriately immobilized.

Once entry to the vehicle has been gained and the victim's injuries assessed, treatment can begin simultaneously with preparation for disentangling the victim. The most important point to remember is that the vehicle is removed from around the victim and not the reverse. Various parts of the vehicle may trap the occupants such as steering wheels, seats, pedals, and dash boards (Figure 7.38). The situation should be assessed with the

victim's safety foremost in the rescuer's mind. Familiarity with all the rescue tools available and the ability to improvise will keep the rescuer from being "locked in" on the use of only one tool or technique.

Before rescuers can begin removing or transferring a victim, the team member responsible for patient care must be sure the patient is ready. If the victim is conscious, coherent, and only slightly injured, then the victim may not need to be packaged for removal to the ambulance.

"Packaging" is a term that means that wounds have been dressed and bandaged, fractures have been splinted, and the victim's body has been immobilized to reduce the possibility of further injury. Proper packaging protects the victim and aids and facilitates the victim's removal.

When the victim has been properly packaged and is ready for removal, cover sharp edges that can cut rescuers and victims. Openings should be widened and edges padded with blankets or fire hose that has been split and prepared beforehand. Openings should be wide enough so that the victim can be removed as smoothly as possible with no jerking or sudden movements.

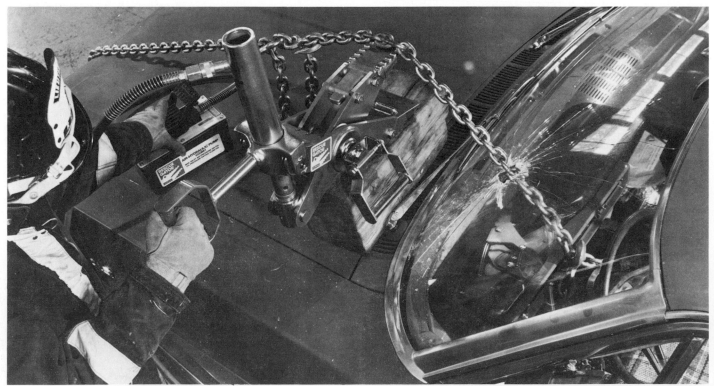

Figure 7.38 Carefully extricate the victim using safe techniques to move steering wheels, seats, dash boards, or other parts of the vehicle. *Courtesy of Bill Maggi.*

EMERGENCY EVACUATION OF ELEVATORS
Elevator Access
ELEVATOR CAR DOORS

Elevator doors are powered by an electric motor that rests on the top of the elevator car. The car door does not have any locks and can be pushed open by hand; however, electrical interlocks are provided to stop car movement whenever the doors open. Elevator car doors are designed to open and close automatically when the car receives instructions from the controller to stop at a particular floor landing. The car door, powered by the motor is also the means employed to open the hoistway door. When the elevator stops at a desired level, the hoistway doors are unlocked by a driving vane which is attached to the car door (Figure 7.39). As the car door opens, the vane strikes a roller which releases the hoistway door lock and the car door pushes the hoistway doors completely open. For closing, as a weight forces the hoistway closed, the driving vane moves away from the roller and the hoistway doors are relocked. This action takes place only when the doors are supposed to open and does not happen if the car stops somewhere other than the landing.

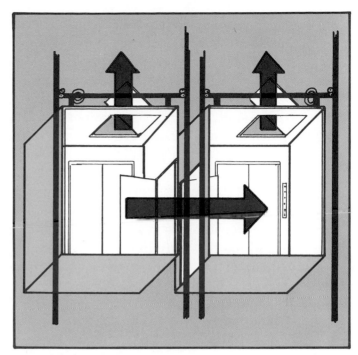

Figure 7.40 Elevator access panels are provided in addition to the main car doors.

Top Exit

A top exit is provided on all electric traction elevators. On hydraulic elevators, a top exit may or may not be provided as the system is equipped with a manual lowering valve. Top exits are only required if the elevator does not have this valve which permits the lowering of the car in the event of trouble. The top exit panel is normally locked from outside the car and is designed to open in that direction. In addition, some elevator cars are provided with electrical interlocks that prevent car movement while the panel is open.

Side Exit

In installations with multiple hoistways, side exits are normally provided so that travel from one car to another is possible for rescue operations. However, it may not be provided on hydraulic installations that have manual lowering valves. Side exits are required to have electrical interlocks to prevent car movement while the panel is open. The panels may be opened from the inside with a special key or from the outside where a permanent handle is provided.

HOISTWAY DOORS

Hoistway doors are equipped with a locking mechanism mounted on one of the panels to pre-

Figure 7.39 Typical elevator car door assembly.

ACCESS PANELS

In addition to the main doors of elevator cars, additional exits are provided to be used in emergency situations. These are located on the top and at the sides of the elevator car (Figure 7.40).

vent the doors from opening when an elevator car is not at the landing. The latching device is shaped on one end so that it hooks around a stop to prevent the doors from opening. In the unlocked position, the shaped end of the latching device is pushed away from the stop and the door can be opened. The latch is attached to a roller or drive block and when an elevator car comes to a landing, a driving vane mounted on the car door is positioned near the roller or block. If the car is not programmed to stop, the vane passes by the roller without contact. When the car does stop, the vane is pushed against the roller and the hoistway latch is released and the movement of the car door then pushes the hoistway door completely open (Figure 7.41). Hoistway doors are equipped with a weight or spring system that applies a constant force against the door toward the closed position. The force opening the car door must be sufficient to overcome the force of the hoistway door weights. As the electric motor closes the car doors, the weight or spring on the hoistway door allows it to close simultaneously. The driving vane then moves away from the roller or block and the latch resets itself, locking the hoistway doors and permitting the car to move on.

Figure 7.41 When the elevator stops at a desired level, the hoistway doors are unlocked by a driving vane which is attached to the car door.

Elevator Control

An elevator car will automatically proceed to the floor from which it receives a call, regardless of what activated the call button. The effects of heat, smoke, and moisture may cause the elevator control mechanism to activate. Because of this, it is

possible for the elevator car to go to the fire floor, open its doors, and expose the occupants to untenable conditions. However, in emergency situations it is desirable to have the ability to control the elevator car without interference. The two methods available are known as independent service and emergency service. Both services are controlled by keys that may or may not be readily available; therefore, firefighters should be aware of how to obtain the control keys.

INDEPENDENT SERVICE

Independent service is a feature available on most elevator installations that allows the car to be controlled manually. The car will respond only to those instructions issued from the operator inside the car and it will not respond to signals initiated from hall landings. This is an advantage to firefighters because they can move the car at will without having it stop for calls at landings or the fire floor. The service is activated by a switch usually located inside the car in a locked panel and identified by the letters I.S. (Figure 7.42). On some in-

Figure 7.42 The switch to activate independent service is usually located inside a locked panel on the interior of the elevator car.

stallations, the switch may be located on the control panel in the main lobby. After the car has been switched to independent service, movement and direction of travel is set by pushing a floor button. When the floor is reached, the door will open automatically.

EMERGENCY SERVICE

A more recent innovation is known as the emergency service mode of operation. The operating device for this is located on the car control panel and is operated by a key that turns it on and off. Once on, it will only respond to signals initiated from within the car and disregard those from hall landings. In addition, the doors must be opened and closed by pressing the appropriate button.

Both of the operations described above require the use of keys in order to make them function. Independent service requires a key to unlock the panel that protects the toggle switch while emergency service requires a key to turn it on.

Evacuation of Elevators

Upon arrival at the scene of an elevator emergency, firefighters should have an elevator mechanic dispatched to the scene. The mechanic is trained to make mechanical adjustment to the elevator that may enable passengers to exit from the elevator car in a normal manner. There may be times when a mechanic is unavailable or when the time of arrival will be extremely long. Also, there may be passengers aboard who are in critical medical condition requiring the fire department to evacuate the car immediately. However, elevator rescue requires training in the use of proper rescue techniques and fire departments should not attempt rescue with untrained personnel. Under no circumstances should firefighters alter the elevator's mechanical system in an attempt to drift the elevator. Adjustments to the mechanical system of the elevator installation should be done only by the elevator mechanic.

DETERMINE CAR LOCATION

After it has been decided that evacuation of the elevator car is necessary, the exact location of the car must be determined. Its position in the hoistway can be located by observing the position

indicator panel in the lobby or by opening the hoistway doors at the lobby and looking into the hoistway.

ESTABLISH COMMUNICATION

Communication must be established with the passengers to assure them of their safety and that work is being done to release them. If a telephone or intercom is not available, shouting through the door near the stall location may be sufficient for passing messages back and forth. Communication with the passengers is essential for their morale and mental state and should be established regardless of whether the fire department is going to rescue or wait for the elevator mechanic.

SECURE ELEVATOR

Rescue procedures will vary depending upon where the elevator car is located in relation to the nearest landing. Elevators may be stalled near the landing (several inches above or below), below the landing, or above the landing. Each situation requires the use of different techniques for passenger removal.

Firefighters must make certain that the stalled elevator car does not begin to move once rescue operations have begun. The elevator can be secured by opening the main power circuit to the elevator drive motor. The disconnect switch is in the machine room and it immobilizes the elevator by removing power and applying the brakes. As a further backup, the car's emergency stop switch should be activated as this also removes power from the drive motor. Passengers can be instructed to activate the switch and when the firefighters enter the car, they should also check to make sure the switch has been activated.

OPENING HOISTWAY DOORS

The opening of the hoistway door can easily be accomplished by the use of formed emergency keys, or if necessary, by forcing the doors open. There are three types of keys found in use for most elevator installations identified as: moon-shaped, T-shaped, and drop keys. The moon-shaped key is inserted into the opening a few inches and then pulled downward. The T-shaped is pushed straight into the opening to operate the locking mechanism.

The drop key is constructed of two sections hinged together. When the key is inserted into the opening far enough, the front section drops to form a 90° angle with the rear section. The key is then rotated to unlock the door (Figure 7.43). After the keys have released the locking mechanism, the hoistway doors can then be pushed open by hand.

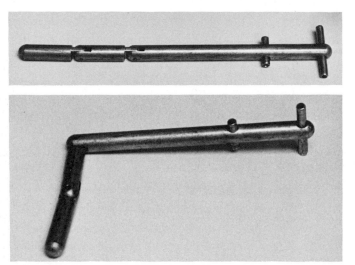

Figure 7.43 Hoistway door keys are inserted into the door and rotated to release the latch.

Alternate methods must be used when elevator keys are not available or are unusable. If the elevator is stalled near or within the landing zone, the passengers can be instructed to push open the elevator car door. At this time, if the car is close enough to the landing, the hoistway door will also open. If it is not close enough to the landing, opening the car door will only expose the closed hoistway doors. However, the locking mechanism may also be exposed and the passengers may be able to release it.

Another method that can be used in multiple hoistways involves lining up another elevator next to the hoistway door of the stalled elevator. A slender pole or slat can be slipped between the car and wall of the operating elevator, and is used to reach the door lock of the stalled elevator. This technique is known as poling.

A less desirable method of opening hoistway doors is to force the door open using forcible entry tools. There is a good possibility of only bending or crimping the door and not opening it. When doors are forced it should be done at the top, near the lock, and in the direction that the door opens.

PASSENGER REMOVAL
Elevator Car Stalled Near Landing

When an elevator car stalls within the landing zone, the car is close enough to the landing to release the locking mechanism on the hoistway door. Evacuation of the elevator car is simply a matter of pushing the doors open and assisting the passengers from the car.

When the elevator is stalled above or below the landing level, the passengers will be forced to climb up or down to reach the landing. Once the doors are open, a firefighter should enter the car to make sure the emergency stop switch is activated. A hazard that exists when the car is stalled above the landing is the opening below the car into the hoistway. Precautions should be taken to guard the hole with a ladder to avoid the possibility of someone falling into the hoistway.

Top Exit Removal

Access to the top of a stalled car is accomplished by opening the hoistway door above it. Firefighters lowered by lifelines can then open the top panel, which can only be opened from the outside. Care should be taken so that the elevator ceiling panels do not fall on the passengers. Once the exit panel is removed, a ladder is placed down into the car and a firefighter should enter the car to check the emergency stop switch. A lifeline should be tied to each passenger as they are assisted through the opening to the top of the car. A firefighter located on the landing will then assist the passenger onto the landing and untie the lifeline. If the top of the elevator is not even with the landing, a ladder may be needed.

Side Exit Removal

Side exit removal of passengers is an effective method of rescue if the elevator installation is in a multiple hoistway. To accomplish a rescue, a rescue elevator car (on independent service mode) is moved even with the stalled elevator car. The side exit doors of both cars are opened and rescuers enter the stalled car to check the emergency stop switch. A walking plank is laid between the two cars and a pike pole is used to act as a handrail while the passengers are guided into the rescue car.

NFPA STANDARD 1001
WATER SUPPLY
Fire Fighter II

4-13 Water Supplies

4-13.1 The fire fighter shall identify the water distribution system, and other water sources in the local community.

4-13.2 The fire fighter shall identify the following parts of a water distribution system:
- (a) Distributors
- (b) Primary Feeders
- (c) Secondary Feeders.

4-13.3 The fire fighter shall identify a:
- (a) Dry-barrel hydrant
- (b) Wet-barrel hydrant.

4-13.4 The fire fighter shall identify the following:
- (a) Normal operating pressure of a water distribution system.
- (b) Residual pressure of a water distribution system.
- (c) The flow pressure from an opening that is flowing water.

4-13.5 The fire fighter shall identify the following types of water main valves:
- (a) Indicating
- (b) Nonindicating
- (c) Post Indicator
- (d) Outside Screw and Yoke.

4-13.6 The fire fighter shall identify hydrant usability by:
- (a) Obstructions to use of hydrant
- (b) Direction of hydrant outlets to suitability of use
- (c) Mechanical above-ground damage
- (d) Condition of paint for rust and corrosion
- (e) The flow by fully opening the hydrant
- (f) Ability to drain.*

*Reprinted by permission from NFPA Standard No. 1001, *Standard for Fire Fighter Professional Qualifications*. Copyright © 1981, National Fire Protection Association, Boston, MA.

IFSTA's Water Supply Transparencies are designed to complement this chapter.

Chapter 8
Water Supply

Technology keeps advancing new methods and materials for extinguishing fires but water still remains the primary extinguisher because of its universal abundance and ability to absorb heat. Water can be conveyed long distances, pumps are adaptable for automatic control to increase pressure, or it can be stored. Therefore, it is important that a firefighter be familiar with water supply.

The intricate working parts of a water system are many and varied. Basically, the system can be described by the following fundamental components:

- The source of supply
- The processing or treatment facilities
- The mechanical or other means of moving water
- The distribution system, including storage

Some of the physical structures necessary for efficient operation of these basic components of a waterworks system are dams which form impounding reservoirs; aqueducts and pipelines; river and lake intakes; pumping stations; filters; water treatment plants; distribution piping; and distribution reservoirs. For efficient operations, each component must be planned so that it will function adequately in relation to the complete water system (Figure 8.1).

Public and/or private water systems provide the methods for supplying water to the more populated areas. As rural areas increase in population, rural communities seek to improve efficient water distribution systems from a reliable source.

The water department is a separate city utility. Its principal function is to provide water safe

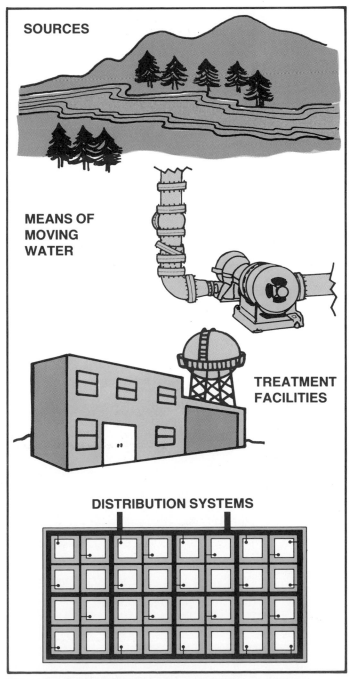

SOURCES

MEANS OF MOVING WATER

TREATMENT FACILITIES

DISTRIBUTION SYSTEMS

Figure 8.1 There are many components included in a water system.

for human use. They should be considered the experts in water supply problems. The fire department must work with the water department in planning fire protection. Water department officials should realize that fire departments are vitally concerned with water supply and should work with them on supply needs, location, and types of fire hydrants.

SOURCES OF WATER SUPPLY

The primary water supply can be obtained from either surface waters or ground waters. Although most water systems are supplied from only one source, there are instances where both sources are used. Two examples of surface water supply are river supply and lake supply. Ground water supply can be water wells or water producing springs.

The amount of water that a community may need can be determined by an engineering estimate. This is the total amount of water it must furnish and is the sum of the water required for domestic or industrial use and the water required for fire service (fire flow).

In small towns the requirements for fire protection exceed other requirements. There are three types of systems for getting water distributed:

- Direct pumping
- Gravity system
- Combination system

Direct Pumping

One or more pumps take the water from the primary source and discharge it through the filtration and treatment processes. From here a series of pumps force the water into the distribution system. If purification of the water is not needed, the water can be pumped directly into the distribution system. Failures in supply lines and pumps can usually be overcome by duplicating these units and providing a secondary power source (Figure 8.2).

Gravity System

This is a primary water source that is located at a higher elevation than the distribution system. The gravity flow from the higher elevation provides the pressure (Figure 8.3).

Combination System

Most communities use the combination system. This system uses a combination of the direct pumping and gravity system. In most cases the gravity flow is supplied by elevated storage tanks. These tanks serve as emergency storage and can

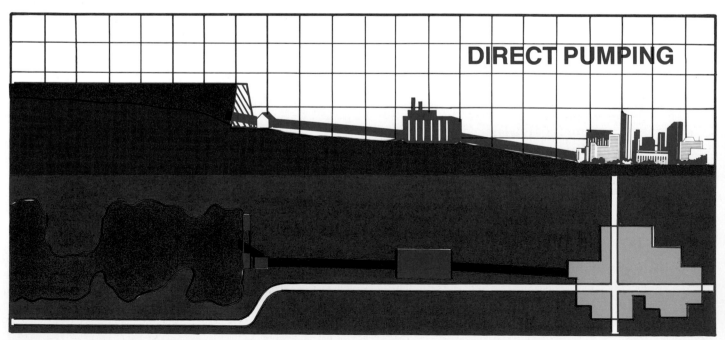

Figure 8.2 A direct pumping system is used where the water source does not have sufficient elevation to create adequate pressure to deliver ample quantities.

provide the necessary pressure because of gravity flow. When the system pressure is high during periods of low use, automatic valves open and allow the storage containers to fill with the water not consumed by customers. When the pressure drops during periods of heavy consumption, the storage containers provide extra water by feeding back into the distribution system. Providing a good combination system involves reliable and duplicated equipment and storage containers of proper size strategically located (Figure 8.4).

The storage of water in elevated reservoirs can also assure water supply when the system becomes inoperative. According to the engineers of the American Insurance Association (AIA), storage should be sufficient, during any period of five days maximum consumption, to provide domestic and industrial demands plus fire flow of two to ten

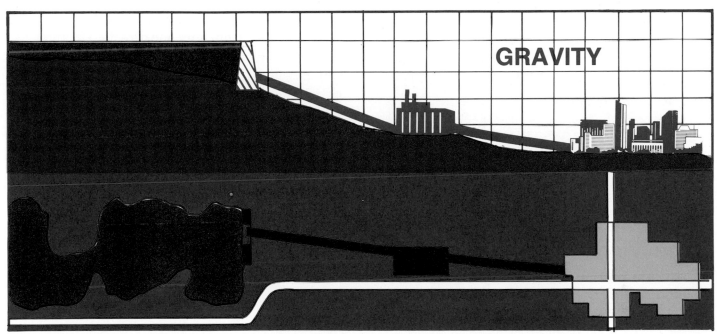

Figure 8.3 A gravity system is used where the water source is elevated.

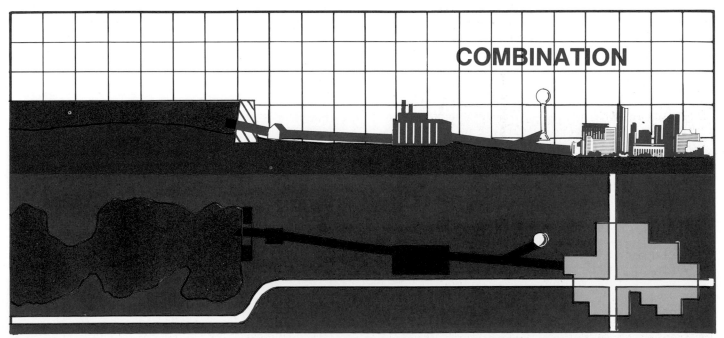

Figure 8.4 A combination of direct pumping and gravity is used to allow storing of water during low demand for use when consumption exceeds pump capacity.

hours depending on the size and character of the community. Such storage should be sufficient to permit making most of the repairs, alterations, or additions to the system. Location of the storage and the capacity of the mains leading from this storage are also important factors.

Many industries provide their own private systems, such as elevated storage tanks, which are available to the fire department. Water may be available to some communities from storage systems such as cisterns, which are considered a part of the distribution system. The fire department pumper removes this water by draft and provides pressure by the pump.

THE DISTRIBUTION SYSTEM

The distribution system of water utility is that part which receives the water from the pumping station and delivers it throughout the area to be served. Fire hydrants, gate valves, elevated stor-age, reservoirs, and standpipes are supplementary parts of the distribution system.

The ability of a water system to deliver an adequate quantity of water relies upon the carrying capacity of the system's network of pipes. When water flows through pipes, its movement causes friction which results in reduction of pressure. There is much less pressure loss in a water distribution system when fire hydrants are supplied from two or more directions. A fire hydrant which receives water from only one direction is known as a dead-end hydrant. When a fire hydrant receives water from two or more directions, it is said to have circulating feed. A distribution system that provides circulating feed from several mains constitutes a grid system. A grid system (Figure 8.5) should consist of the following:

- Primary feeders consist of large pipes (mains) with relatively widespread spacing which convey large quantities of water to

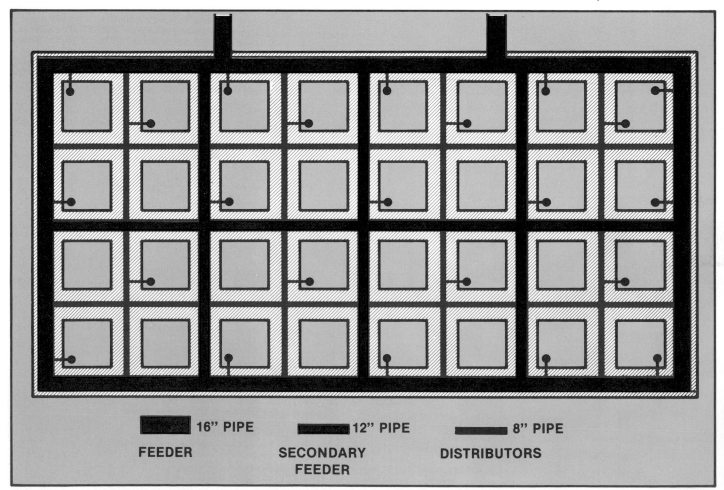

Figure 8.5 A well-gridded water distribution system consists of primary and secondary feeders and distribution mains.

various points of the system for local distribution to the smaller mains.

- Secondary feeders for the network of pipes of intermediate size which reinforce the grid within the various loops of the primary feeder system, and aid the concentration of the required fire flow at any point.

- Distributors consist of a grid arrangement of smaller mains serving the individual fire hydrant and blocks of consumers.

To assure sufficient water, two or more primary feeders should run by separate routes from the source of supply to the high value and industrial districts of the community. Similarly, secondary feeders should be arranged as far as possible in loops so as to give two directions of supply to any point. This practice increases the capacity of the supply at any given point and assures that a break in a feeder main will not completely cut off the supply.

In residential areas the recommended fire hydrant supply main should be at least 8 inches (20 cm) in diameter and no smaller than 6 inches (15 cm). These should be closely gridironed by 8-inch (20 cm) cross-connecting mains at intervals of not more than 600 feet (180 m).

In the high-value and industrial districts, the minimum size is an 8-inch (20 cm) main with cross-connecting mains every 600 feet (180 m) (Figure 8.6). Twelve-inch (30 cm) mains should be used on all principal streets and for long mains not cross-connected at frequent intervals.

Water Main Valves

The function of a valve in a water distribution system is to provide a means for controlling the flow of water through the distribution piping. Valves should be located at frequent intervals in the grid system, so that only small districts will need to be cut off when necessary to stop the flow at specified points. Valves should be operated at least once a year to keep them in good condition. The actual need for valve operation in a water system rarely occurs, sometimes not for many years. Valve spacing should be such that a minimum length of pipe will be out of service at one time. The

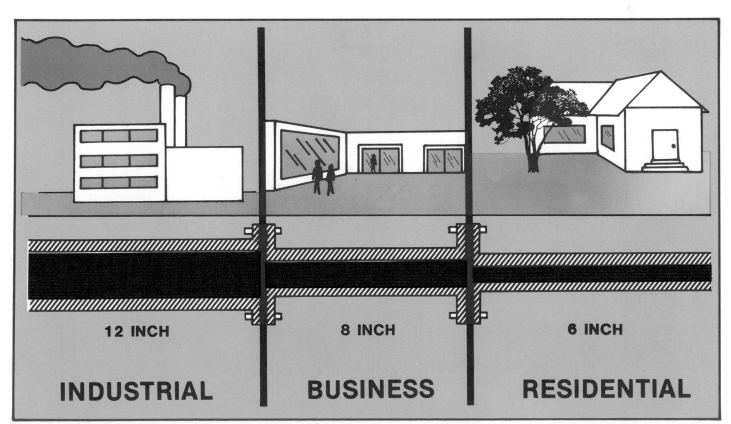

12 INCH   **8 INCH**   **6 INCH**

INDUSTRIAL   **BUSINESS**   **RESIDENTIAL**

Figure 8.6 Main sizes will depend on the flows required and should be frequently cross-connected.

maximum lengths should be 500 feet (150 m) in high value districts and 800 feet (240 m) in other areas, as recommended by ISO engineers.

One of the most important factors in a water supply system is the utilities' ability to promptly operate the valves during an emergency or breakdown of equipment. A well-run water utility has records of the location of all valves and the procedure for inspecting, maintaining, and operating each valve not less than once a year.

If each fire department company is informed of the location of valves in the distribution system, their condition and accessibility can be noted during fire hydrant inspection and the water department informed of needed attention.

Valves for water systems are broadly divided into indicating and nonindicating types (Figure 8.7). An indicating valve visually shows the position of the gate or valve seat, whether open or closed or partially closed. Valves in private fire protection systems are usually of the indicating type. Except for possibly a few valves in treatment plants and pump stations, valves in public water systems are of the nonindicating type.

Two common indicator valves are the post indicator valve (PIV) and the outside screw and yoke valve (OS&Y). The post indicator valve is a hollow metal post that is attached to the valve housing. The valve stem is inside of this post, and on the stem the words "OPEN" and "SHUT" are printed so that position of the valve is shown. The OS&Y valve has a yoke on the outside with a threaded stem which controls the gate's opening or closing. The threaded portion of the stem is out of the yoke when the valve is open and inside the yoke when the valve is closed.

Nonindicating valves in a water distribution system are normally buried or installed in manholes. If a buried valve is properly installed, a way to operate the valve from above ground through a valve box can be provided. A special socket wrench on the end of a reach rod is available for this pur-

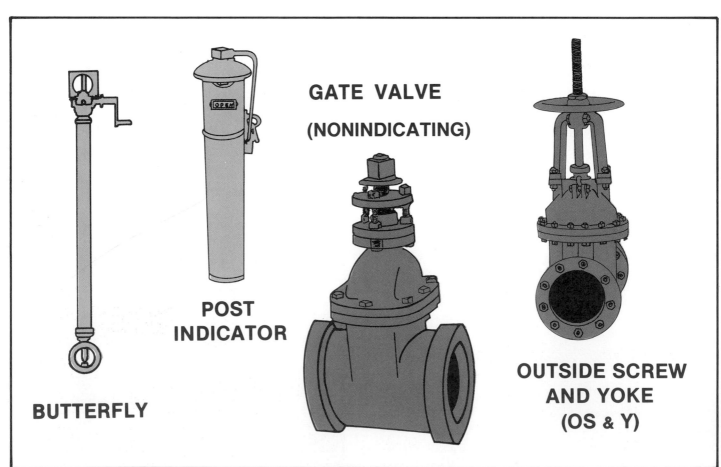

GATE VALVE

(NONINDICATING)

POST INDICATOR

BUTTERFLY

OUTSIDE SCREW AND YOKE (OS & Y)

Figure 8.7 A variety of indicating and nonindicating valves are used in water mains.

pose. Control valves in water distribution systems may be either gate valves or butterfly valves. Gate valves are usually of the nonrising stem type, and as the valve nut is turned by the valve key (wrench) the gate either rises or lowers to control the water flow. Butterfly valves are tight closing and they usually have rubber or a rubber composition seat which is bonded to the valve body. The valve disk rotates 90 degrees from the fully open to the tight shut position. The nonindicating type also requires a valve key. Its principle of operation provides satisfactory water control after long periods of inactivity. Both valves can be of the indicating type which will be discussed under sprinkler and standpipe systems.

The advantages of proper valving in a distribution system are readily apparent. If valves are installed according to established standards, it will normally be necessary to close off only one or perhaps two fire hydrants from service while a single break is being repaired. The advantage of proper valving will, however, be reduced if all valves are not properly maintained and kept fully open. High friction loss is caused by valves that are only partially open. When valves are closed or partially closed, the condition may not be noticeable during ordinary domestic flows of water. As a result, the impairment will not be known until a fire occurs or at a time when a detailed inspection and a fire flow test is made. A fire department will experience difficulty in obtaining water in areas where there are closed or partially closed valves in the distribution system.

Water Pipe

Water pipe, which is used underground, is generally made of either cast iron, ductile iron, asbestos cement, steel, plastic, or concrete. Whenever pipe is installed, it should be of the proper type for the soil conditions and pressures to which it will be subjected. When water mains are installed in unstable or corrosive soils or in difficult access areas, steel or reinforced concrete pipe may be used to give the strength needed. Some conditions which may require extra protection include beneath railroad tracks and highways, areas close to heavy industrial machinery, earthquake areas, or in rugged terrain.

The internal surface of the pipe, regardless of the material from which it is made, offers resistance to water flow. Some materials, however, have considerably less resistance to water flow than others. The engineering division of a water department should determine the type best suited for the conditions at hand.

Kinds of Pressure

The speed by which a fluid travels through a pipe is developed by pressure upon that fluid. The speed of travel is often referred to as velocity, and the pressure should be identified as to the kind of pressure because the word "pressure" in connection with fluids has a very broad meaning. Pressure, in a normal sense, may be defined as a force or energy that may be measured in pounds per square inch (psi) (kPa). An understanding of the terms static pressure, normal operating pressure, residual pressure, and flow pressure is essential (Figure 8.8). These terms identify the kinds of pressure.

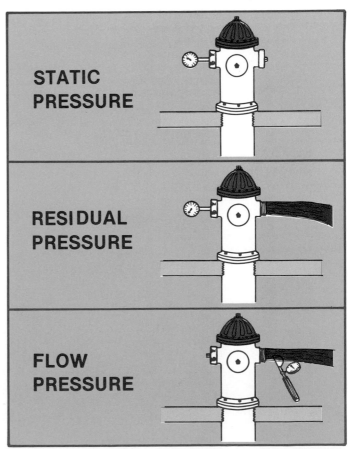

Figure 8.8 The different types of pressure are used in calculating water flows and determining the remaining amount of water when pumping at fires.

STATIC PRESSURE

If the water is not moving, the pressure exerted is static. A water flow definition of static pressure could be as follows: *"Static pressure is stored potential energy that is available to force water through pipe, fittings, fire hose, and adapters."*

NORMAL OPERATING PRESSURE

The flow of water through a distribution system to supply consumers continuously fluctuates during the day and night. A water flow definition of normal operating pressure can be as follows: *"Normal operating pressure is that pressure which is found on a water distribution system during normal consumption demands."*

RESIDUAL PRESSURE

The term residual represents the pressure which is left in a distribution system, at a specific location, when a quantity of water is flowing. A water flow definition of residual pressure could be as follows: *"Residual pressure is that part of the total available pressure that is not used to overcome friction or gravity while forcing water through pipe, fittings, fire hose, and adapters."*

FLOW PRESSURE

The forward velocity of a water stream exerts a pressure that can be read on a Pitot tube and gauge. A water flow definition of flow pressure can be as follows: *"Flow pressure is that forward velocity pressure at a discharge opening while water is flowing."*

FIRE HYDRANTS

The two main types of fire hydrants are dry-barrel or wet-barrel (Figure 8.9). The dry-barrel hydrant, utilized in climates where freezing is ex-

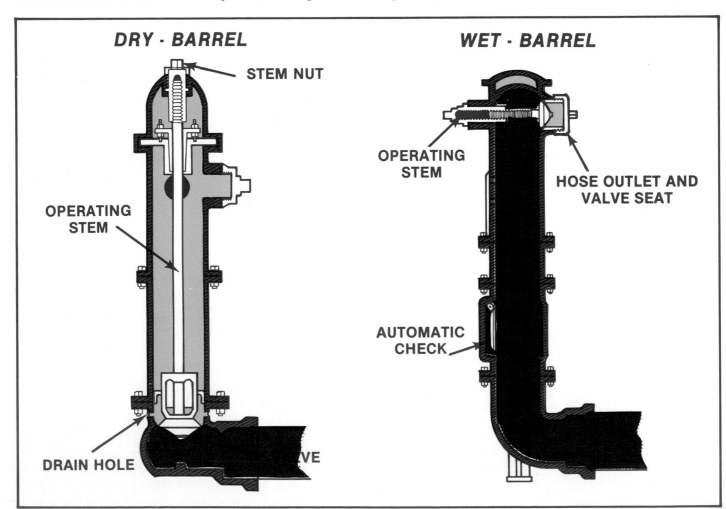

Figure 8.9 Dry-barrel hydrants are used in freezing climates with the valve below the freeze line. Wet-barrel hydrants are used in mild climates and have the valves at the outlets.

pected, is usually classified as a compression, gate, or knuckle-joint type which either opens with the pressure or against pressure. Any water that may remain in a closed dry-barrel hydrant will drain through a small valve at the bottom. Wet-barrel hydrants do not have this feature. This drain valve opens as the main valve approaches the closed position. Wet-barrel hydrants usually have a compression type valve at each outlet, or they may have only one valve that is located in the bonnet which controls the flow of water to all outlets. In general, hydrant bonnets, barrels, and foot pieces are made of cast iron. The important working parts, however, are usually made of bronze, but valve facings may be made of rubber, leather, or composition.

The flow of a hydrant will vary for several reasons already mentioned such as feeder main location and size. Incrustations and deposits within the distribution system may increase the resistance to water flow. Firefighters can make better decisions affecting a fire attack if they know at least the relative available water flow of different hydrants in the vicinity. To aid them, a system was developed by NFPA of coloring hydrants to indicate a range of water flow. With the system, hydrants are classified as follows:

Color of Tops	Flow
Green	1000 GPM (3800 L/min) or greater
Orange or Yellow	500-1000 GPM (1900-3800 L/min)
Red	Less than 500 GPM (1900 L/min)

Local variation of coloring may be found, but the main intent of any color scheme is simplicity.

Location of Fire Hydrants

Although the installation of fire hydrants is usually performed by water works personnel, the location, spacing, and distribution of fire hydrants should be the responsibility of the fire chief. In general, fire hydrants should not be spaced more than 300 feet (90 m) apart in high value districts. A basic rule to follow is to place one hydrant near each street intersection and to place intermediate hydrants where distance between intersections exceeds 350 to 400 feet (105 to 120 m). This basic rule represents a minimum requirement and should be regarded only as a guide for spacing hydrants. Other factors more pertinent to the situation include types of construction, types of occupancy, congestion, sizes of water mains, fire flows, and pumping.

Fire Hydrant Maintenance

In most cities, repair and maintenance of fire hydrants is a responsibility of the water department since they are in a better position to do this work than any other agency.

In addition to the pressure and drain tests prescribed for local conditions, a hydrant should be inspected for the following:

- Have obstructions such as sign posts, utility poles, or fences been erected too near the hydrant?

- Do the outlets face the proper direction and is there sufficient clearance from them to the ground?

- Has the hydrant been damaged by traffic?

- Is the hydrant rusting or corroded?

OTHER SOURCES OF WATER

Fire departments should not limit their study of water supply to just the piped public distribution system. Areas outside the public system should be studied for available water. Water supply can be provided where an industry has its own private water system. With today's modern pumper, water can be drawn from many natural sources such as the ocean, lakes, ponds, and rivers. Water can also sometimes be found in farm stock tanks and swimming pools. A good method of providing water for fire protection is to construct storage tanks. The fire department pumper could get water from a connection on the tank or by dropping a hard suction sleeve into the tank.

Fire department pumpers carry water for initial attack streams. Many fire departments also have one or more tank vehicles to carry water for fire fighting in locations outside the public water system and some make use of portable folding tanks.

NFPA STANDARD 1001
FIRE STREAMS
Fire Fighter I

3-7 Fire Hose, Nozzles, and Appliances
3-7.2   The fire fighter shall demonstrate the use of nozzles, hose adaptors, and hose appliances carried on a pumper.

3-7.5   The fire fighter shall demonstrate the techniques for cleaning fire hose, couplings, and nozzles; and inspecting for damage.

3-8 Fire Streams
3-8.1   The fire fighter shall define a fire stream.

3-8.3   The fire fighter shall define water hammer and at least one method for its prevention.

3-8.4   The fire fighter shall demonstrate how to open and close a nozzle.*

Fire Fighter II
4-7 Fire Hose, Nozzles, and Appliances
4-7.1   The fire fighter shall identify, select, and demonstrate the use of any nozzle.

4-7.3   The fire fighter shall demonstrate inspection and maintenance of fire hose, couplings, and nozzles, and recommend replacement or repair as needed.

4-8 Fire Streams
4-8.3   The fire fighter shall identify characteristics of all types of fire streams.

4-8.5   The fire fighter shall identify three conditions that result in pressure losses in a hose line.

4-8.6   The fire fighter shall identify four special stream nozzles and demonstrate at least two uses or applications for each.

4-8.7   The fire fighter shall identify and define foam making appliances, and shall demonstrate a foam stream from each.

4-8.8   The fire fighter shall identify three observable results that are obtained when the proper application of a fire stream is accomplished.

4-8.9   The fire fighter shall identify and define those items required to develop three types of fire streams, and shall demonstrate each.*

*Reprinted by permission from NFPA Standard No. 1001, *Standard for Fire Fighter Professional Qualifications*. Copyright © 1981, National Fire Protection Association, Boston, MA.

IFSTA's Fire Streams Transparencies are designed to complement this chapter.

Chapter 9
Fire Streams

A fire stream can be defined as a stream of water, or other extinguishing agent, after it leaves the fire hose and nozzle until it reaches the desired point in the proper configuration. The perfect fire stream can no longer be sharply defined since individual desires and extinguishing requirements vary. During the time a stream of water passes through space, it is influenced by its velocity, gravity, wind, and friction with the air. The condition of the stream when it leaves the nozzle is influenced by operating pressures, nozzle design, nozzle adjustment, and the condition of the nozzle orifice.

Fire streams are intended to reduce temperatures and provide protection by one of the following methods:

- By applying water directly to the burning material

- By reducing high atmospheric temperature and by absorbing and/or dispersing hot smoke and fire gases from a heated area in a confined space

- By reducing the temperature over an open fire and thus permit a closer proximity with hand hoselines to effect extinguishment

- By protecting firefighters and property from heat through the use of fire streams as a water curtain

THE EXTINGUISHING PROPERTIES OF WATER

Water has the ability to absorb large quantities of heat. Quantity of heat is measured in terms of British thermal units (Btu's). A British thermal unit is defined as the amount of heat needed to raise the temperature of one pound of water one degree Fahrenheit.

At sea level, water boils and vaporizes into steam at 212°F (100°C). Complete vaporization, however, does not happen the instant the water reaches the boiling point, because each pound of water will then require approximately 970 Btu's of additional heat to completely turn it into steam. When a fire stream is broken into small particles, it will absorb heat and be converted into steam more rapidly than it would in a compact form because more of the water surface is exposed to the heat. For example, if a one square inch cube of ice is dropped into a glass of water, it will take quite some time for the ice cube to absorb its capacity of heat because a surface area of only 6 square inches (15 square cm) of the ice is exposed to the water, but if that cube of ice is divided into ⅛-inch (3 mm) cubes and dropped into the water, a surface area of 48 square inches (121 square cm) of the ice is exposed to the water. The finely divided particles of ice will absorb heat more rapidly. This same principle applies to water in the liquid state.

Another characteristic of water which is an aid to fire fighting is its expansion when converted into steam. When water changes into steam, it expands. The amount of expansion varies with the temperatures of the fire area. At 212°F (100°C), a cubic foot of water expands approximately 1,700 times its original volume. The greater the temperature, the higher the amount of expansion. For example, at 500°F (260°C), the expansion ratio is approximately 2,400 times and at 1,200°F (649°C), the ratio is approximately 4,200 times.

It might be well to visualize a nozzle discharging 75 gallons (284 L) of water fog every minute into a heated area at approximately 212°F, (100°C) and being converted into steam. During one minute of operation, ten cubic feet of water will have been vaporized, and the ten cubic feet of water will have expanded to approximately 17,000 cubic feet (518 m) of steam. This is enough steam to fill a room approximately 10 feet high (3 m), 25 feet (8 m) wide, and 68 feet (20 m) long, as shown in Figure 9.1. In extremely hot atmospheres, steam will further expand to greater volumes. Steam expansion is not gradual but rapid, but if the room is already full of smoke and gases, the steam that is generated will displace these gases. As the room cools, the steam condenses and allows the room to refill with cooler air. The steam produced can also be an aid in fire extinguishment by smothering when certain types of materials burn. Smothering is accomplished when the expansion of steam reduces oxygen in a confined space.

Water may be used as a smothering agent when it will float on liquids that are heavier than water, such as carbon disulfide. If the material is water soluble, such as alcohol, the smothering action is not likely. Water may also cause a smothering action on fires by the formation of an emulsion over the surface of certain flammable liquids.

When a spray of water agitates the surface of certain flammable liquids which are burning, the agitation can cause the water to be temporarily suspended in an emulsion bubble on the surface and thereby smother the fire. This action can only take place when the flammable liquid has sufficient viscosity. Viscosity is the tendency of a liquid to possess internal resistance to flow (e.g. — Water is low, molasses is high). A light fuel oil (No. 2 grade) will retain an emulsion for only a short time while a heavier fuel oil (No. 6 grade) will retain an emulsified surface for a longer period of time. The water which is present in the emulsion absorbs heat from the oil adjacent to it, reduces the oil temperature, and decreases the amount of flammable vapors that are emitted.

Summing up the extinguishing properties of water, several characteristics which are extremely valuable from the fire extinguishing point of view can be listed.

- Water has a greater heat absorbing capacity than other common extinguishing agents.

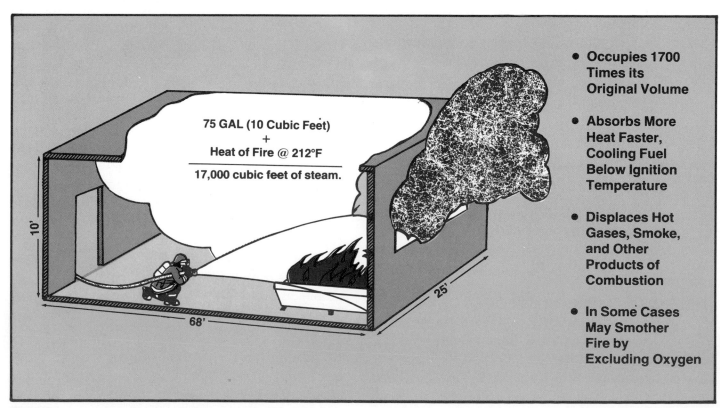

75 GAL (10 Cubic Feet) + Heat of Fire @ 212°F

17,000 cubic feet of steam.

- Occupies 1700 Times its Original Volume

- Absorbs More Heat Faster, Cooling Fuel Below Ignition Temperature

- Displaces Hot Gases, Smoke, and Other Products of Combustion

- In Some Cases May Smother Fire by Excluding Oxygen

10'

68'

25'

Figure 9.1

- A relatively large amount of heat is required to change water into steam.

- The greater the surface area of the water that is exposed, the more rapidly heat will be absorbed.

- Water converted into steam occupies several hundred times its original volume.

TYPES OF PRESSURE

The speed by which a fluid travels through hose is developed by pressure upon that fluid. The speed of travel is often referred to as velocity, and the pressure should be identified as to the kind of pressure because the word "pressure" in connection with fluids has a very broad meaning. Pressure, in a normal sense, may be defined as a force or energy that may be measured in pounds per square inch (psi) (kilo paschals - kPa). An understanding of the terms static pressure, normal operating pressure, residual pressure, and flow pressure is essential. These terms identify the kinds of pressure.

Static Pressure

If the water is not moving, the pressure exerted is static. A water flow definition of static pressure could be as follows: *"STATIC PRESSURE IS STORED POTENTIAL ENERGY THAT IS AVAILABLE TO FORCE WATER THROUGH PIPE, FITTINGS, FIRE HOSE, AND ADAPTERS."*

Normal Operating Pressure

The flow of water through a distribution system to supply consumers continuously fluctuates during the day and night. A water flow definition of normal operating pressure can be as follows: *"NORMAL OPERATING PRESSURE IS THAT PRESSURE WHICH IS FOUND ON A WATER DISTRIBUTION SYSTEM DURING NORMAL CONSUMPTION DEMANDS."*

Residual Pressure

The term residual represents the pressure which is left in a distribution system, at a specific location, when a quantity of water is flowing. A water flow definition of residual pressure could be as follows: *"RESIDUAL PRESSURE IS THAT PART OF THE TOTAL AVAILABLE PRESSURE THAT IS NOT USED TO OVERCOME FRICTION OR GRAVITY WHILE FORCING WATER THROUGH PIPE, FITTINGS, FIRE HOSE, AND ADAPTERS."*

Flow Pressure

The forward velocity of a water stream exerts a pressure that can be read on a pitot tube and gauge. A water flow definition of flow pressure can be as follows: *"FLOW PRESSURE IS THAT FORWARD VELOCITY PRESSURE AT A DISCHARGE OPENING WHILE WATER IS FLOWING."*

FRICTION LOSS

If an opening is made in a closed system of piping or fire hose, a difference of pressure will exist between the internal pressure and the atmospheric pressure outside the pipe or hose. This difference in pressure causes the water to flow toward the lesser pressure. This loss of pressure is usually called friction loss or loss because of friction. The only pressure available to overcome this resistance is the total pressure. A fire stream definition of friction loss could be as follows: *"FRICTION LOSS IS THAT PART OF TOTAL PRESSURE THAT IS LOST WHILE FORCING WATER THROUGH PIPE, FITTINGS, FIRE HOSE, AND ADAPTERS."* The difference in pressure on a hoseline between a nozzle and a pumper is a good example of friction loss. Friction loss can be measured by inserting in-line gauges in a hoseline. The difference in the residual pressures between gauges when water is flowing through the hose will be the friction loss for the length of hose between the gauges for that rate of flow.

One point that should be considered in applying pressure on water in a hoseline is that there is a limit to the velocity or speed at which the stream can travel. If the velocity is increased beyond these limits, the friction becomes so great that the entire stream is agitated by resistance. This agitation causes a point of turbulence, called "critical velocity." Beyond this point, it becomes necessary to parallel or siamese hoselines in order to increase the flow and reduce friction.

Certain characteristics of hose layouts affect friction loss (Figure 9.2). In order to reduce pressure loss due to friction, consider the following.

- Check for rough linings in old hose.
- Replace crushed couplings.
- Eliminate sharp bends.
- Use adapters only when necessary.
- Keep nozzles and valves fully open when possible.
- Use proper size gaskets.
- Use short lines as much as possible.
- When flow must be increased, use larger hose or multiple lines.

WATER HAMMER

When the flow of water through fire hose or pipe is suddenly stopped, the resulting surge is referred to as water hammer. Water hammer can often be heard as a distinct sharp clank, very much like a hammer striking a pipe. This sudden stop-page results in a change in the direction of energy and this energy is instantaneously multiplied many times. These excessive pressures can cause considerable damage to water mains, plumbing, fire hose, and fire pumps. Nozzle controls, hydrants, valves, and hose clamps should be operated slowly to prevent water hammer (Figure 9.3).

TYPES OF FIRE STREAMS

Fire streams may be defined in terms of size or volume. A small stream can be considered a low volume stream which discharges less than 40 gpm (151 L) including those fed by booster hoselines. Handline streams may be supplied by 1½-inch (38 mm) to 2½-inch (65 mm) hose. Streams supplied by 1½-inch (38 mm) hose will have capacities ranging from 40 to 125 gpm (151-455 L) while streams fed by 2½-inch (65 mm) hose will discharge from 125 to 300 gpm (455-1153 L). Nozzles with flows in excess of 300 gpm (1153 L) are not recommended for handlines. A master stream is a large volume stream which discharges more than 300 gpm (1153 L), and is usually fed by two or more hoselines

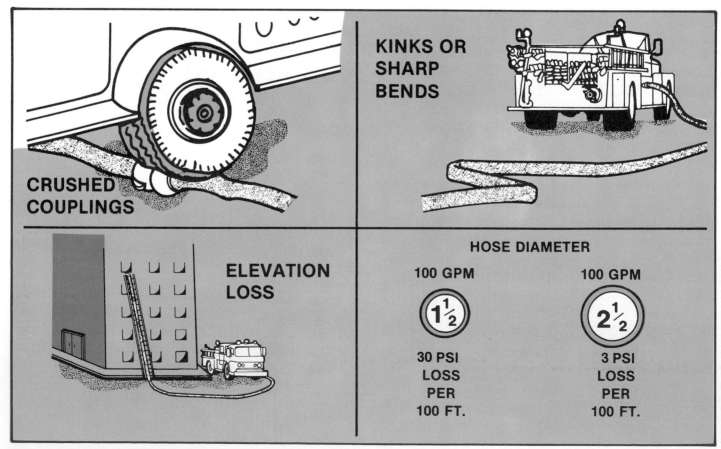

Figure 9.2 Pressure loss in hose at fires is caused by elevation or several sources of friction.

WATER HAMMER HITS EVERYTHING

PUMP

PIPING

HOSE

HYDRANT

COUPLING-

MAINS

Open And Close All Nozzles And Valves Slowly

Figure 9.3 A water hammer can have damaging effects throughout the water supply system.

siamesed into a master stream device. In any classification, the stream may be either solid or spray pattern.

Solid Streams

The fire service requires a fire stream nozzle to produce a specific pattern of water jet in order to perform a specific function (Figure 9.4). The solid stream nozzle is designed to produce a stream as compact as possible with little shower or spray. Its ability to reach areas that might not be reached by other mediums has established the solid stream as necessary to the fire service. It is recommended that when solid stream nozzles are used on a handline, that 50 psi (350 kPa) nozzle pressure be used. If the nozzle is to be used on a master stream device, 80 psi (550 kPa) is recommended. Modern constant flow or constant gallonage fog nozzles will, however, produce a straight stream which is substantially equivalent to a solid stream.

SOLID
- *Little Shower or Spray*
- *Good Reach and Penetration*

Figure 9.4

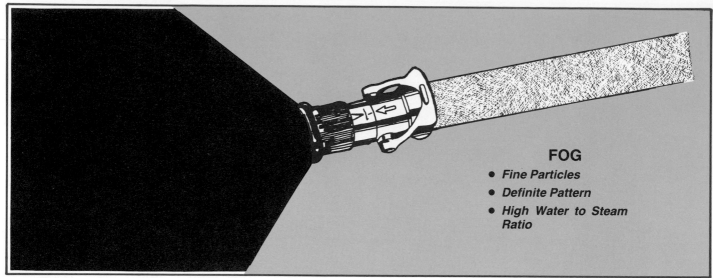

Figure 9.5

Fog Streams

The term *fog stream* is commonly used in the fire service to describe a patterned stream composed of droplets, and will be used in this manual (Figure 9.5).* It should be understood that a *straight stream* is a pattern of the adjustable fog nozzle whereas a *solid stream* is discharged from a smooth-orifice nozzle.

Broken Streams

A broken fire stream may be a solid stream that has been broken into coarsely divided drops (Figure 9.6). A solid stream may become a broken stream because of reacting forces, and it may be desirable that a stream break near the fire to sprinkle the burning material. Other means may be employed to produce broken streams, such as rotary distributor nozzles, water curtain nozzles, or by directing two solid streams together in midair. The droplets are larger than those of a fog stream and have great penetration, so the broken stream can be useful where neither a fog stream nor a solid stream would be as effective in situations requiring these stream qualities.

CHARACTERISTICS OF SOLID STREAMS

The extreme limit at which a solid stream of water can be classified as a good stream cannot be

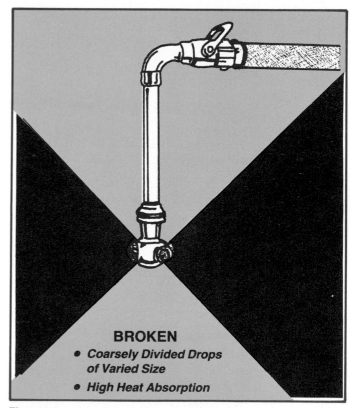

Figure 9.6

sharply defined and is, to a considerable extent, a matter of judgement. It is difficult to say within five feet (2 m), or in some instances even within ten feet (3 m), just exactly where the stream ceases to be good. Any observations or recordings that may

---

*Here is where the difference between a fog stream and a broken stream lies: the fog stream has a definite pattern, whereas the broken stream does not always have a definite pattern and is usually composed of larger drops.

be made to classify solid streams in turn reflect the characteristics of the condition of the nozzle. Observations and tests, covering the effective range of fire streams, have classed as good those streams which at the point of breakover have the following required physical characteristics:

- A stream that at the point of breakover has not lost continuity by breaking into showers of spray.

- A stream that up to the point of breakover appears to shoot nine-tenths of the whole volume of water inside a circle 15 inches (38 cm) in diameter and three-quarters of it inside a 10-inch (25 cm) circle, as nearly as can be judged by the eye.

- A stream that would probably be stiff enough to attain in fair condition the height required even though a fresh breeze is blowing.

Water from a nozzle producing a jet to conform with these characteristics would be thrown farther than the breaking point, but beyond this point, it tends to be in the form of a heavy rain which is easily carried away. These characteristics and the point of breakover are illustrated in Figure 9.7.

Flow Capacity of Solid Streams

The rate of discharge of a fire stream is measured in gallons per minute (gpm) or liters per minute (L/min). Flow from a nozzle is largely determined by the velocity of the stream and the size of the discharge opening. These two factors are influenced by other forces and, for this reason, they will be considered separately.

Effective Reach of a Solid Stream

Just as a solid fire stream will deliver a definite discharge under given conditions, it will also travel a predetermined distance from the nozzle. The reach of a solid fire stream is the distance which a stream can be effectively thrown from a nozzle. Two forces, gravity and friction of the air, tend to decrease this reach. These opposing forces must be overcome by the forward velocity and the volume of the stream. Since the stream is beyond mechanical control after it leaves the nozzle, its speed while passing through space and the distance it will travel are determined by these two factors.

If the stream is encased within a pipe and put under 50 psi (350 kPa) pressure at the base of the pipe, the water will fill to a height of 115 feet (35 m). In other words, if one psi (7 kPa) will raise water in the pipe 2.3 feet (70 cm), then 50 psi (350 kPa) will raise the water 115 feet (35 m). If the same stream of water is not encased in a pipe but is directed vertically into the air, the performance changes. After leaving the tip, the stream tends to hold its shape for a considerable distance and then begins to flare out and break into drops. The broken drops of water soon lose their momentum and fall to the ground. This breaking away continues until only the inner part of the column reaches the maximum height. The friction of the air on the outer surface of the water column breaks the

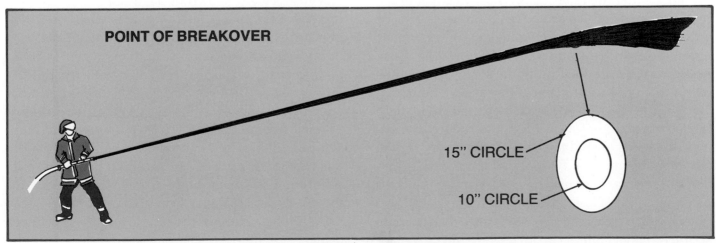

POINT OF BREAKOVER

15" CIRCLE

10" CIRCLE

Figure 9.7 The point of breakover is where the solid stream breaks apart into heavy rain droplets.

stream, and the force of gravity pulls the drops to the ground.

CHARACTERISTICS OF FOG STREAMS

Factors that should be considered when selecting a fog nozzle are the type of jet, the size and weight of the nozzle, and the simplicity of its operation. The design of the fog nozzle controls certain factors which influence the fog stream. A fog nozzle may be considered to have a perfect design which meets all requirements, yet without the required pressure at the nozzle it may prove to be inadequate. Low nozzle pressure will adversely affect the quality of water fog.

One question which has been of great concern to the fire service is the matter of operating pressures. When the design of the nozzle and the pressure at the nozzle are considered alone, it is easy to overemphasize operating pressures. As pressures are increased, factors appear which influence the development of the fog stream pattern. The velocity of the jet, the size of the water particles, and the volume of water discharged, are more important than any excess in pressure which may be necessary to produce a stream that will reach a desired objective.

Although nozzle designs differ, the water pattern which is controlled by the nozzle setting usually affects the ease with which this particular nozzle may be handled. Fire stream nozzles, in general, are not considered to be easily handled, especially nozzles of the solid stream type. This difficulty is due to the fact that water from an ordinary solid stream nozzle is projected directly away from the nozzle and the reaction is equally strong in the opposite direction. Thus, the reaction caused by the velocity and mass of a straight stream, which acts against the nozzle and the curves in the hose, makes the nozzle hard to handle. If water can be made to travel at angles to the direct line of discharge, the reaction forces may be made to more or less counterbalance each other and reduce the nozzle reaction. This balancing of forces is the reason why fog streams can be handled more easily. The desired performance of fog stream nozzles is judged by the amount of heat that a fog stream will absorb and the rate by which the water is converted into steam or vapor. These two factors should be considered when one fog stream nozzle is compared with another because it is from these two performances that a decrease in temperature, the displacement of smoke and gas, and a minimum of water damage are obtained.

Velocity of the Jet

Velocity may be defined as the rate of motion of a particle in a given direction. Direction and rate of motion are both essential parts of velocity and may be measured and expressed pictorially or graphically. Speed is the rate of travel irrespective of direction. The exact measurement of speed of the water towards its objective is of little value to firefighters. It does not matter, from a fire extinguishment standpoint, whether it takes one or five seconds for a particle of water to travel from the nozzle to the fire, as long as the stream reaches the objective in the desired pattern. When a stream is driven against an obstacle, the forward velocity is partially reduced in proportion to the shape of the obstacle and the angle at which it hits. If the surface of the obstacle is at a right angle to the stream, the entire forward velocity will be lost, but if it is less than a right angle, the loss will be in proportion to the angle. As soon as the water is discharged from a nozzle, it is carried forward by its momentum and downward by the force of gravity, and its motion is retarded by friction with the air. Obviously, the faster the water is traveling forward, the farther it will travel before it is pulled to earth by gravity. Forward velocity, therefore, is one factor which governs the reach of a fog stream.

The effect that nozzle design or adjustments has upon a fog stream may be illustrated by operating two identical adjustable nozzles at identical pressures. Since the pressures are equal at both nozzles, it can be assumed that the water is being discharged from both at the same velocity, but the difference in the adjustment of these nozzles demonstrates a marked difference in the two streams. The fog stream in Figure 9.8 appears to have considerably less reach and forward velocity than the one in Figure 9.9. The design of the nozzle permits different adjustment of the fog head to produce different stream patterns from the two nozzles. This adjustment changes the angle of the discharge orifice and increases the forward velocity of the

Figure 9.8 A wide angle fog pattern has less velocity and thus less reach.

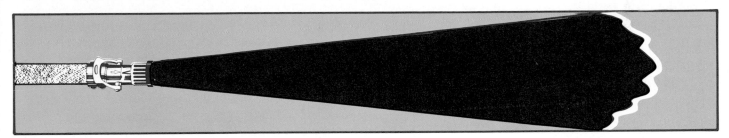

Figure 9.9 A narrow angle fog pattern has more velocity and greater reach.

stream. Fog streams should be evenly divided into a fine spray with a uniform discharge pattern around the "cone" regardless of the nozzle setting.

A short-reach fog has little forward velocity. A long-reach fog has considerable forward velocity, and its reach will vary in proportion to the pressure applied. Unsatisfactory long-reach fog discharge patterns are often caused by insufficient nozzle pressures. There is a maximum reach, however, to any long-reach fog pattern, and once the nozzle pressure is sufficient to produce a satisfactory reach, further increases in the nozzle pressure will have relatively little effect upon the stream except to increase the volume.

Size of the Water Particles

Any jet of water is retarded in its path through the air because of the friction between the outer surface of the jet and the air. A fog stream is composed of droplets of water surrounded by air. The air is entrained in the stream and becomes part of the stream. A fog stream is retarded by the friction between the outer surface of the fog cone and the surrounding air, turbulence within the fog stream,

and the process of entraining air. If two fog streams are of the same diameter and the water droplets had the same initial velocity, the one containing the most water will have the greatest reach because it will have a greater momentum. The size of the droplets will have some effect on the reach of the fog stream because of the amount of entrained air. Since fog streams have a larger diameter in comparison with straight streams, there is more area subject to friction with the air, and its forward velocity is retarded more rapidly.

Water may be divided so finely that the air appears to be saturated with a fine mist, and the small water particles appear to be suspended in the air. Water particles in this form are so fine that they may be carried away from the fire by air currents and never penetrate the fire area. The result is that they do not have an opportunity to absorb heat with sufficient rapidity and quantity to be effective. In order to maintain momentum and obtain desired reach and penetration, a spray of water may be desirable and effective. Fog patterns should be sufficiently heavy to work into a moderate wind without being destroyed.

Volume of Water Discharged

Another factor that is determined by the design of the nozzle and the pressure at the nozzle is the volume of water discharged. Fog streams are superior to solid streams when their heat absorption qualities are compared, but fog streams must have sufficient volume to penetrate the heated area. Water cannot absorb its full capacity of heat unless it is completely converted into steam. If a low volume nozzle producing finely divided particles is used where the heat is generated faster than it is absorbed, the fire might be checked to a certain extent but extinguishment probably would not be accomplished. It is essential for a fog stream to deliver a volume of water which is sufficient to absorb heat more rapidly than it is generated.

Reach of the Fog Stream

Before the introduction of a "shut off" to fog nozzles, they were considered likely to cause unnecessary water damage. They also did not have what was then considered to be the desired quality of penetrating the actual seat of the fire. Another serious objection to these fog nozzles was the poor reach of the stream. The process of constructing a nozzle to produce fog at ordinary working pressures caused a considerable reduction in the velocity of the jet. This reduction in velocity also caused a serious reduction in reach which required a very close approach to the fire. These objections were not well-founded, since the space occupied by the fog stream was given little consideration at that time. The reach of a fog stream and the ease of handling a fog stream were then considered to be of prime importance because fire extinguishment was to be obtained by direct application of water to the seat of the fire. The application of water fog into the heated area, to absorb heat and displace smoke, had not yet been considered. Application of water fog for this purpose demands that the fog stream have both sufficient reach and volume.

CHARACTERISTICS OF BROKEN STREAMS

Broken streams are produced by broken stream nozzles. They are especially useful for fires in walls, attics, and basements. The effectiveness of the nozzle will be determined by the magnitude of the fire, its location and fuel type.

When used on basement fires, a cellar nozzle or distributor is commonly used. They should be lowered through holes cut in the floor or through some other suitable opening. Unseen obstructions may inhibit the effectiveness of the nozzle used through the hole.

Since these nozzles do not have shutoffs on them, an inline shutoff valve should be placed at a convenient location back from the nozzle. Lower the nozzle until it hits something solid, as a check for obstructions. If the drop is far enough to indicate that there are no obstructions between the floor and ceiling of the basement, raise the nozzle a little higher than halfway and apply the stream.

A water curtain is a fire stream which takes, as nearly as possible, the shape of a curtain or fan between a fire and other combustible material. A water curtain is sometimes desirable to protect firefighters from heat. One who has had the experience of following a fog nozzle into a heated room or against a fire can readily appreciate this use. The curtain must cover a wide area and be reasonably heavy to fill this requirement. Tests have indicated that a water curtain between the fire and combustible material is not as effective as the same amount of water flowing over the surface of the combustible material. It is, therefore, better to direct fire streams upon exposed surfaces rather than to establish a water curtain between a fire and an exposure.

FIRE HOSE NOZZLES

Fire hose nozzles are grouped according to the type of stream desired and the purpose for which they are to be used (Figure 9.10). The most common types of nozzles are:

- Solid Stream
- Fog Stream
- Broken Stream
- Master Stream
- Special Stream

The solid stream nozzle is designed to produce a stream as compact as possible with little shower or spray. Its ability to reach areas that might not be reached by other nozzles has established the solid stream as necessary for the fire service.

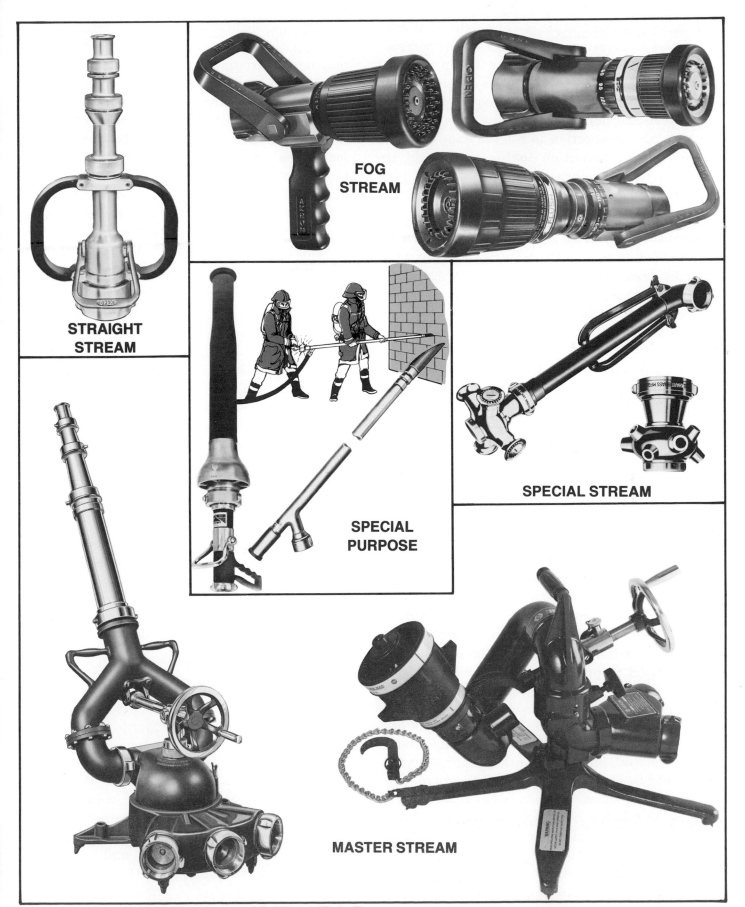

STRAIGHT STREAM

FOG STREAM

SPECIAL PURPOSE

SPECIAL STREAM

MASTER STREAM

Figure 9.10 Nozzles have been designed to develop solid, fog, broken, master, and special streams. *Courtesy of Akron Brass and Elkhart Brass.*

Many designs and shapes of nozzles that produce solid streams are available to the fire service, and manufacturers are constantly striving for better designs. Although these nozzles differ in shape and size, they all conform to the required physical characteristics for solid stream nozzles. The shutoff valve connection to the playpipe and the shutoff valve connection to the tip should be threaded with 1½-inch (38 mm) National Hose thread.

The adjustable fog nozzle is the most popular method of making a fog stream. These nozzles may be adjusted from a wide-angle fog pattern to a straight stream pattern. Remember that a *straight stream* is a pattern of the adjustable fog nozzle whereas a *solid stream* is discharged from a smooth-orifice nozzle.

It is difficult to draw an exact line between broken stream and fog stream nozzles except that streams from fog nozzles are usually produced in definite patterns and in finer particles. A broken stream should cover the most space possible and yet afford enough heat absorption to extinguish fire. If it is possible to get the nozzle close to the fire, the breaking of the stream may be done at the nozzle. These broken streams are useful when fighting a basement fire through the floor above or when fighting fire in a partition. The volume of a broken stream should be comparable to that of a solid stream. Where broken streams are used, water damage is not generally given as much concern as extinguishing the fire, since considerable water is projected over the general area. Regardless of the method used to produce a broken stream, such streams have a definite purpose in fire fighting.

The term *master stream* is applied to any fire stream that is too large to be controlled without mechanical aid. Safety precautions, such as tying off, should be employed. A master stream may be either solid or fog and is produced by using special nozzles, paralleled hoselines and large-capacity pumps. A master solid stream requires a large, smooth nozzle; a master fog stream requires a fog nozzle of a size sufficient to deliver an amount of water fog comparable in volume to that of the solid stream.

Because master stream devices are used from fixed positions most of them have some means for moving the stream in either a vertical or horizontal plane, or both. To permit such adjustments the water must pass through one or more sharp bends near the throat of the nozzle. On some larger master stream devices there are two bends that form a loop in the shape of a ram's horns. Some other master stream devices have the shape of a lateral, angular *U*. The amount of friction loss will vary from device to device. Although the friction loss in a master stream device may be assumed to be some fixed figure, such as 10 or 25 psi (70 or 170 kPa), each department must determine the friction loss in the devices it has available.

Special stream nozzles are designed for special purposes to meet specific needs. The degree to which such nozzles are needed in the fire service depends upon the particular situation that exists or upon the particular fuel that has set up the situation. The widespread use of hazardous industrial processes, the storage and transportation of flammable liquids, the testing of aviation fuels and engines, and the construction of buildings in general are some of the activities that have increased the demand for special-purpose nozzles for fire fighting.

Nozzles should be inspected periodically to make sure they are in proper working condition. This inspection should include:

- Checking the gasket.
- Checking for external damage.
- Checking for internal damage and debris.
- Check the ease of operation.

The nozzles should be thoroughly cleaned with soap and water, using a soft bristle brush.

FOAMS

Fire fighting foams are created by the mechanical mixing of foam concentrate, water, and air. Foams are well suited for use against flammable liquids because of their light density, high water content, blanketing tendencies, and resistance to rapid breakdown. These properties allow foam to be floated over burning flammable liquids, smothering the fire out, and cooling hot objects in and near

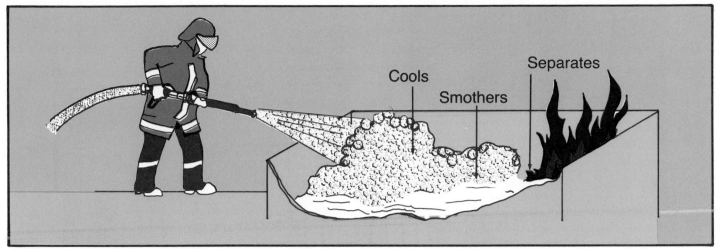

Figure 9.11 Foams extinguish by cooling, excluding the oxygen, and separating the vapors from the heat source.

flammable liquids (Figure 9.11). Foams are not suitable for water-reactive or three dimensional fires, such as leaking flanges.

There are three basic types of foam agents. These are chemical foams, protein foams, and synthetic based foams. Of these three the protein and synthetic are generally called mechanical foams to distinguish them from the chemical-type foams. All three foams can be utilized with fresh or salt water.

Chemical Foams

Chemical foams are made when an acid salt and an alkaline salt come into contact in solution. The acid salt is usually termed the "A" ingredient and the alkaline salt called the "B" ingredient.

Systems using chemical foam have inherent maintenance problems due to the corrosive nature of the "A" and "B" powders. Powder mixing often results in clogging problems, and the foams produced have serious drawbacks because of their tendency to "bake" hard and crack. This results in fissures where the flammable liquids can vaporize and burn. For these reasons, and because other suitable foams are readily available, chemical foams are considered obsolete, and are included in this section primarily to address remaining installations.

Protein Foams

REGULAR PROTEIN FOAMS

Regular protein foams are chemically broken-down natural protein solids. The end product of this chemical digestion is protein polymers, which have excellent elasticity, water retention capabilities and high strength. These characteristics make protein foam excellent for use as a fire fighting agent. Basic protein foams are nontoxic and work within a 100°F (38°C) temperature range from 20°F to 120°F (7°C to 49°C). These foams are not well-suited for use on polar solvents, for extremely cold temperatures, for subsurface injection, or for use with dry-chemical powders. However, certain organic solvents can be added to the foam to improve the suitability for use in these situations.

Regular protein foams, as well as most other foams, are marketed to be mixed into 3 percent and 6 percent foam solutions. This means that in a 3 percent foam solution for every one hundred parts of foam solution (foam concentrate and water mixed) there are 3 parts of concentrate to 97 parts water. For 6 percent solution there are 6 parts concentrate and 94 parts water. The main reason for this difference in concentration is due to technological advancements in foam concentrate production (Table 9.1). Initially, the best concentrate available was 6 percent. After a time, procedures were improved to allow a suitable 3 percent foam concentrate to be marketed. The 3 percent gave the advantage of costing less to ship per area of coverage, and requiring only one-half the storage space of 6 percent foam concentrates. Six percent foam concentrates are still marketed to meet the needs of departments that cannot afford to convert their systems to a 3 percent foam concentrate.

TABLE 9.1
Low-Expansion Foam Concentrate Required for
10 Minutes of Application

Nozzle GPM Flow of Foam/Water Concentrate	Gallons of Solution Used for 10 min. Application	Gallons of Foam Concentrate		Gallons of Water		Gallons of Aerated Foam
		3%	6%	3%	6%	
60 (227.12 L)	600 (2271.24 L)	18 (68.13 L)	36 (136.27 L)	582 (2203.10 L)	564 (2134.97 L)	4800 (18,169 L)
95 (359.61 L)	950 (3596.14 L)	28.5 (107.88 L)	57 (215.76 L)	921.5 (3488.25 L)	893 (3380.37 L)	7600 (28769.12 L)

These are minimums for each nozzle. Backup supply should be equal to minimum requirement. With good application technique, one-third less AFFF could be used for spill fires.

NOTE: Since 1 gallon (3.79 L) of solution produces 8 gallons (30.28 L) of aerated foam, 5 gallons (18.92 L) of 6 percent foam concentrate should yield 664 gallons (2513.5 L) of aerated foam; 5 gallons (18.92 L) of 3 percent concentrate should yield 1328 gallons (5027.0 L) of aerated foam.

Low-temperature foams are protein based foams with added nonflammable antifreeze solutions. These foams can be used at air temperatures down to -20°F (-7°C). As with normal protein foam, they are marketed in 3 percent and 6 percent solution.

FLUOROPROTEIN FOAMS

Fluoroprotein foams are basically protein foams fortified with fluoronated solvents. The addition of these solvents gives fluoroprotein foams several advantages in that these tend to separate from the flammable liquids they are mixed into. This quality makes fluoroprotein foams excellent for subsurface injection, or surface applications in which the foam becomes agitated with the flammable liquid. Other advantages are that fluoroprotein foams tend to be more stable than regular protein foams and more compatible for use with dry-chemical agents.

Fluoroproteins have the same temperature characteristics as regular protein foams. It is also marketed in 3 percent and 6 percent concentrates.

ALCOHOL FOAMS

Alcohol foams have been developed for use on polar solvents, such as alcohol, lacquer thinner, acetone, and ketones. Polar solvents are miscible in water. This characteristic causes regular foams to break down rapidly. The alcohol foams were developed because regular foams are very miscible in polar solvents, and tended to melt into the burning liquid without extinguishing the fire. Additionally, regular hydrocarbon liquids mixed with even small amounts of polar solvents tend to destroy the effectiveness of regular foam products.

There is one type of protein-based alcohol foam. This foam is derived from regular protein foam mixed with heavy-metal salts suspended in organic solvents. Protein-based alcohol foams must, to be effective, be applied gently to the burning surface, and must be applied immediately after eduction into the water. Protein-based alcohol foams when pumped through hose or piping lose effectiveness. This loss is in proportion to the distance it is pumped. For this reason, protein-based alcohol foams are usually mixed into the water stream at or near the application nozzle. Application temperature is from 35°F to 120°F (20°C to 49°C).

Synthetic Foams
ALCOHOL FOAMS

There are two types of synthetic alcohol foams. Both of these foams have the advantage over protein-based alcohol foams in that they do not have to be applied gently to the surface, and can be pumped long distances without loss of effectiveness. Both are marketed in 3 percent and 6 percent solutions and have the same temperature range as protein alcohol foam.

The first type of synthetic alcohol foam is catalytic alcohol foam. It is the result of the mixing of a two-part solution, a polymer and catalyst. The resulting foam is a very stable alcohol-resistant foam. This foam can be pumped long distances without loss of effectiveness.

The second synthetic alcohol foam is sometimes referred to as multi-use foam. This is because it can be used effectively on either hydrocarbon or polar liquids, and works effectively when applied through most foam-mixing devices. These foams are synthetically produced concentrates in a single-component foam. The proportioning is very similar to that of regular protein foams, and they may be pumped long distances without loss of effectiveness.

DETERGENT FOAMS

Several synthetically produced detergent foams are available. These foams have a high foam yield, but are generally less stable than other foams. They have a low resistance to heat or physical destruction, and require high application rates to be effective.

Detergent foams cause a very low surface tension of water, and may be thought of as wetting agents. This makes their use on Class "A" combustibles effective and will sometimes form a frothy emulsion on top of flammable liquids. However, this characteristic causes detergent foams to break down other foams.

AQUEOUS FILM FORMING FOAMS

Aqueous film forming foams (AFFF) are dual-action synthetic foams. The first action of AFFF is similar to the detergent foam's air-entrapping action. This action forms a blanket of strong foam which spreads over the burning liquid surface. This blanket smothers the fire and retards vaporization of the flammable liquid to below flammable limits. The second action is the forming of a film of aqueous solution void of bubbles across the surface of the liquid. This aqueous film smothers the fire and prevents vaporization just as the foam film. The aqueous film is self-healing, and will re-cover open areas caused by agitation of the flammable liquid surface (Figure 9.12). The result of foam ac-

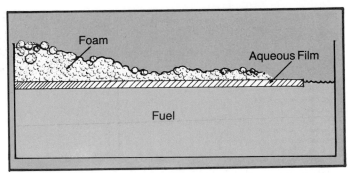

Figure 9.12 AFFF has a film that floats on the surface of the liquid restricting vaporization.

tion and the film forming action make AFFF one of the most dependable and versatile foams available. Use of AFFF foams has the added advantages of not requiring special application devices and being compatible with dry chemical.

AFFF foams are generally strengthened by the addition of fluoronated solvents like those used to stabilize protein foams. These foams can be used on flammable liquids, and under some conditions on polar solvents. Often AFFF foams are used on flammable liquid spills to prevent ignition of the flammable liquid. AFFF foams are marketed in 3 percent and 6 percent solutions, and have temperature characteristics like alcohol foams.

HIGH EXPANSION FOAMS

High expansion foams are foam concentrates mixed into a 2 percent solution and then mixed with air to form high-air-content, good-quality foams. High expansion foam is different from other foams in the respect that it is useful to fight fires in indoor structures and inaccessible places through a total flooding application. It has been shown to be an effective agent when used with water sprinklers. Outdoor use of high expansion foam can be effective, dependent upon the wind conditions.

When using high expansion foams it is necessary to provide ventilation to allow the foam to flow properly. If air becomes trapped ahead of the foam it will not flow adequately.

High expansion foam acts through oxygen displacement in a confined area. It also provides a good insulating blanket which serves as exposure protection. It is nontoxic, but entry into the foam is hazardous because of lack of visibility. Also, foams generated using air contaminated with products of combustion will have the toxic characteristics of the contaminant, and entry will therefore require breathing apparatus. Additionally, the products of combustion cause the foam to break down rapidly.

Foam Application

As in all fire fighting, the development of a good technique of applying a foam blanket requires much training and practice. Since there is a definite aging and deterioration of foam concentrates, a program should be set up to use the oldest foam for training with a definite schedule of replacing the foam stock over a period of time. Protein foam has a shelf life of only five years.

Laboratory tests on mechanical or protein foams are run at a rate of application of one gallon per minute (L/min) water rate expanded to the normal 10 to 1 expansion, per square foot (30 square cm) of burning surface of flammable liquid. This is a good rate of application for practical purposes. This means with a 60 gpm (227 L/min) nozzle, a fire in a normal hydrocarbon fire of 600 square feet (180 square m) should be controlled. The rate of application of alcohol foams on alcohol fires is normally about twice this amount and some of the other surfactants require a slightly higher than one gallon per minute (L/min) per square foot (30 square cm) application. When fighting interior fires, the rate of foam application is very important. With a limited water supply a more rapid rate of application is often better than extending the time of application with smaller rate of application.

The most successful method of applying the foam to a liquid surface is to bounce the stream off a solid surface. If the fire is a spill on a level surface, the stream should be directed at the ground in front of the fire to start a blanket of foam. While sweeping the stream the width of the fire the nozzle is raised to push a blanket of foam over the surface. Care should be taken not to apply the stream directly into an existing foam blanket and thus blow foam from the surface. The movement of the nozzle should not be violent but slow and deliberate. The stream should be directed to the side until foam appears and then turn it to the fire. If foam should

stop while in the process of extinguishing the fire, the stream should immediately be removed from the foam blanket so as not to wash it away. Always use the maximum reach of the stream and let the foam do the work (Figure 9.13).

Figure 9.13 Either roll the foam onto the fire from the front to control a spreading surface fire or bank it off a wall or other solid object to cover the fire without splattering the fuel.

If the fire is in a tank or located so that the above method cannot be used, the stream should be directed against the side of the tank or other obstacle. This method uses the object as a splash board which permits the foam to form a blanket to slide down and across the surface. An attempt to lob or drop the foam on to the surface should be avoided. This method is very ineffective since the foam must fall through the heat where it will break up and fall on the surface in a rather scattered form. Normally, the foam blanket should be formed about three inches thick.

FOAM PROPORTIONERS

For effective foam generation the foam concentrate must be mixed with the proper ratio of water and then aerated (except those chemical foams and catalytic foams which do not require aeration). It is important that the ratio of concentrate to water is maintained. Too much concentrate wastes the product, and foams lean on concentrate are ineffective.

FOAM NOZZLE EDUCTOR

This type of eductor utilizes a venturi action to draft concentrate. It can draft concentrate up to six feet (1.8 m). Immediately after the venturi proportionment the foam-making solution is aerated by the nozzle. It is well-suited for use with protein alcohol foams which need to be applied soon after mixing. The foam nozzle eduction is a one location application appliance because of the trouble of moving the nozzle about along with the concentrate container.

IN-LINE EDUCTOR

The in-line eductor utilizes the same venturi action as the foam nozzle eductor. However, the foam solution can be pumped to a remote location for aeration. The in-line eductor can be connected directly to a foam nozzle to form a foam nozzle eductor, or can be put on the hoseline or even on the discharge of the pump. Eductor placement should follow the manufacturer's recommendations to allow for the back pressure that will be developed. One eductor can be used to supply more than one line. Caution is necessary when using an in-line eductor in this fashion because this type of device is usually designed to flow a certain amount of water at a given pressure. With multiple nozzles, this design criterion may be exceeded and the foam produced becomes ineffective. Care must be taken to match the nozzle or nozzle combination GPM (L/min) with the eductor design criterion. Some fixed systems utilize in-line eductors, but most systems depend on more accurate and dependable proportioner systems.

Foam Nozzles

Foam nozzles are foam application devices. These nozzles can be adjusted or moved to change the foam application pattern. One type of foam nozzle has already been discussed, and this was the foam nozzle eductor. This particular nozzle is special as it has the eductor built into the nozzle. Other foam nozzles require a separate eductor. The remainder of the foam nozzles fall into three categories: foam sprinklers, foam-aspirating nozzles, and high expansion foam generators.

FOAM SPRINKLERS

Foam sprinklers are found on foam deluge and foam water systems. Foam sprinklers resemble aspirating nozzles in that they utilize a venturi velocity to mix air into the foam-making solution. However, some systems utilize standard sprinklers to form a low-grade foam through simple turbulence of the water droplets falling through the air. These systems generally use AFFF foams. Foam sprinklers come in upright and pendant designs. The deflector on them must be adapted to meet the specific installation requirements.

FOAM-ASPIRATING NOZZLES

Foam nozzles come in two major categories: the fog-foam nozzle and the variable-pattern fog nozzle. These nozzles are employed on fixed systems for manual application of foam in conjunction with the system application. They give the system the added versatility of using portable nozzles. The fog-foam-type nozzle is marketed with two-stream-shaping foam adapters. The basic nozzle breaks the foam-making solution into small streams, plus inducts air through a venturi action. The attachments serve to give the foam application qualities different than the standard nozzle. The cone-shaped attachment allows the nozzle extra reach, and the screen produces a more homogeneous high-air-content for gentle applications.

The water fog nozzle, as used with foam solution, produces a low-quality short-standing foam. This nozzle breaks the foam solution into tiny droplets and utilizes the agitation of water droplets moving through air to achieve the foaming action. Its best application is when coupled with AFFF, which, due to their filming characteristics, do not require a high-quality foam to be effective. Water fog nozzles have found a growing acceptance because of the capability to be used as both a normal fire fighting nozzle and a foam nozzle.

HIGH BACK PRESSURE ASPIRATORS

The high back pressure aspirator, or forcing foam aspirator, is an in-line aspirator utilized in situations requiring foam to be delivered under pressure. These devices are best suited for subsurface injection, but are used in conjunction with many fixed-piping systems. High back pressure aspirators operate through venturi action. This action typically produces a low-air-content but homogeneous and stable foam.

HIGH EXPANSION FOAM GENERATORS

High expansion foam generators produce a high-air-content, stable foam. The air content ranges from 100 parts air to one part foam solution (100x) to 1,000 parts air to one part foam solution (1,000x). There are two basic types of high expansion foam generators: the mechanical blower and the water aspirating. The water aspirating is very similar to the other foam-producing nozzles except it is much larger and longer. The back of the nozzle

is open to allow airflow. The foam solution is pumped through the nozzle in a fine spray which mixes with air to form a moderate expansion foam. The end of the nozzle has a screen, or series of screens, which breaks the foam up and further mixes it with air. These nozzles typically produce a lower-air-volume foam than do mechanical blower generators.

Mechanical blower generators resemble smoke ejectors in appearance (Figure 9.14). They operate along the same principle as the water-aspirating nozzle except the air is forced through the foam spray instead of being pulled through by the water movement. These devices produce a higher-air-content foam, and are typically associated with total flooding applications.

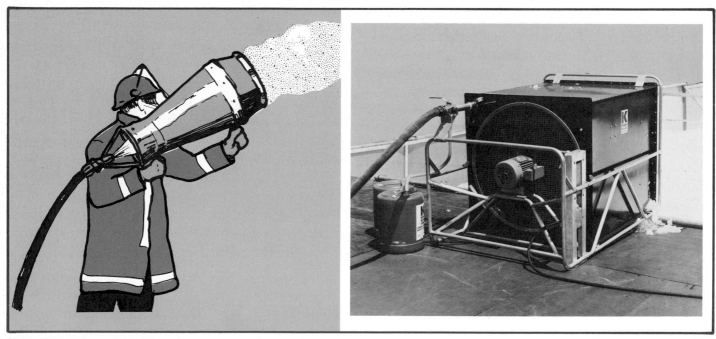

Figure 9.14 High expansion foam generators may be portable or hand held. The portable units may look similar to mechanical smoke ejectors. *Courtesy of Walter Kidde & Co., Inc.*

**NFPA STANDARD 1001
FIRE HOSE
Fire Fighter I**

3-7 Fire Hose, Nozzles, and Appliances

3-7.1 The fire fighter shall identify the sizes, types, amounts, and use of hose carried on a pumper.

3-7.2 The fire fighter shall demonstrate the use of nozzles, hose adapters, and hose appliances carried on a pumper.

3-7.3 The fire fighter, given the necessary equipment and operating as an individual and as a member of a team, shall advance dry hose lines of two different sizes, both of which shall be 1½ in. or larger, from a pumper:

  (a) Into a structure
  (b) Up a ladder into an upper floor window
  (c) Up an inside stairway to an upper floor
  (d) Up an outside stairway to an upper floor
  (e) Down an inside stairway to a lower floor
  (f) Down an outside stairway to a lower floor
  (g) To an upper floor by hoisting.

3-7.4 The fire fighter, given the necessary equipment and operating as an individual and as a member of a team, shall advance charged attack lines of two different sizes, both of which shall be 1½ in. or larger, from a pumper:

  (a) Into a structure
  (b) Up a ladder into an upper floor window
  (c) Up an inside stairway to an upper floor
  (d) Up an outside stairway to an upper floor
  (e) Down an inside stairway to a lower floor
  (f) Down an outside stairway to a lower floor
  (g) To an upper floor by hoisting.

3-7.5 The fire fighter shall demonstrate the techniques for cleaning fire hose, couplings, and nozzles; and inspecting for damage.

3-7.6 The fire fighter shall connect a fire hose to a hydrant, and fully open and close the hydrant.

3-7.7 The fire fighter shall demonstrate the loading of fire hose on fire apparatus and identify the purpose of at least three types of hose loads and finishes.

3-7.8 The fire fighter shall demonstrate three types of hose rolls.

3-7.9 The fire fighter shall demonstrate two types of hose carries.

3-7.10 The fire fighter shall demonstrate coupling and uncoupling fire hose.

3-7.11 The fire fighter shall work from a ladder with a charged attack line, which shall be 1½ in. or larger.

3-7.12 The fire fighter shall demonstrate the techniques of carrying hose into a building to be connected to a standpipe, and of advancing a hose line from a standpipe.

3-7.13 The fire fighter shall demonstrate the methods for extending a hose line.

3-7.14 The fire fighter shall demonstrate replacing a burst section of hose line.*

Fire Fighter II

4-7 Fire Hose, Nozzles, and Appliances

4-7.2   The fire fighter shall demonstrate all hand hose lays.

4-7.3   The fire fighter shall demonstrate inspection and mainte-nance of fire hose, couplings, and nozzles, and recommend replacement or repair as needed.

4-7.4   The fire fighter shall demonstrate all hydrant to pumper hose connections.

4-7.5   The fire fighter shall select adapters and appliances to be used in a specific fire ground situation.*

*Reprinted by permission from NFPA Standard No. 1001, *Standard for Fire Fighter Pro-fessional Qualifications.* Copyright © 1981, National Fire Protection Association, Boston, MA.

IFSTA's two Hose Transparency Sets are designed to complement this chapter.

CONSTRUCTION AND CARE OF HOSE

Fire hose is classified by its size (inside diameter) and by material from which it is constructed. Present-day fire hose is made of many materials which may be susceptible to deterioration and wear, and it can be made in several grades and degrees of quality. Fire hose is manufactured by three specific basic construction methods: (1) braided, (2) wrapped, and (3) woven. Some of the major fibers being used in the construction of the outer jacket of fire hose are cotton, nylon, rayon, vinyl, and polyester fibers. It is important to remember that fire hose must withstand relatively high pressures, be able to transport water with a minimum loss in pressure, and be sufficiently flexible to permit loading into a hose compartment without occupying excessive space (Figure 10.1).

Types and Sizes

The term "fire hose" identifies a type of flexible tube used by firefighters to carry water under pressure from the source of supply to a point where it is

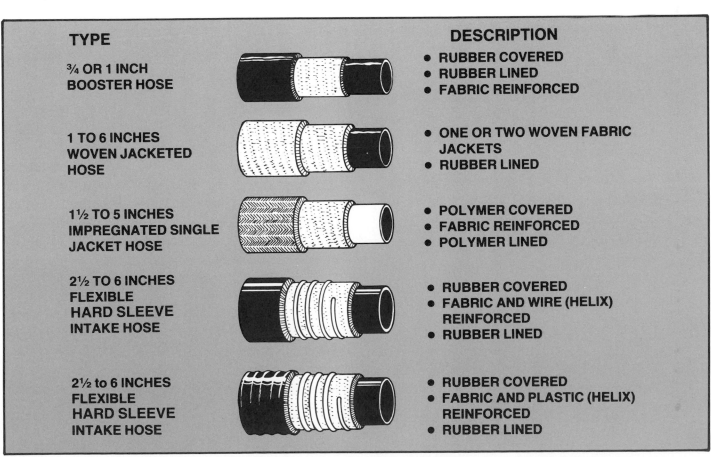

TYPE	DESCRIPTION
¾ OR 1 INCH BOOSTER HOSE	• RUBBER COVERED • RUBBER LINED • FABRIC REINFORCED
1 TO 6 INCHES WOVEN JACKETED HOSE	• ONE OR TWO WOVEN FABRIC JACKETS • RUBBER LINED
1½ TO 5 INCHES IMPREGNATED SINGLE JACKET HOSE	• POLYMER COVERED • FABRIC REINFORCED • POLYMER LINED
2½ TO 6 INCHES FLEXIBLE HARD SLEEVE INTAKE HOSE	• RUBBER COVERED • FABRIC AND WIRE (HELIX) REINFORCED • RUBBER LINED
2½ to 6 INCHES FLEXIBLE HARD SLEEVE INTAKE HOSE	• RUBBER COVERED • FABRIC AND PLASTIC (HELIX) REINFORCED • RUBBER LINED

Figure 10.1 A variety of constructions are used to produce the different types and sizes of fire hose.

discharged. In order to be reliable, fire hose should be constructed of the best materials and it should not be used for purposes other than fire fighting. Fire hose is the most used item in the fire service and the way in which it is used requires it to be flexible, watertight, have a smooth lining, and a durable covering. The manner in which fire hose is used with and applied to other appliances has associated its use with other essential functions of fire fighting.

The many different sizes of fire hose used by fire departments are all designed for a specific purpose. When reference is made to the diameter of fire hose, the dimensions which are stated are the inside diameter of the hose. Fire hose is usually cut and coupled into lengths of fifty feet for convenience of handling and replacement, but lengths greater than fifty feet (15 m) may be obtained. These lengths are commonly referred to as sections and they must be coupled together to produce a continuous hoseline. The National Fire Protection Association (NFPA) Standard No. 1961 lists specifications for fire hose, and their Standard No. 1963 lists specifications for fire hose couplings and screw threads. The following list includes the commonly used sizes of fire hose:

- The ¾- or 1-inch (16-25 mm) rubber-covered, rubber-lined hose, equipped with 1-inch couplings, is commonly called booster or chemical hose.

- The 1-inch (25 mm) woven-jacketed rubber-lined hose, equipped with 1-inch (25 mm) couplings, sometimes used on booster pumps.

- The 1-inch (25 mm) woven-jacketed rubber-lined single jacket hose with 1-inch (25 mm) couplings, commonly called forestry hose.

- The 1½-inch (38 mm) woven-jacketed rubber-lined hose, single or double jacket, equipped with 1½-inch (38 mm) couplings.

- The 1¾-inch (44 mm) rubber cover, single, or double woven-jacketed rubber-lined hose, equipped with 1½-inch ((38 mm) couplings.

- The 2-inch (50 mm) woven-jacketed rubber-lined hose, single or double jackets, equipped with 2½-inch (65 mm) couplings.

- The 2½-inch (65 mm) woven-jacketed rubber-lined hose, single or double jacket, equipped with 2½-inch (65 mm) couplings.

- The 2¾-inch (69 mm) woven-jacketed rubber-lined hose, single or double jacket, equipped with 2½-inch (65 mm) couplings.

- The 3-inch (77 mm) woven-jacketed rubber-lined hose, single or double jacket, equipped with 2½-inch (65 mm) reducing couplings or 3-inch (77 mm) couplings.

- The 3½-inch (90 mm) woven-jacketed rubber-lined hose, single or double jacket, equipped with 2½-inch (65 mm) or 3-inch (77 mm) reducing couplings, or 3½-inch (90 mm) couplings.

- The 4-, 4½-, 5-, and 6-inch (100-, 115-, 130, and 160 mm) woven-jacketed, or plastic covered, rubber-lined synthetic intake or supply hose.

- The 2½-, 3-, 4-, 4½-, 5-, and 6-inch (65-, 77-, 100-, 115-, 130, and 160 mm) hard sleeve intake hose.

Care and Maintenance

Since fire hose is a device or a tool to use during fire fighting, it is only natural that it will be subjected to all sorts of situations. Little can be done at fires to provide safe usage and to protect it from injury except in certain instances. Probably the most important factor relating to the life of fire hose is the care it gets after fires, in storage, and on the fire apparatus. Fire hose should be selected with caution to assure its lasting qualities, but regardless of its good materials it cannot stand up under mechanical injury, heat, mildew and mold, chemical contacts, and pressures which may exceed those for which it was tested. The life of fire hose is, however, considerably dependent upon how well the hose is protected against these five destructive causes.

MECHANICAL INJURY

Fire hose may be damaged in a variety of ways when being used at fires. Some common mechanical injuries are worn places, rips, abrasions, crushed or damaged couplings, and cracked interlinings. To prevent these damages the following good practices are recommended:

- Avoid laying hose over rough, sharp corners.
- Provide warning lights and use hose bridges in traffic lanes.
- Prevent vehicles from running over fire hose.
- Avoid closing the nozzle abruptly to prevent a water hammer.
- Change position of bends in hose when reloading.
- Provide chafing blocks to prevent abrasion to hose when it vibrates near the pumper.
- Avoid excessive pump pressure on hoselines.

HEAT

The exposure of hose to excessive heat or its contact with fire will char, melt, or weaken the fabric and dry the rubber lining. A similar drying effect may occur to interlinings when hose is hung in a drying tower for a longer period of time in high temperatures. To prevent this damage, firefighters should conform to the following recommended good practices:

- Protect hose from excessive heat or fire when possible.
- Do not allow hose to remain in any heated area after it is dry.
- Use moderate temperature for drying. A current of warm air is much better than hot air.
- Keep the outside woven jacket of the hose dry.
- Hose that has not been used for some time should have water run through it to prolong its life.
- It is not good practice to dry fire hose on hot pavement.

MILDEW AND MOLD

Mildew and mold may occur on the woven jacket when moisture is allowed to remain on the outer surfaces. This condition will cause rot or decay and the consequent deterioration of the hose. Some methods of preventing mildew and mold are as follows:

- Remove all wet hose from the apparatus after a fire and replace it with dry hose.
- If hose has not been unloaded from the apparatus during a period of thirty days, it should be removed, inspected, swept, and reloaded.
- Some fire hose has been chemically treated to resist mildew and mold but such treatment is not always 100 percent effective. Regardless of this, hose should be exercised every thirty days, and water should be run through it every ninety days to prevent drying and cracking of the rubber lining.

CHEMICAL CONTACT

Chemicals and chemical vapors will damage the rubber lining and often cause the lining and jacket to separate. When hose is exposed to petroleum products, paints, acids or alkalies, it may be weakened to the point of bursting (Figure 10.2).

Figure 10.2 Chemicals will damage the hose jacket and rubber lining.

After being exposed to chemicals or chemial vapors, hose should be cleaned as soon as practical. Some recommended good practices are as follows:

- Thoroughly scrub and brush all traces of acid contacts with a solution of bicarbonate of soda and water. The soda will neutralize the acids.
- Guard againt spilling gasoline on hose when filling the fuel tank.

- Remove hose from the apparatus periodically, wash it with plain water, and dry thoroughly.

- If the least suspicion of injury exists, the hose should be properly tested.

- Avoid laying hose in the gutter or where automobiles have been parked next to the curb, because parked cars drop oil from the mechanical components and acid from the battery.

- Runoff water from the fire may also carry various foreign materials that can damage fire hose.

WASHING, DRYING, AND STORAGE

The importance of reliable fire hose is sometimes not fully appreciated and hose which has been neglected is more likely to fail when needed. The replacement of a defective section of hose results in loss of time which may permit a fire to reach proportions beyond control. Undependable fire hose can also be responsible for serious injury to firefighters and other persons. The techniques of washing and drying fire hose and the provisions for storage are very important functions in the care of fire hose.

After fire hose has been in use at fires, the usual accumulation of dust and dirt should be thoroughly brushed from it. If the dirt cannot be removed by brushing, the hose should be washed and scrubbed with clear water. When fire hose has been exposed to oil it should be washed with a mild soap or detergent, making sure that all oil is completely removed. The hose should then be properly rinsed. If a commercial type hose washing machine is not available, common scrub brushes or brooms can be used with streams of water from a hoseline and nozzle. One type of hose washer which will wash almost any size of fire hose up to 3 inches (77 mm) in diameter is shown in Figure 10.3. The flow of water into this device can be adjusted as desired and the movement of the water assists in propelling the hose through the device. The hoseline which supplies the washer with water can be connected to a pumper or used directly from a hydrant. It is obvious that higher pressure will give better results.

Figure 10.3 This water jet type hose washer helps propel the hose while cleaning it.

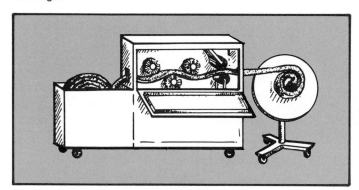

Figure 10.4 This hose washing machine scrubs the hose with brushes, rinses, drains, and rolls the hose.

A hose washing machine of the cabinet type which washes, rinses, and drains fire hose is shown in Figure 10.4. This type of washer can be operated by one person, is self-propelled, and can be used with or without detergents. A hose washing machine is a very important appliance in the care and maintenance of fire hose. Its cost can be justified by comparing the value and cost of the best grade of fire hose. Hose washing machines will, however, not clean hose couplings sufficiently when the coupling swivel becomes stiff or sluggish with dirt or other foreign matter. The swivel part should be submerged in a container of warm soapy water and worked forward and backward to thoroughly clean the swivel (Figure 10.5). The male threads should be cleaned with a suitable brush, and if clogged by tar, asphalt, or other foreign matter a wire brush may be necessary.

Once a fire hose has been thoroughly washed, it should be either hung to dry in a hose tower, placed on an inclined drying rack, or placed in a cabinet type hose dryer (Figure 10.6). Where hose drying towers are available they are usually built into and as a part of the fire station. Tower drying

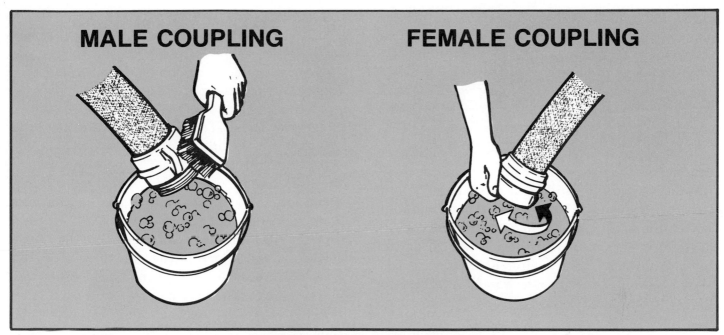

MALE COUPLING

FEMALE COUPLING

Figure 10.5 A brush is used to clean male threads, while the female swivel is rotated in warm, soapy water.

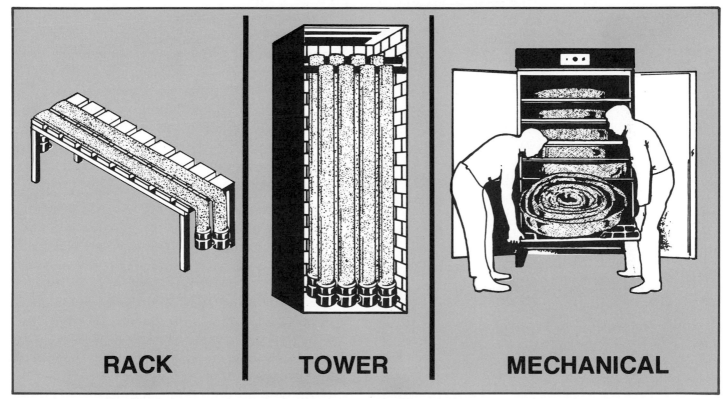

RACK

TOWER

MECHANICAL

Figure 10.6 Hose drying after cleaning is done on racks, in towers, or mechanical dryers.

has proved to be quite successful in most installations. It is, however, important to make sure that the tower is adequately ventilated and that excessive temperatures or direct sunlight are not permitted to reach the hose.

Most authorities feel that it is best for the inside of the rubber tube to remain damp or moist while in storage or on the fire apparatus. Since tower drying provides complete draining and oftentimes drying of the rubber tube, it may be desirable to dampen the inner tube after the outer jacket has dried.

Hose drying racks can be constructed with wood or metal framework and covered with wood slats. Drying racks are more adaptable to long spacious interior walls which are free from window openings. Wall openings may permit the sun's rays to shine upon the outstretched fire hose. Each deck should be about 52 feet (16 m) long and an overall space of 60 feet (18 m) should be provided to permit adequate working space at each end.

Mechanical cabinet dryers may be purchased commercially or they may be built into the structure of a building. Built-in dryers should be provided with adequate fans for forced air circulation within the drying space. Most mechanical cabinet dryers are equipped with thermostats to control temperature and fans to circulate the warm air. When possible, provisions to vent the damp air to the outside should also be made. Racks, upon which the hose is placed, are usually removable so that the sections of hose can be arranged for complete drying.

After fire hose has been adequately brushed, washed, dried, and rolled, it should be stored in suitable racks. These racks should be in the fire station and readily accessible to fire apparatus on the apparatus floor. Here again these racks may be purchased commercially but many successful hose storage racks have been built in fire department shops. Some fire stations provide a hose storage room with racks built along the wall. Some hose storage racks are made of wood, while others are made of metal or pipe (Figure 10.7).

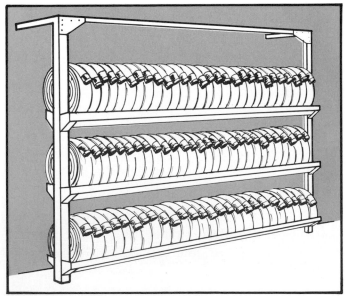

Figure 10.7 Suitable storage racks readily accessible to the fire apparatus should be provided for the cleaned hose.

COUPLINGS AND APPLIANCES

Since the early use of leather and canvas hose to convey water to fires, some means has been necessary to connect sections of the hose together. Stories of early usage reveal that hose couplings were even then made of metal. Fire hose couplings of today are made of durable material and so designed that it is possible to couple and uncouple in a short time with little effort. The materials used for fire hose couplings are generally alloys of brass, aluminum, or magnesium in varied percentages. Such materials lend themselves to the methods of attaching the couplings to the hose and they do not corrode easily. Much of the efficiency of the fire hose operation depends upon the condition and maintenance of its couplings, and firefighters should be knowledgeable of couplings with which they work.

Types of Hose Couplings

There are several types of hose couplings used in the fire service, and some of the more common ones are shown in the following illustrations. The part of the coupling into which the hose is attached may be known by several different names. Some of these identifying terms are tailpiece, bowl, shell, and shank. In order to provide clarity of description, and to maintain consistency, this manual will refer to this part of the coupling as the shank.

The three-piece coupling consists of a male threaded shank, a female threaded swivel, and a non-threaded shank to which the female swivel is attached. One good way to identify the male and female parts of the coupling when a connection is made is to remember that the male shank is equipped with lugs, while the female shank does not usually have lugs. The physical structure of fire hose couplings is many and varied. Some of the popular materials are brass, aluminum alloy, and aluminum alloy with hard coating. Each material may be formed into a coupling by several methods. Cast couplings are probably the weakest and they have the least resistance to internal and external forces or other forms of abuse. Extruded couplings are somewhat stronger than cast couplings and they have a greater resistance to forces. They usually have a smoother surface and lack swivel protection rings. Drop-forged couplings are harder

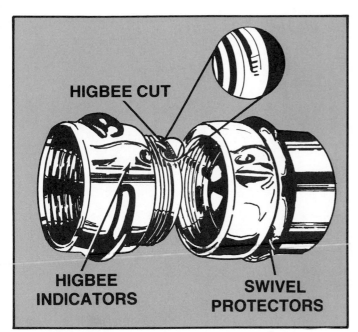

Figure 10.8 Drop-forged coupling with swivel protectors, Higbee indicators, and Higbee cut.

and more desirable for strength on the fireground as compared to cast or extruded couplings. The embossed ridges on the shanks of a drop-forged coupling are known as swivel protectors (Figure 10.8). These protectors strike the ground first when a coupling is dropped or when hose is being pulled from the apparatus.

An added feature that may be otained with screw type couplings is the Higbee cut and indicators. The Higbee cut is a special type of thread arrangement that is cut at the beginning of the thread to provide positive indentification of the first thread and it tends to eliminate cross threading. The Higbee cut is on the thread itself and is shown in Figure 10.8 Higbee indicators are notches or grooves that are cut into the coupling lugs which identify from touch or sight the exact location of the Higbee cut. When the groove on the female swivel lug is in line with the groove on the female coupling lug, the threads are in line for connection.

The materials used in fire hose couplings should be considered on the basis of use and reuse by the local department. Cast brass couplings may not lend themselves to recoupling without permanent expansion of the shank because of their softness. Drop-forged brass may, however, be reused with less swelling during the expansion process. Aluminum alloy (pyrolite) couplings with hard

coating show considerable merit, and they are stronger than drop-forged brass with about one-half the weight. The hard coating protects them from galling, corrosion, and electrolysis which have been common among aluminum alloy materials. Another highly desirable feature is their ability to be reused without excessive stretching of the shank. Intermixing of brass and aluminum alloy theaded connections should be avoided due to the chance of electrolysis taking place over a period of time thus causing the connection to corrode and "freeze" tight.

The three-piece and five-piece reducing coupling shown in Figure 10.9 is used to fit hose larger than 2½-inch (65 mm) and is either reduced in the coupling or by means of separate fittings so that standard 2½-inch (65 mm) threads may be used. This coupling has little effect upon any increase in friction loss because of the jet effect at each coupling. The reducing coupling may be used where 3-inch (77 mm) or larger hose is used in connection with the standard 2½-inch (65 mm) fire hose fittings.

Other methods of connecting couplings include the snap and quarter turn types. The snap

Figure 10.9 Examples of three-piece and five-piece reducing couplings. *Courtesy of Akron Brass.*

Figure 10.10 Quick connect couplings may be either snap, quarter, or one-third turn.

coupling has spring-loaded clips or lugs on a female coupling which clamp over the ring on a male coupling. Couplings are broken by disengaging the clips. A quarter turn coupling has double lugs on the couplings at each end of the hose. They are connected by twisting a quarter turn, which causes the double lugs to interlock. Both of these types can be connected and disconnected quickly, but the advantage of being able to couple with another fire department's hose may be lost. Examples of snap and quarter turn couplings are shown in Figure 10.10.

Hose couplings for the various sizes of intake hose are equipped with extended lugs which afford convenient handles for attaching intake hose to a hydrant or pump intake. The general three-piece construction of fire hose couplings is also used for intake hose couplings as shown in Figure 10.11.

Figure 10.11 Three-piece coupling with extended lugs for intake hose.

Lugs are provided on couplings so that a grip can be obtained with spanner wrenches to assist in tightening and breaking couplings. Most hose purchased today comes equipped with rocker lugs, to help the coupling slide over obstructions when the hose is moved on the ground or around objects. Hose couplings may be obtained with either two or three rocker lugs. The two-lug design has less projection to catch on objects, but the three-lug design is easier to grip with a spanner when breaking couplings. They both give equally satisfactory service. Among the various types of lugs used on couplings are the rocker, recessed, and pin lug. These three types are illustrated in Figure 10.12. The rocker lug is more frequently used because of its improved design.

Care of Fire Hose Couplings

Whether the male or female part of a fire hose coupling is more susceptible to damage is controversial. The male threads on a coupling are exposed when not connected and are, therefore, subject to damage. The female threads are not exposed but the swivel is subject to bending and damage. When the female and male coupling parts are connected together, there is less danger of damage to either part during common usage. Connected couplings seldom receive injury from being dropped or dragged. It is the disconnected separated ends that require the most protection. Connected couplings can be bent or crushed when they are run over by heavy vehicles, which is reason enough to prohibit vehi-

Figure 10.12 There are three types of lugs used on couplings to obtain a grip with spanner wrenches.

cles from running over fire hose. Some simple rules for the care of fire hose couplings are as follows:

- Avoid dropping and/or dragging couplings.
- Do not permit vehicles to run over fire hose.
- Examine couplings when hose is washed and dried.
- Remove gasket and free swivel in warm soapy water.
- Clean threads of tar, dirt, gravel, and oil.
- Inspect gasket and replace if cracked or creased.

There are two types of gaskets used with fire hose couplings. The fire hose swivel gasket is used to make the connection watertight when a female and male end are connected. The fire hose expansion ring gasket is used at the end of the hose where it is expanded into the shank of the coupling. These two gaskets are not interchangeable: the difference in them lies between their thickness and width. A comparison of these two gaskets reveals that the swivel gasket is relatively wide in structure with a larger outside diameter than the expansion ring gasket. The inside diameters of the two gaskets may be equal. The expansion ring gasket has a smaller outside diameter, but is thicker in body and cannot be used as a swivel gasket. Swivel gaskets should occasionally be removed from the coupling and checked for cracks, creases, and general elastic deterioration. The gasket inspection can be made by simply pinching the gasket together between the thumb and index finger as illustrated in Figure 10.13. This method usually discloses any defects and demonstrates the inabil-

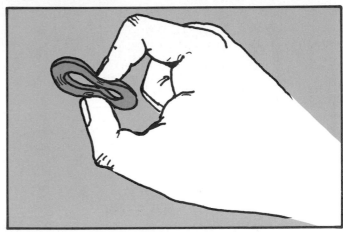

Figure 10.13 Pinching will disclose the condition of the gasket.

ity of the gasket to return to normal shape. One method of holding a gasket that is to be placed into a coupling swivel is to hold the gasket between the middle finger and thumb with the index finger resting on the inside rim of the gasket (Figure 10.14).

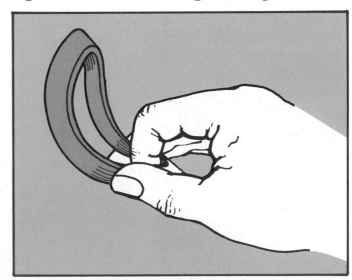

Figure 10.14 Grasp the gasket and pull with the index finger to fold the outer ring of the gasket upward.

Then fold the outer ring of the gasket upward by pulling on the index finger and place the gasket into the swivel by permitting the large loop of the gasket to enter into the coupling swivel at the place provided for the gasket. Allow the small loop to fall into place by releasing your grip on the gasket (Figure 10.15). If a swivel has become stiff or sluggish, place the swivel into a container of warm soapy water and work the swivel forward and backward in the solution.

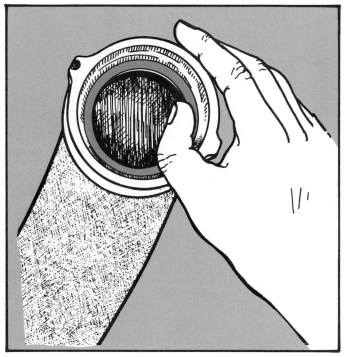

Figure 10.15 Place the large loop of the gasket inside the coupling swivel, then the small loop which will fall into place as the grip is released.

When a coupling has been severely bent (egg-shaped), the usual solution is to replace the coupling. There are cases, however, when slightly bent or egg-shaped couplings can sometimes be straightened by using the expander tool to help round the shape of the coupling. Tapping the coupling with a small hammer while under slight expansion will sometimes help. After a coupling has been straightened, the threads should be rerun by a tap or die threading tool. Male threads can sometimes be repaired by using a small three-cornered file for this purpose.

Hose Appliances and Tools

A complete hose layout for fire fighting purposes includes one end of the hose attached to a source of water supply and the other to a nozzle or similar discharge device. There are various devices other than hose couplings and nozzles used with fire hose to complete such an arrangement. These devices are usually grouped into one category called appliances. Such appliances as connections, adapters, or systems are devices through which water passes when connected to fire hose. Devices that are used in connection with fire hose through which water does not pass are known as hose tools. The process of maneuvering fire hose at the fire involves the use of a variety of hand tools, such as hose clamps, spanners, bridges, ropes, straps, and others. In practically every hose maneuver, both appliances and tools are used with fire hose to accomplish a complete procedure.

WYE APPLIANCES

Certain occasions make it desirable to divide a line of hose into two or more hoselines. Various types of wye connections are used for this purpose. The 2½-inch (65 mm) to 1½-inch (38 mm) wye is used for direct connection to 2½-inch (65 mm) hose to supply to 1½-inch (38 mm) hoselines (Figure 10.16). The 2½-inch (65 mm) wye is used to divide one 2½-inch (65 mm) or larger hoseline into two 2½-inch (65 mm) lines and is shown in Figure 10.17. Wye appliances are often gated so that water being fed into the hoselines may be controlled at the gate.

WATER THIEF APPLIANCES

The water thief is a variation of the wye adapter. It is regularly made with quarter-turn gate valves on 1½-inch (38 mm) hose outlets. The 2½-inch (65 mm) hose outlet may also be provided with a quarter-turn gate valve. The water thief is intended to be used on a 2½-inch (65 mm) or larger hoseline, usually near the nozzle, so that 2½-inch (65 mm) and 1½-inch (38 mm) hoselines may be used as desired from the same layout (Figure 10.16). The water thief can be obtained for 1½-inch (38 mm) hose.

SIAMESE APPLIANCES

The siamese and wye adapters are often confused because of their close resemblance. Siamese fire hose layouts consist of two or more hoselines which are brought into one hoseline or device. It

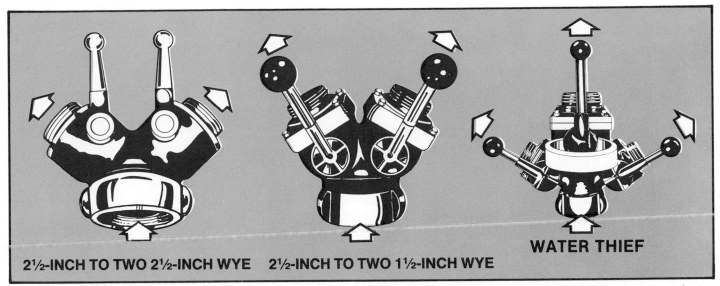

2½-INCH TO TWO 2½-INCH WYE   2½-INCH TO TWO 1½-INCH WYE   WATER THIEF

Figure 10.16 A single line is divided into two lines by a wye and three lines by a water thief.

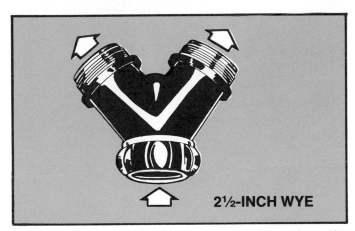

2½-INCH WYE

Figure 10.17 This ungated straight wye can be used to supply two 2½-inch (65 mm) handlines from a 3-inch (76 mm) line.

Figure 10.18 A straight siamese has two female and one male connection.

can be noticed that the siamese adapter, as shown in Figure 10.18, has two female connections and one male connection, while the wye adapter has one female connection and two male connections. All siamese adapters can be purchased with or without clapper valves, but it is usually advisable to specify them. The siamese appliances that are 2½-inch (65 mm) or larger are usually equipped with clapper valves so that one or more hoselines may be used in any hook-up, such as two hoselines into a pump (Figure 10.19). The deluge set, or multiversal master stream adapter often employs a siamese to merge several hoselines into one master fire stream.

SPECIAL HOSE APPLIANCES

A variety of special hose appliances which are sometimes used in hose layouts are shown in Fig-

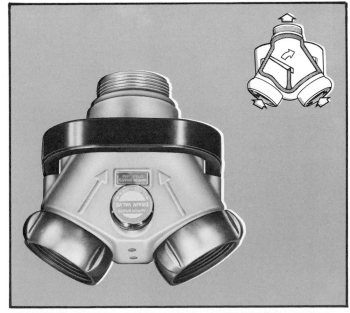

Figure 10.19 A clappered siamese allows charging the first hoseline before the second one is connected. *Courtesy of Akron Brass.*

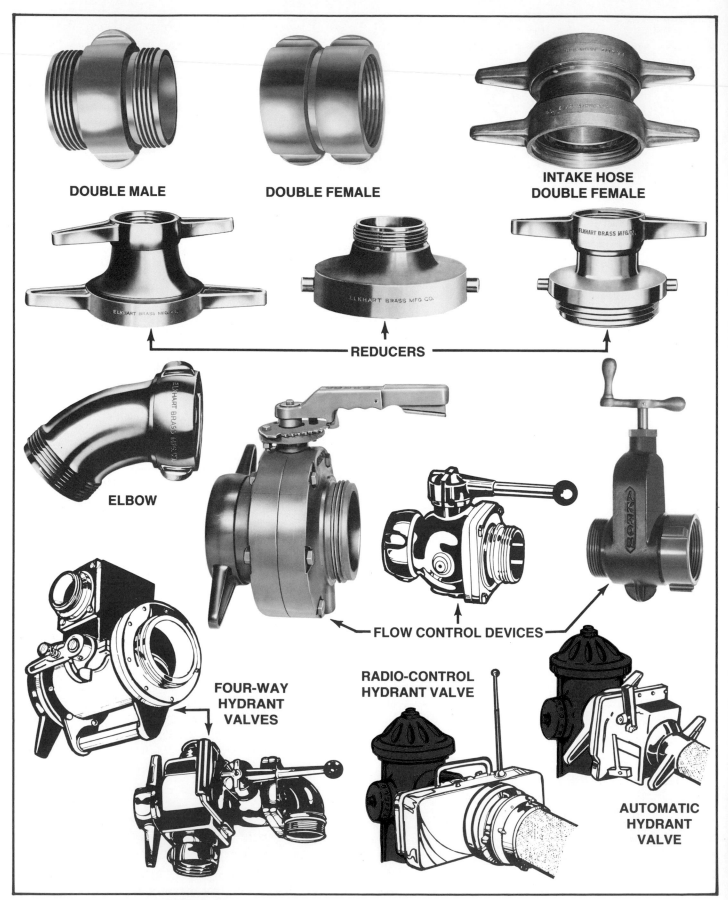

DOUBLE MALE

DOUBLE FEMALE

INTAKE HOSE DOUBLE FEMALE

REDUCERS

ELBOW

FLOW CONTROL DEVICES

FOUR-WAY HYDRANT VALVES

RADIO-CONTROL HYDRANT VALVE

AUTOMATIC HYDRANT VALVE

Figure 10.20 Various hose appliances usually carried on a pumper. *Courtesy of Akron Brass and Elkhart Brass.*

ure 10.20. The double male and double female adapters are probably used more than any other special hose appliance. These two appliances may be purchased or they may be made by welding together old hose couplings of the same size.

Reducers range from 1½-inch (38 mm) to 1-inch (25 mm), 2½-inch (65 mm) to 1½-inch (38 mm), 4- to 6-inch (115-160 mm) intake couplings to 2½-inch (65 mm). Reducers and double female appliances for intake hose couplings are usually preferred with long extended lugs. Likewise, discharge hose elbows are appliances that may be used for pump connections.

A hose cap can be attached to the male thread to cap off a hoseline or pump outlet. A hose plug can be attached to the female thread to plug pump suction auxiliary intakes or to plug female hose couplings.

FLOW CONTROL DEVICES

The flow of water is controlled by various devices in hoselines, at hydrants, and at pumpers. These include ball valves, gate valves, four-way hydrant valves, and butterfly valves. Generally, ball valves are used in a hoseline, gate valves and four-way hydrant valves to control the flow from a hydrant, and butterfly valves on large pump intakes. Ball valves are open when the handle is in line with the hose and closed when it is at a right angle to the hose. Gate valves have a baffle that moves up and down by a handle and screw arrangement. Four-way hydrant valves are described under Hose Lays. A butterfly valve uses a flat baffle operated by a quarter turn handle. The baffle is in the center of the waterway when the valve is open. There are also mechanical and radio operated valves now available to control the flow from a hydrant into a supply line.

HOSE HOIST TOOLS

A hose hoist is a device over which rope or hose may be pulled to hoist or lower equipment when firefighters are operating in buildings above the ground level. It consists of a metal frame, curved so that it will fit over a windowsill or the edge of the roof. It contains two or more rollers over which rope or hose may be drawn. This device reduces the pos-

sibility of cutting the hose on the sharp edge of the wall, cornice, or roof while it is being raised or lowered.

During hose operations the hose hoist is placed over a windowsill, cornice, or edge of the roof and is made secure to the roof's support or other projections by means of an attached short rope or a screw clamp. The hose can be hoisted with a rope handline and pulled over the roller. The roller and the method of using it over the edge of a roof is shown in Figure 10.21. Its use is similar over a windowsill or cornice.

Figure 10.21 The hose roller or hoist protects hose from sharp edges while it is being raised or lowered.

HOSE JACKET TOOLS

The name "Hose Jacket" has been applied to a tool that is used to seal small cuts or breaks which may occur in fire hose or to connect mismated or damaged couplings of the same size. The name of this device is easily confused with the woven-jacket or fabric rubber-lined fire hose. Although, in a

Figure 10.22 Place couplings or burst hose in bottom of hose jacket.

Figure 10.23 Stomp on the hose jacket to lock both sides together. *10.23b Courtesy of Akron Brass.*

sense, water passes through the hose jacket, it is considered as a hose tool rather than an appliance. The method of connecting two mismated couplings is illustrated in Figures 10.22-10.23. With the top open, place one side under the hose and couplings. Place the upper half into position but do not lock. Stomp sharply with one foot to lock both sides together. The hose jacket is equipped with a flexible rubber seal that fits snugly around the hose. Care should be taken to place the edges of flattened hose in the jacket so that they do not interfere with snapping the tool closed. As soon as pressure is applied to the hose, the rubber flanges are sealed together around the hose to make a watertight joint.

HOSE CLAMP TOOLS

The hose clamp is a tool to shut off the water in hoselines when other control valves are not applicable. It is used to replace a burst section of hose, to extend lines, or to hold water back for line advancement without shutting off the sources of supply. The clamp may also be used on a hoseline which has been laid from a hydrant to the fire. Applying the clamp near the fire permits the person at the hydrant to operate the hydrant without a signal

from the company officer at the fire. The hose clamp can be removed when all dry lines have been advanced and thus charge the hose with water.

The methods of applying hose clamps to a line of hose may vary according to the kind of clamp used. The clamping jaws of most hose clamps are designed to prevent excessive pressure on the hose which might cause injury to the jacket and the lining. Some clamps are adjustable to fit several sizes of hose. The general types of hose clamps may be classified by the method in which they are operated. These methods are press down lever clamp, pull up lever clamp, screw down clamp, and hydraulic press clamp as shown in Figure 10.24. Some general rules which apply to all portable clamps are as follows:

- Place hose clamp at least 20 feet (6 m) to the rear of fire apparatus.

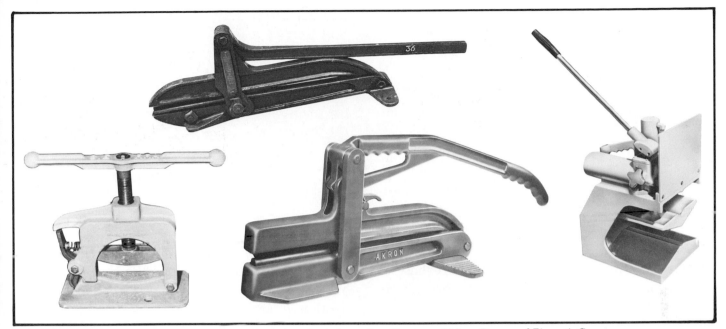

Figure 10.24 Hose clamps are generally classified by their method of operation. *Courtesy of Akron Brass and Ziamatic Corp.*

- Place hose clamp not closer than 6 feet (2 m) to a coupling and on the supply side.

- Stand to one side when applying or removing a lever-operated clamp.

- Place the hose evenly in the jaws to avoid pinching. The hose must be placed all the way inside the jaw to assure complete stoppage of water flow.

- Close and open the clamp slowly to avoid water hammer.

HOSE SPANNERS AND HYDRANT WRENCHES

The principal use of the hose spanner wrench is to tighten or loosen hose couplings, but this versatile tool can also be used to close utility cocks, pry, and hammer. Several styles are illustrated in Figure 10.25. Some hose couplings of special designs require special spanners. The hydrant wrench is usually equipped with a pentagon opening in its head which will fit most standard fire hydrant operating nuts. The lever handle may be threaded into the operating head to make it adjustable or the head and handle may be of the ratchet type. The head may also be equipped with a spanner. Although the rubber mallet is not a wrench, it may be used to tighten and loosen intake hose couplings. It is sometimes difficult to get a completely airtight connection with intake hose couplings even though these couplings may be equipped with long operating lugs. A hard rubber or rawhide mallet similar to the one shown may be carried on each pumper for this purpose.

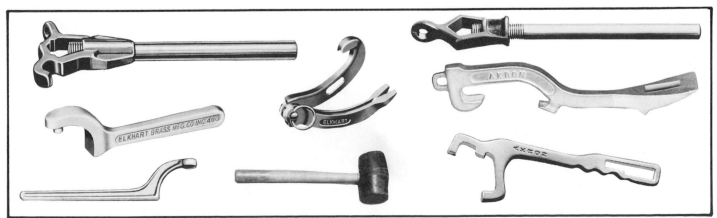

Figure 10.25 Hose spanner and hydrant wrenches of various styles are useful for multiple purposes. *Courtesy of Akron Brass and Elkhart Brass.*

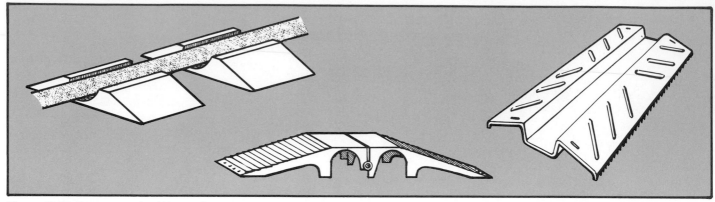

Figure 10.26 Hose ramps should be used to prevent traffic damage to hose.

HOSE RAMPS OR BRIDGES

One common cause of injury to fire hose is crossing hoselines by vehicles. When a hoseline is charged and under proper pressure, the damage will not be as great as when the hose is flat without pressure. However, hose couplings are always subject to damage when they are run over by a vehicle. Not only will a vehicle cause damage to the hose as it crosses, firefighters at the nozzles will receive sudden jerks each time a wheel cuts off the water momentarily. Damage to the hose jacket may also be caused by skidding the hose on the pavement when a vehicle tries to cross. Whenever traffic must be moved across fire hose, hose ramps or bridges similar to that shown in Figure 10.26 should be used. A very versatile hose ramp is one made of heavy gauge sheet metal. Another hose bridge is made of wood to form an incline for each wheel.

CHAFING BLOCKS

Chafing blocks are devices that are made to fit around fire hose where the hose is to be subjected to rubbing from vibrations. Chafing blocks are particularly useful where intake hose comes in contact with pavement or curb steps. At these points, wear on intake hose is most likely, since pumper vibrations may be keeping the intake hose in constant motion. Chafing blocks may be made of wood, leather, or sections from old truck tires.

ROPE, STRAP, AND CHAIN HOSE TOOLS

Numerous rope, strap, and chain hose tools are available which aid in the maneuvering of hoselines at fires. Some of their uses are carrying, securing, dragging, and assisting in hoseline operation in connection with ladders, fire escapes, stairways, and hoisting. Some of the more common types are shown in Figure 10.27.

FIRE HOSE ROLLS
Straight Roll

The basic hose roll consists of starting at one end, usually at the male couplings, and rolling the hose toward the other end to complete the roll. When the roll is completed, the female end is exposed, with the male end protected in the center of the roll. This "straight" roll is commonly used prior to placing hose in storage (Figure 10.28).

A variation of this method is to begin the roll at the female coupling so that when the roll is completed, the male coupling is exposed. This is often

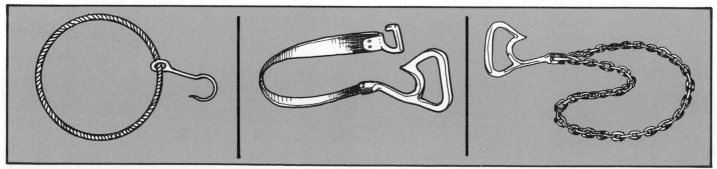

Figure 10.27 These are the common types of rope, strap, and chain hose tools.

Figure 10.28 A straight roll starts at the male end and rolls to the female.

done to denote a damaged coupling or piece of hose. A tag is usually attached to the male coupling indicating the type and location of the damage.

The Donut-Roll

The donut-roll has certain advantages that the straight roll does not possess. Two main advantages are that both ends are available on the outside of the roll and that the hose is less likely to spiral or kink when unrolled. Its adaptation to service makes it a preferred hose roll in many instances. When a section of fire hose needs to be rolled into a donut-roll, one or two firefighters may perform the task and any one of three methods may be used. The following describes two methods used to make the donut-roll (Figures 10.29-10.35).

Method No. 1

Step 1: Lay the section of hose flat and in a straight line. Start the roll from a point 5 or 6 feet (2 m) off center toward the male coupling and roll toward the female end. Leave sufficient space at the center loop to insert the hand for carrying (Figure 10.29).

Step 2: Near the completion of the roll, the male coupling will be enclosed within the roll as the hose is rolled over it. Use the short length of hose at the female end to protect the male threads (Figure 10.30).

Step 3: Continue rolling the hose over the female end until this end is off the ground and behind the operator. With the roll in this position, kneel alongside the roll with the inside knee. Place the inside arm over the top of the roll and insert the fingers of the hand in the center loop for lifting (Figure 10.31).

Figure 10.29 Lay the male coupling on the hose, and proceed to a point 6 feet (1.82 m) short of the center to initiate the donut.

Figure 10.30 Near completion of the donut, the male coupling will be protected by the remaining hose.

Figure 10.31 Complete the donut by rolling past the female. To carry: kneel, place the hand in the center loop, and lift.

Method No. 2

Step 1: Grasp either coupling end, carry it to the opposite end, and cause the looped section to lie flat, straight, and without twist (Figure 10.32).

Step 2: Face the coupling ends, start the roll on the male coupling side about 2½-feet (65 mm) from the bend (1½-feet [38 mm] for 1½-inch [38 mm] hose) and roll toward the male coupling (Figure 10.33).

Step 3: If the hose behind the roll becomes tight during the roll, pull the female side back a short distance to relieve the tension (Figure 10.34).

Step 4: As the roll approaches the male coupling, lay the roll flat on the ground and draw the female coupling end around the male coupling to complete the roll (Figure 10.35).

Figure 10.33 Start the donut on the male coupling side about 2½-feet (.65 mm).

Figure 10.34 When the donut becomes tight, pull the female coupling side back.

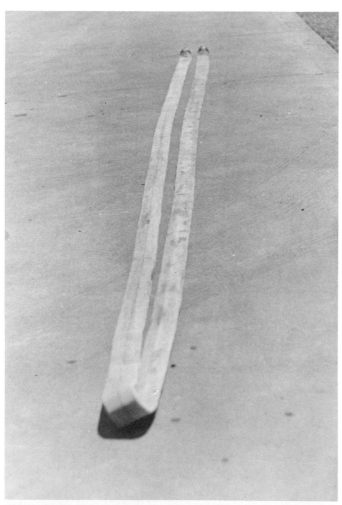

Figure 10.32 Form a loop as the hose section is halved with both sides straight and flat.

Figure 10.35 As the donut approaches the male coupling, lay it flat, and complete the donut by bringing the female coupling around the male coupling.

Twin Donut-Roll

The twin donut-roll is more adaptable to 1½-inch (38 mm) hose, although 2½- or 3-inch (65 mm-77 mm) hose can be used. Its purpose is to arrange a compact roll which may be transported and carried for special applications. The following steps describe how the twin donut-roll can be made (Figures 10.36-10.39).

Step 1: Place the male and female couplings together and lay the hose flat without twist to form two parallel lines from the loop end to the couplings (Figure 10.36).

Step 2: Fold the loop end over and upon the two lines to start the roll (Figure 10.37).

Step 3: Continue to roll both lines simultaneously toward the coupling ends, which forms a twin roll with a decreased diameter (Figure 10.38).

Step 4: The twin donut-roll may be carried in the same manner as the standard donut-roll, or a short piece of strap or rope may be looped through the roll and tied with a quick-releasing hitch for storage on fire apparatus fireground operations (Figure 10.39).

Figure 10.37 Fold the loop over upon the two parallel lines to start the donut.

Figure 10.38 Roll both lines simultaneously toward the couplings.

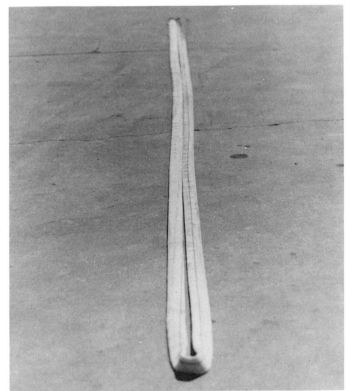
Figure 10.36 Place the hose flat in two parallel lines with a tight loop.

Figure 10.39 The smaller diameter donut can be tied for storing and transporting.

Self-Locking Hose Roll

The self-locking hose roll is a means by which fire hose can be rolled into a twin roll and secured together by a portion of the hose itself. The self-locking loop can be regulated to various lengths as might be preferred by local requirements. The following steps describe how to make the roll (Figures 10.40-10.46).

Step 1: Place the male and female couplings together and lay the hose flat without twist to form parallel lines from the loop end to the couplings (Figure 10.40).

Step 2: Move one side of the hose up and over 2½- to 3 feet (65 mm to 77 mm) to the opposite side without turning. This lay-over method prevents a twist in the hose at the big loop. The size of this loop (butterfly) determines the length of the shoulder loop for carrying (Figure 10.41).

Step 3: Face the coupling ends, bring the back side of the loop forward toward the couplings and place it on top of where the hose crosses. This action forms a loop on each side without twist (Figure 10.42).

Step 4: Start rolling toward the coupling ends and form two rolls side-by-side (Figure 10.43).

Figure 10.42. Place the back side of the loop on top of where the hose crosses.

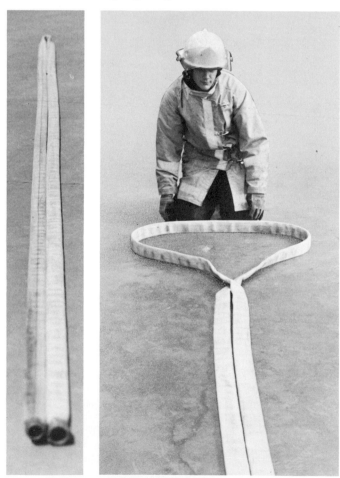

Figure 10.40 (left) Place the hose flat in two parallel lines.

Figure 10.41 (right) Cross one side of the hose over the other while keeping both flat. Form a large loop of adequate size for carrying the completed donut from the shoulder.

Figure 10.43 Start rolling both lines toward the couplings.

Figure 10.44 Complete the donut with both couplings on top. Pull one side to adjust the loops.

Step 5: When the rolls are completed, allow the couplings to lie across the top of each roll and adjust the loops, one short and one long (Figure 10.44), by pulling only one side of the loop through.

Step 6: Place the long loop through the short loop just behind the couplings and tighten snugly. The loop forms a shoulder sling (Figure 10.45).

Step 7: The coupling ends may be carried in front or to the rear (Figure 10.46).

Figure 10.46 The couplings can be carried in either direction with the shoulder loop.

Figure 10.45 Place the long loop through the short one and tighten.

Special Hose Assemblies

With the advent of high-rise buildings having standpipe risers, special hose assemblies are needed for use within the building. These special assemblies can vary in a number of ways, but the principle of having a prearranged hose assembly ready for instant use is the same. Hose assemblies that include hose appliances and tools and can be carried by one firefighter are illustrated in Figures 10.47 - 10.51.

In the event it is necessary for the firefighter to wear protective breathing equipment, provision can be made to carry the hose assembly by one hand and alongside the body (Figure 10.52).

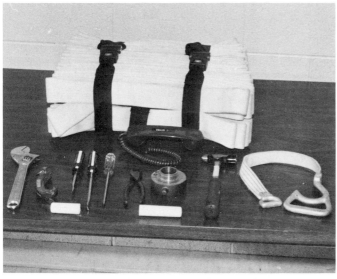

Figure 10.47 Hose, a nozzle, and various tools and adapters are assembled for standpipe operations. *Courtesy of Southfield Fire Department.*

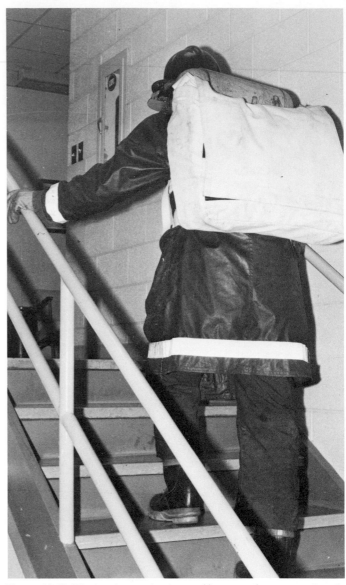

Figure 10.49 The bag is easy to carry and has room for a spare air tank on top. *Courtesy of Southfield Fire Department.*

Figure 10.48 The standpipe operating equipment can be conveniently placed in a canvas bas. *Courtesy of Southfield Fire Department.*

Figure 10.50 Alternate carrying methods are a strap pack and a back pack.

Figure 10.53 One firefighter positioning for coupling hose by the foot-tilt method.

Figure 10.51 (left) These packs can also be easily carried.

Figure 10.52 (right) The bag or either pack can be hand carried if self-contained breathing apparatus is required.

COUPLING AND UNCOUPLING HOSE

The process of coupling hose is, for the most part, a simple process of fastening together the male and female hose couplings. The need for speed and accuracy under emergency conditions requires that specific techniques for coupling hose be developed. When coupling two sections of hose keep the flat sides of the hose in the same plane. This practice makes it easier to handle and load the hose. Nozzles may be attached to the hose by using the same methods as when coupling two sections of hose together.

Coupling Hose (Foot-Tilt Method)

Stand facing the two couplings so that one foot is near the male end. Place the foot directly behind the male coupling and apply pressure to tilt it upward. Position the feet well apart for balance, grasp the female end with one hand behind the coupling and place the other hand on the coupling swivel. Bring the two couplings together and turn the swivel with the thumb of the hand to make the connection (Figure 10.53).

Coupling Hose (Over-The-Hip Method)

Grasp the female coupling with one hand on the swivel. The hose is brought across the hip with the feet spread comfortably apart. Allow the female coupling to hang about ten inches (254 mm) over the hip toward the ground. Pick up the male coupling with the other hand, align the couplings, and turn the swivel to engage the threads (Figure 10.54).

Figure 10.54 One firefighter positioning for coupling hose by the over-the-hip method.

Coupling Hose (Two-Firefighter Method)

Two firefighters holding opposite couplings, stand facing each other. The firefighter with the male coupling will grasp the coupling with both hands, bend the hose directly behind the coupling and hold the coupling and hose tightly against the upper thigh or mid-section with the male threads pointed outward. The alignment of the hose must be done by the firefighter with the female coupling. It is important for the firefighter with the male coupling to look off in another direction in order to prevent trying to help align the couplings. The firefighter with the female coupling grasps the coupling with both hands, brings the two couplings together, aligns their positions, and then turns the swivel clockwise to complete the connection. The positions are illustrated in Figure 10.55.

Figure 10.55 When two firefighters couple a hose, the operator of the female coupling swivel aligns the hose.

Breaking a Tight Coupling

It may sometimes be necessary to break a tight coupling when spanner wrenches are not available. A method known as the "Knee-Press," is most effective under normal conditions. The principle of operation is to compress the hose gasket just enough to permit a free turn of the swivel. An alter-nate two-firefighter method, known as the "Stiff-Arm" may also be used for quick release.

KNEE-PRESS METHOD

Grasp the hose behind the female coupling and stand the connection on end with the male coupling below. Best results will be obtained if the hose is bent close to the male coupling. Set the feet well apart for balance, and place the right knee upon the hose and shank of the female coupling. Keep the thigh in a vertical plane with the couplings and apply body weight to the connected coupling. As this weight is applied, quickly snap the swivel in a counterclockwise direction (Figure 10.56).

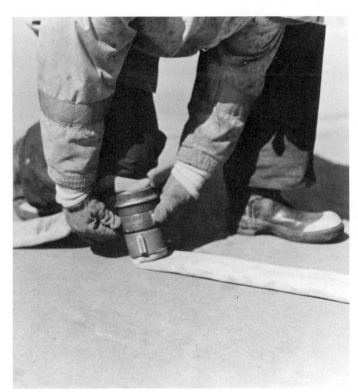

Figure 10.56 Applying body weight to compress the coupling gasket is an effective method of breaking a tight coupling. The female coupling should be on the top for ease of operation.

STIFF-ARM METHOD

Two firefighters face each other with the hose coupling between them. Each firefighter takes a firm two-handed grip on their respective coupling and presses the coupling toward each other thereby compressing the gasket in the coupling. Keeping their arms stiff, and elbows out, both firefighters use the weight of their bodies to turn the hose coupling counterclockwise (Figure 10.57).

Fire Hose Compartments

Hose compartments are generally referred to as "Hose Beds." The size of the compartment restricts the amount of hose that can be carried and its shape determines the type of load that can be used. Certain general characteristics are common in any hose bed, such as the open slats in the bottom that enable air to circulate throughout the hose load. Without this feature the hose will likely mildew and rot in a very short time. The width of the bed may be only as wide as the distance between the rear wheels of the apparatus or it may extend over the rear wheels to the maximum width allowed by the law.

A fire hose compartment (hose bed) may be divided or separated at some point in order for the compartment to hold two or more separate loads of hose. The divider or separator is known as a "baffleboard" although some are made of metal. The baffleboard may be used for special adaptations of hose loads to fulfill a particular need.

Any one of the three basic hose loads may be used for a divided hose compartment load. The one used for this example is the flat load. Each load is started next to the baffleboard at the rear of the hose bed. The male couplings are left exposed on each side of the baffleboard as illustrated in Figure 10.58. After the two loads are completed, two lines of hose can be laid simultaneously. The female end from one load can, however, be brought over and connected to the opposite exposed male coupling as

Figure 10.58 A divided hose compartment has the advantage of laying two parallel lines simultaneously.

Figure 10.57 With a firm grip and arms stiff, the body weight can be used to break a tight coupling as both firefighters turn it counterclockwise.

HOSE LOADS

Before fire hose can be systematically loaded for accurately planned layouts, certain requirements must be known. The space provided for hose on the apparatus and the construction of the hose compartment are basic standard requirements that have been established. These are factors that specific requirements must govern. There are two basic layout requirements in common practice which are known as forward and reverse lays. The manner in which hose is loaded in the hose compartment does not necessarily determine whether the lay will be forward or reverse. Fire apparatus that is expected to make the initial lay at fires from the source of water supply to the fire is generally loaded with the female coupling to come off first. Likewise, apparatus that is expected to make the initial lay from the fire to the source of water supply is generally loaded with the male coupling to come off the load first with the nozzle attached. However, nothing prevents the apparatus from laying hose either forward or reverse if double male and female adapters are properly used. These two layout requirements are discussed separately.

illustrated in Figure 10.59. One continuous hoseline can then be laid if so desired. This provides a very versatile combination and has proved to be effective for certain conditions. In some cases, the baffleboard may be situated at one side of the hose bed to form a smaller compartment for 1½ or 1¾-inch (38 or 44 mm) hose.

Figure 10.59 If desired, the hose beds can be joined and a single long hoseline laid.

Hose Loading Factors

Although the loading of fire hose on fire apparatus is not an emergency operation, it is a very vital operation that must be done correctly. When fire hose is needed at a fire, the proper hose load permits efficient and effective operations to be carried out. It is not intended for this manual to favor any particular load or technique. It is intended, however, to present some of the more commonly used techniques so that one or several can be selected. Several factors should be considered when determining the hose load to use. If only one fire apparatus is available, a load should be selected that will most nearly fill all requirements. If, however, a department's apparatus includes one or more pumpers, a squad, and a ladder truck, each apparatus may be loaded to fill a specific need. After studying and practicing the different loads and load finishes suggested, the department needs may be determined (Figure 10.60).

The most common terminology used to describe a fire hose compartment is "hose bed." Hose beds vary in size and shape and they are sometimes built for specific needs. In order to clarify terminology of hose bed positions in this manual, the front of the hose bed is designated as that part of the compartment toward the front of the apparatus.

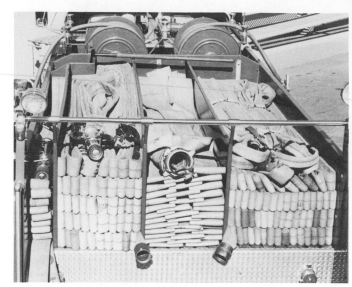

Figure 10.60 An example of hose load designed to meet a variety of local needs.

The rear of the hose bed is designated as that part of the compartment toward the rear of the apparatus. When loading any hose, firefighters should observe certain basic rules to enable the hose to lie evenly in the bed and to prevent difficulty in disconnecting the couplings. Some of the basic rules are:

- Before connecting any coupling, check gaskets and swivel.

- When two sections of hose are connected, keep the flat sides of the hose in the same plane. The alignment of the lugs on the couplings is not important.

- When two sections of hose are connected, the couplings should be made hand tight. Do not use wrenches or undue force.

- When fire hose must be bent to form a loop in the hose bed, all wrinkles should be removed by pressing with the fingers so that the inside of the bend is smoothly folded. During the loading process, a coupling will frequently come in position so that it must turn around to be pulled out. To avoid this situation, make a short fold or reverse bend in the hose as shown in Figure 10.61. This practice is commonly referred to as a "dutchman." The dutchman serves two purposes: one to change the direction of the hose and the other to change the location of a coupling.

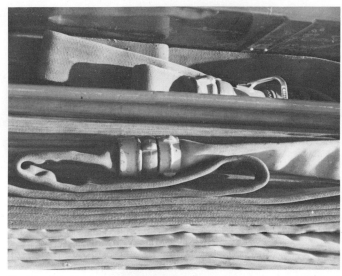

Figure 10.61 The dutchman changes the position and direction of the coupling so it will not have to make a turn in the hose bed.

The Accordion Load

The accordion load may be started in any corner of the hose bed, and either end of the hose may be used, depending upon the need for a forward or a reverse lay. The load places bends at each end of the hose bed, but it has an added feature of having all flakes the same length. This feature is a distinct advantage in hoseline advancement.

A minimum of three firefighters should be used for loading, one in the hose bed and two at the rear of the hose bed. The following steps are illustrated in Figures 10.62 - 10.64.

Step 1: The load is started in a rear corner with the hose on edge and extend to the front of the bed where it is folded back toward the front. Make each succeeding fold the length of the bed to make an even load (Figure 10.62).

Step 2: As each folded layer of hose is placed into the accordion load, the remaining space in the hose bed becomes narrower. During the process, a coupling will frequently come into position so that it must turn around to be pulled out. To avoid this situation, make a short fold or reverse bend in the hose as previously described.

Step 3: When the first tier is completed, gradually raise the last fold from the front toward the rear along the side of the hose bed as the hose approaches the tailboard.

The second tier is started by simply bending the hose toward the open bed instead of toward the side of the bed. Each fold is then placed as previously explained and the load is made toward the opposite side (Figure 10.63).

Figure 10.62 Start an accordion load with the coupling next to baffleboard at the rear. With the hose on edge proceed to the front of the hose bed and continue making folds.

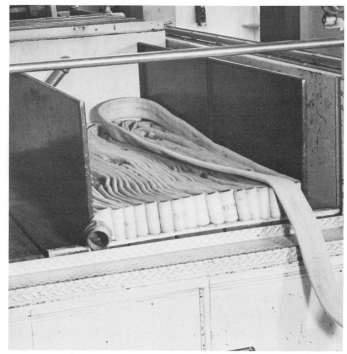

Figure 10.63 The next tier is started by gradually raising the last fold and bending the hose toward the open bed.

Alternate Method: Bend the last fold of the first tier toward the open hose bed instead of folding it toward the side of the bed and gradually raise this fold along the side of the bed as it approaches the front. This pactice prevents the hose from kinking when it is pulled from the bed at this point. Extend the hose to the front of the hose bed, cross it over to the opposite side, and start a new tier (Figure 10.64). Crossing over to the opposite side permits the hose to always pay out from the same side of the hose bed.

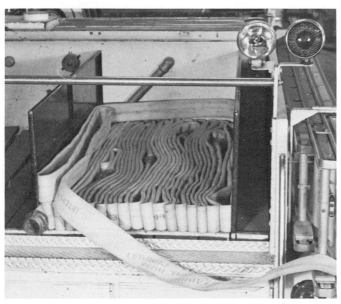

Figure 10.64 Beginning the next tier by going to the front of the hose bed and crossing over to the opposite side allows the hose to always come out from the same side.

The Flat Load

The flat load, as the name implies, consists of folding the hose back and forth on its flat sides and lengthwise in the hose compartment. Either end of the hose may be used to start the load and it can be started in any corner of the hose bed. The load places approximately the same number of bends in the hose as the accordion load and it has the same added feature of having all the flakes the same length.

A minimum of three firefighters should be used for loading the flat load, one in the hose bed and two at the rear of the hose bed. The following steps for loading are illustrated in Figures 10.65 - 10.66.

Step 1: The hose is started at an inside corner of the rear of the hose bed, leaving the coupling exposed at the tailboard. Lay the hose flat along one side to the front of the hose bed, fold it back toward the rear, and lay it at a slight diagonal in order to start the second flake alongside the first (Figure 10.65).

Step 2: Make each succeeding flake in the same manner until the tier is full. Start a second tier by simply folding back to the beginning side. The folded ends may be staggered at each tier to permit the center of the load to fill evenly with folded ends (Figure 10.66).

Figure 10.65 Start a flat load at an inside rear corner and lay the hose flat to the front bringing it back at a slight diagonal. *Courtesy of Rose Fire Company.*

Figure 10.66 Continue additional tiers by simply folding the hose back to the initial side. Stagger tier folds to obtain even loading. *Courtesy of Rose Fire Company.*

The Horseshoe Load

The horseshoe load may be started at a corner at either end of the hose bed and either end of the hose may be used to start the load, depending upon the need for a forward or reverse lay. This load places fewer bends in the hose than the accordion or flat loads, but the flakes are considerably long and are unequal in length.

A minimum of three firefighters should be used for loading the horseshoe load, one in the hose bed and two at the rear of the bed. The following steps for loading are illustrated in Figures 10.67 - 10.69.

Step 1: The load is started in a rear corner, with the hose on edge and extend to the front of the bed where it is folded back to the front. Place the hose across the front of the bed and bring it along the opposite side to the rear of the bed. Fold the hose back toward the front and continue to load the hose in a horseshoe fashion (Figure 10.67).

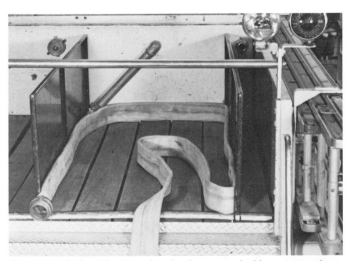

Figure 10.67 Start the hoseshoe load at a rear inside corner and proceed on edge to the front of the hose bed, cross the front, go down the opposite side, and return.

Step 2: As the tier fills toward the center, a coupling may appear at this point. This coupling may be in a position that makes it necessary for it to turn around when it pays out. To remedy this situation, place a "dutchman" in the load as previously described. The folded ends of the load at the rear of the bed may be staggered or placed evenly, depending upon departmental policy (Figure 10.68).

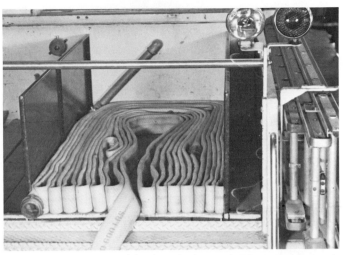

Figure 10.68 As the remaining hose bed space becomes narrower, additional short folds to create dutchmen will be necessary. Staggered or even folds can be used.

Step 3: To make the crossover to the second tier at the front of the hose bed, place the last fold from the center flat on the load and extend it toward a front corner. At this corner, turn the hose back on edge, bring the hose across the front of the bed, and resume loading as previously described (Figure 10.69). (The crossover to succeeding tiers should be made in alternate corners.)

Figure 10.69 To rise to the next tier, lay the fold flat to a front corner, return to the edge, lay the hose across the front of the bed, and resume as on the tier below or bring the last fold to the rear of the bed and fold the hose around the ends of one-half of the tier.

Alternate Method: Another method to make the crossover to the next tier of a horseshoe load is to bring the last fold from the center to the rear of the bed and fold the hose around the ends of one-half of the tier. At this point, place the hose between the tier and the side of the bed and pull snugly. Form a short fold at this point by bending the hose toward the side of the bed instead of toward the open bed. This short fold prevents the forming of a kink when the hose is pulled from the bed.

Large Diameter Hose

Due to the materials used in the construction of large diameter hose and its size, it should be placed on fire apparatus using the flat load method. This enables easy removal from the hose bed and prevents chafing that occurs when it is placed on edge. Large diameter hose may also be loaded on hose reels.

Hose Load Finishes

Hose load finishes are added to the basic hose load to increase the versatility of the load. Finishes are normally loaded for two purposes: to provide enough hose to make a hydrant connection, and to provide a working line at the fire scene.

STRAIGHT FINISH

A straight finish consists of the last length or two flaked loosely back and forth across the top of the hose load. This finish is normally associated with forward lay operation. A hydrant wrench, gate valve, and any necessary adapters should be strapped at or near the female coupling (Figure 10.70).

REVERSE HORSESHOE LOAD FINISH

A reverse horsehoe load is what its name implies, a horseshoe load with its ends reversed in the hose bed. Thus, the large loop of a single tier is at the rear of the hose bed (near tailboard) instead of at the front. This finish consists of two such loads, one on each side resting on top of a basic hose load.

This load finish can be started at either the center or outside of the space it will occupy. If initiated at the center it will build toward the 2½-(65

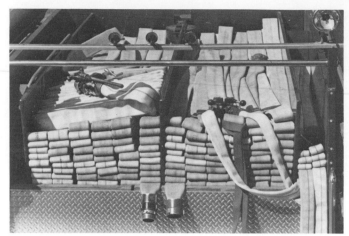

Figure 10.70 The straight finish in the left bed provides hose to reach the hydrant for a forward lay.

mm) by 1½-inch (38 mm) wye at the center of the load, it can also build away from the wye as illustrated in Figures 10.71 - 10.73.

Step 1: Connect the hose to the wye at the rear of the hose bed and proceed to the front, fold the hose against itself and return to the rear, form a loop across the rear of the hose bed, turn the hose as the loop is formed to remove the twist, and proceed along the side of the bed to the front (Figure 10.71).

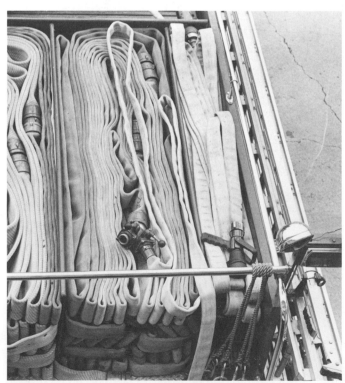

Figure 10.71 From the wye or discharge connection extend a fold to the front of the hose bed, return to the rear and form a loop before proceeding to the front along the side of the hose bed.

Step 2: Fold the hose back against that just loaded, and continue loading to the rear. Stay inside existing loops and continue the pattern until load is complete with desired amount of hose. The twist will have to be removed as each loop is formed (Figure 10.72).

Step 3: Place the nozzle on the line in the center of the hose load. The nozzle may be placed toward the front or the rear of the hose bed depending upon how it will be removed (Figure 10.73).

The reverse horseshoe load can also be used for a preconnected line and loaded in two or three layers. With the nozzle extending to the rear it can be placed over a shoulder and the opposite arm extended through the loops of the layers pulling the hose from the bed for an arm carry. A second preconnect can be bedded below when there is sufficient depth.

Figure 10.73a Place nozzle in center of load. Nozzle can extend toward rear or front.

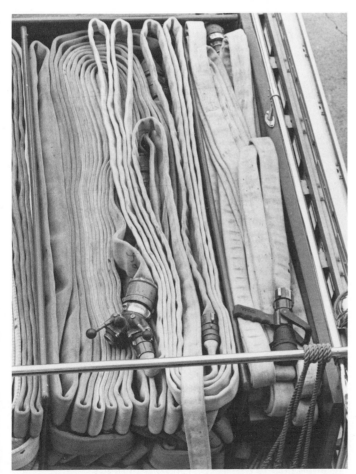

Figure 10.72 Fold the hose back against itself turning the hose as the loop is formed to remove the twist. Continue loading staying inside existing loops.

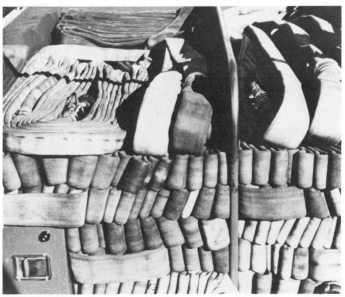

Figure 10.73b The reverse horseshoe load may be placed in the hose bed either flat or on edge. *Courtesy of Skokie Fire Department.*

Preconnected Hose Loads

FLAT LOAD

Any preconnected hose load must start with the female end where it is connected to a pump discharge. A pump discharge connection should be specifically located for a preconnected hoseline. A special hose compartment is also necessary for preconnected loads. The hose compartment for this preconnected load may vary in size and shape but it must be wide enough for free movement of the hose. The following steps illustrate how a preconnected flat load is made (Figures 10.74 - 10.75).

Step 1: Connect the female coupling to a pump discharge connection and lay the hose flat as described for the standard flat load. When approximately one-third of the hose has been loaded, leave a loop extended about 8 inches (203 mm) beyond the rear of the bed toward the tailboard (Figure 10.74).

Step 2: Finish loading the hose in the same manner, attach a nozzle, and place it upon the completed load.

Alternate Method: If the preconnected line must be longer than 150 feet (45 m), two extended loops, one large and one small, may be placed in the load at the tailboard (Figure 10.75).

Figure 10.74 A flat preconnected load is bedded similar to a flat load except a loop should be added at the rear of the bed after one-third of the hose is loaded. Two loops may also be used in the same tier. *Courtesy of Rose Fire Company.*

Figure 10.75 Some departments prefer two loops, one short for the hand, and one long for the shoulder, especially on longer lines. *Courtesy of Rose Fire Company.*

MINUTEMAN LOAD

The minuteman load is a concept that significantly increases the ease and efficiency with which a fast initial fire attack can be made using a minimum of personnel. The minuteman load concept utilizes 1½-inch (38 mm) or larger hose preconnected to the apparatus. The hose bed is arranged and the hose is packed and advanced in a manner that offers substantial improvements over contemporary preconnected hose loads in the areas of efficiency and safety. Some of the main advantages of this load include:

- Permits fast, effective, initial attack with a minimum of personnel.

- The line can be removed from the bed and advanced easily by one person.

- The line peels from the top as it is advanced, and the load on the firefighter gets progressively lighter.

- The possibility of snags in the hose bed or on obstructions is greatly reduced.

- Each minuteman load occupies less than 5% of the total space available in a typical 20" (51 cm) x 73" (185 cm) x 120" (304 cm) hose bed.

- Most existing apparatus hose beds can be easily modified to accept the minuteman load.

- The minuteman load has been tested in actual use for several years and has given excellent results.

The hose compartment for this preconnected load need only be 4¼ x 18 x 120 inches (11 x 18 x 304 cm) deep. The method for making the hose load is illustrated in Figures 10.76-10.78.

Step 1: The female end of the first length of hose is connected to the outlet. One or more layers of hose are placed in the bed, then the remainder of the length is temporarily set aside (Figure 10.76).

Step 2: Place additional lengths of hose in bed with nozzle at the bottom (Figure 10.77).

Step 3: Couple the free ends of the hose and place the remainder of the first length in the bed (Figure 10.78).

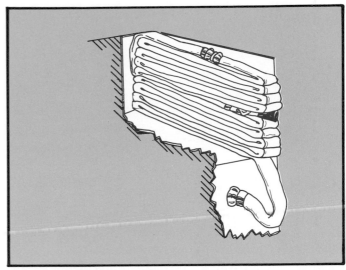

Figure 10.78 Couple the last section to the remainder of the first section, and finish the load.

Figure 10.76 Start the minuteman load by connecting to the discharge outlet, bedding the amount desired and placing the remainder over the front of the hose bed. A loop may be desired at the bottom.

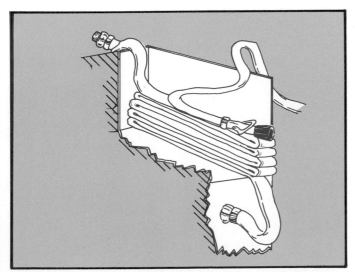

Figure 10.77 Place the nozzle at the rear of the hose bed, and place additional lengths of hose on top of it.

TRIPLE LAYER LOAD

The triple layer load is a variation from the preconnected flat load and it too can be loaded in one or more tiers. The load gets its name from the hose being folded back upon itself to form three layers. All three layers are then loaded as illustrated in Figures 10.79-10.82.

Step 1: Connect the hoseline to a preconnect discharge outlet. Extend the line until it is lying straight back from the rear of the hose bed (Figure 10.79).

Figure 10.79 Start the triple layer load by connecting to the discharge outlet and stretching the hoseline straight out from the rear of the apparatus. *Courtesy of Rose Fire Company.*

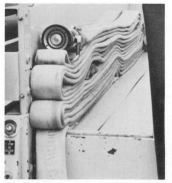

Figure 10.80 Fold the hoseline back on itself to make three even length stacked layers. It is recommended the fold at the nozzle be placed through the handle (bail) or banded in some manner to secure it and make pulling the load easier.

Step 2: At a point, about two-thirds of the distance from the hose bed to the nozzle, fold the line back upon itself and return the fold to the rear of the hose bed. The three layers of hose should be stacked with the nozzle at the end of the extended layers (Figure 10.80).

Step 3: Load the stacked layers of hose flat, keeping the three layers together as though they were a single line (Figure 10.81). If the hose compartment is wider than one hose width, cross to the next side at the front of the hose bed.

Step 4: Place the nozzle and the fold upon which it lies at the rear of the compartment when the load is finished. Finished loads for both narrow and wide compartments are shown in Figure 10.82.

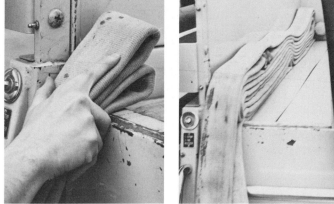

Figure 10.81 Load the stacked layers flat as though a single line. *Courtesy of Rose Fire Company.*

Figure 10.82 The nozzle is placed on top of the load at the rear of the hose bed. *Courtesy of Rose Fire Company.*

Booster Hose Reels

Booster hoselines are preconnected hose which are usually carried coiled upon reels. These booster line reels may be mounted several places upon the fire apparatus according to specified needs and the design of the apparatus. Some booster reels are mounted above the fire pump and behind the apparatus cab. This arrangement provides booster hose which can be unrolled from either side of the apparatus but its advancement above ground level is limited to its length. Hand- and power-operated reels are available.

HOSE LAYS

Successful application of fire streams is largely dependent on the speed and efficiency of engine companies laying hoselines and connecting to a water supply. Hose compartments should be arranged to facilitate the laying of either single or multiple hoselines.

Forward Lay

The first layout practices in the days before the installation of pumps on motor fire apparatus was from the hydrant (or sources of water supply) to the fire and became known as the "Forward Lay." The operation consists of stopping the apparatus at the water supply source and permitting the hydrant person to safely leave the apparatus and secure the hose, after which the apparatus proceeds to the fire laying either a single or dual line of hose (Figure 10.83). An alternate method of wrapping the hydrant is to tie a piece of rope or nylon in the form of a loop to the hose so that it may be dropped over the hydrant.

FOUR-WAY HYDRANT VALVES

The object of a four-way hydrant valve is to provide a means of changing from a direct (unsupported) hydrant supply line to a supply line supplied (supported) by a pumper without interrupting the flow of water. With a four-way valve, an unsupported hydrant line is used by the first pumper until the second pumper arrives at the hydrant and connections are made between the valve and the second pumper. The second pumper then discharges through the valve increasing the pressure and flow in the initial supply line. There are several manufacturers providing four-way valves that have the same basic operating principles. The following steps describe the typical application of a four-way hydrant valve (Figures 10.84-10.86).

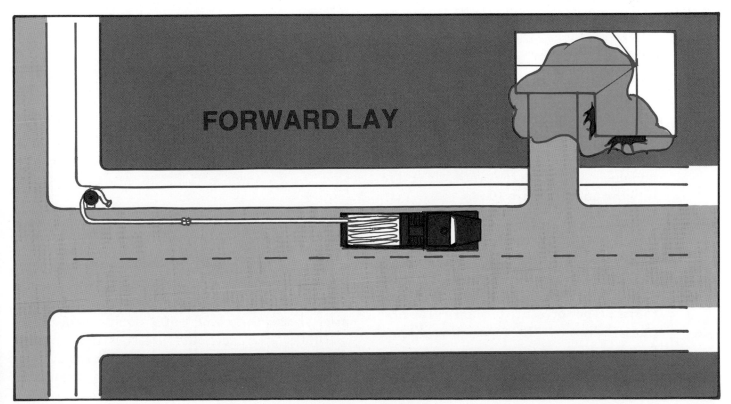

Figure 10.83 A forward lay brings water from the source to the apparatus at the fireground.

Step 1: The first pumper connects the four-way valve to the hydrant and completes a forward lay.

Step 2: Since the supply line is carried attached to the valve, the hydrant can be turned on immediately. Open the hydrant completely (Figure 10.84). Use a hose clamp at the pumper until the hoseline is connected into the pump.

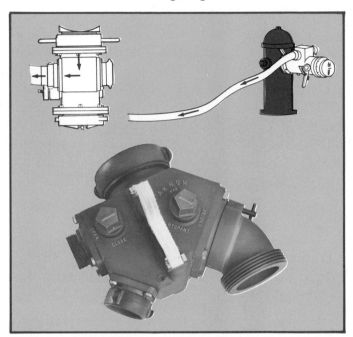

Figure 10.84 The first pumper connects the four-way valve to the hydrant and lays an unsupported supply line to the fire. When the hydrant is opened water flows through the valve to the attack pumper or wagon. Two types are illustrated. *Photo Courtesy of Akron Brass.*

Step 3: The second pumper stops at the hydrant and connects its intake sleeve to the 4½-inch (115 mm) connection on the valve and opens the valve to permit water flow into the pump (Figure 10.85).

Step 4: The second pumper connects a discharge line to the four-way valve inlet. The four-way valve is switched from hydrant to pumper supply on those where this is necessary. The discharge line is then brought to the proper pressure to support the first pumper through the supply line (Figure 10.86).

As discussed the four-way valve allows the first arriving pumper to mount a limited initial attack with preconnected attack lines using water flowing to the pump through the hydrant fed supply line. It also maintains the further use of the remaining available hydrant capacity. The attack can be expanded since the valve provides a means for the second arriving pumper to lay additional hoselines to the same hydrant, hook up, and charge these lines with no interruption of the flow in the initial supply line.

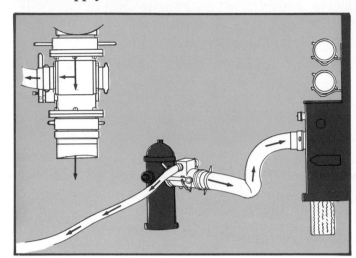

Figure 10.85 The second pumper connects to the 4½-inch (115 mm) male on the valve and opens the valve. Water is now flowing to both pumpers.

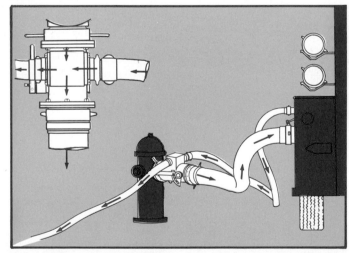

Figure 10.86 The second pumper connects to the hydrant valve 2½-inch (65 mm) inlet (opens the valve waterway if necessary) and begins pumping through the supply line.

Reverse Lay

Sometimes it is necessary to place fire apparatus at the source of water supply to feed the hoselines with adequate pressure and volume. This operation makes it necessary for the pumper to lay hoselines in reverse (from the fire to the water supply source). Because of its nature, it became known as the "Reverse Lay." The operation

Figure 10.87 In a reverse lay the pumper lays hose from the fire to the water source.

consists of stopping the apparatus so that a sufficient amount of hose may be removed from the bed. All tools that may be required at the fire should be unloaded in a safe convenient place before the apparatus leaves the scene since it is to remain at the water supply source (Figure 10.87).

When removing hose from a horseshoe load, pull the folds from the side opposite the fire. This will allow the hose to pay out easily. These folds should be placed next to any already pulled folds on the side away from the fire. When additional folds are being removed from an accordion load, they should be placed next to any previous folds on the side away from the fire (Figure 10.88). Signal the driver to proceed to the hydrant laying either single or multiple lines.

Figure 10.88 Additional hose as it is removed should be placed next to the hose already on the ground on the side away from the fire. *Courtesy of Rose Fire Company.*

Split Lay

The split lay is a hoseline laid in part as a forward lay and in part as a reverse lay. This can be accomplished by one pumper making a forward lay from an intersection or driveway entrance toward the fire. A second pumper can then make a reverse lay to the water supply source from the point where the initial line was laid (Figure 10.89).

Figure 10.89 A split lay is a hoseline laid by two pumpers, one making a forward lay and one making a reverse lay from the same point.

MAKING HYDRANT CONNECTIONS

When making hydrant connections from a forward lay, the following procedures are used:

Step 1: The driver stops the apparatus approximately 10 feet (3 m) past the hydrant. Grasp sufficient amount of hose to reach the hydrant. Step down from the tailboard facing the direction of travel with all of the equipment necessary to make the hydrant connection (Figure 10.90).

Figure 10.90 Remove sufficient hose to reach the hydrant and proceed to the hydrant with a hydrant wrench and all necessary fittings.

Step 2: Approach the hydrant and loop hose around the hydrant, place foot on the "X" created by the hose crossing over itself, and signal the driver to proceed (Figure 10.91). An alternate method of wrapping the hydrant is to tie a piece of rope or nylon in the form of a loop to the hose so that it may be dropped over the hydrant.

Step 3: Place the hydrant wrench on the operating nut. Remove caps from hydrant and place gate valve on outlet away from the

fire, remove hose loop from the hydrant and connect to the outlet nearest the fire. When using large diameter hose a threaded-to-quick-coupling adapter must be placed on the hydrant before the hose can be connected to it (Figure 10.92).

Figure 10.91 Loop hydrant with hose, place foot on hose crossover point, and signal driver before starting to loosen caps.

Figure 10.92 Place hydrant wrench on the operating nut, gate valve on the outlet away from the fire, and the supply line on the fire side outlet. The gate valve can be carried attached to the hoseline and the hose disconnected after the valve is placed on the side away from the fire.

Step 4: Open hydrant fully. When returning to the apparatus tighten leaking couplings and push hose toward the curb. **NOTE:** If multiple lines are laid, follow the same procedures as for a single line.

Step 5: When the apparatus has completed the lay, the hose is uncoupled from the bed (allowing enough hose to reach the pump inlet), and connected to the pump. The signal to charge the line is then given. Signaling to charge the line can be accomplished with hand signals, a hand light, or by using the bell, siren, or air horn.

An alternate method is to place a hose clamp on the supply line 20 feet (6 m) behind the apparatus, as this allows room to remove fire fighting lines, and signal for the line to be charged. The hose can then be uncoupled from the bed and connected to the pump while the line is being charged. When the pump connection is complete, the hose clamp is released.

When hydrant connections are made from a reverse lay the intake hose may be either soft sleeve or hard sleeve. Connecting a pumper to a fire hydrant with this intake hose involves coordination and teamwork to perform the task. More persons are needed to connect a hard sleeve hose than are needed to connect soft sleeve hose. The following steps in making the soft sleeve connection are illustrated in Figures 10.93 - 10.95.

Step 1: Spot the pumper at a convenient angle to the hydrant and within the limits of the length of the intake hose. Check to see if the booster tank valve is closed and remove the pump intake cap.

Step 2: Remove the soft sleeve hose, hydrant wrench, and gate valve from the pumper (Figure 10.93), make pumper connection, and unroll the intake hose. Place the hydrant wrench on the hydrant operating nut with the handle pointing away from the outlet.

Step 3: Remove the hydrant cap and make the hydrant connection, using any adapters which may be necessary for this operation.

Should an adapter be needed, it is usually used at the hydrant connection (Figure 10.94).

Step 4: Open the hydrant slowly and tighten any connection which leaks (Figure 10.95).

Figure 10.93 Remove the soft sleeve, hydrant wrench, and gate valve to make the hydrant connection. Make the pumper connection.

Figure 10.94 Unroll the hose, place the hydrant wrench and gate valve on the hydrant, and make the connection.

Figure 10.95 Slowly open the hydrant completely and tighten any leaking connections.

Figure 10.96 Short dual lines can be used to make pump connections from double hydrants.

Some departments carry the soft sleeve hose preconnected to the pump intake. This arrangement requires a gated intake valve at the pump to prevent leakage. Two 2½-or 3-inch (65 or 77 mm) hoselines are sometimes used for hydrants equipped with two 2½-inch (65 mm) outlets and can be connected to a siamese on the pump (Figure 10.96).

It is more efficient to connect a 4½-inch (115 mm) or larger soft intake hose to a hydrant with only 2½-inch (65 mm) outlets. Such a connection is made by using a 4½-inch (115 mm) to 2½-inch (65 mm) reducer coupling. These reducers should have male threads on the 4½-inch (115 mm) side when

the soft intake hose is equipped with two 4½-inch (115 mm) female couplings. Such a connection is illustrated in Figure 10.97.

Various conditions determine the angle at which a fire department pumper should be positioned at the curb or hydrant. No definite rule can be given to determine the distance to spot the apparatus from the curb or hydrant, for not all hydrants are at the same distance from the curb, and the hydrant outlet may not directly face the street. Another determining factor is that while most apparatus have the pump intakes on both sides, others may also have one at the front or rear. It is considered a good policy to stop the apparatus with

Figure 10.97 A large soft sleeve can also be connected to a double hydrant by using a reducer.

the intake just short of the hydrant outlet and the necessary curves can be made in the intake hose. The procedure for making the hard sleeve connection is illustrated in Figures 10.98-10.101.

Step 1: Spot the pumper at a convenient angle to the hydrant and within the limits of the length of the intake hose. Operator checks to see if the booster tank valve is closed and removes the pump intake cap. The other person secures the hydrant wrench and adapter, removes the hydrant outlet cap, and places the hydrant wrench on the hydrant operating nut with the handle pointing away from the outlet (Figure 10.98).

Step 2: Place the adapter on the 4½-inch (115 mm) outlet.

Step 3: Remove the hard sleeve intake hose from the apparatus and connect it to the hydrant (Figure 10.99). Some departments prefer to connect to the pump intake first (Figure 10.100).

Step 4: Connect the remaining end to the pump intake (Figure 10.101).

Step 5: Open the hydrant and set the pump control.

Figure 10.99 After placing the adapter on the hydrant, remove the hard sleeve and connect it to the hydrant.

Figure 10.100 The hard sleeve can be connected to the pumper first, if preferred.

Figure 10.98 As the operator checks the tank valve and removes the pump intake cap, another firefighter proceeds to the hydrant with wrench and adapter to remove the hydrant cap.

Figure 10.101 Connect the hard sleeve to the pump intake.

HANDLING HOSELINES

To effectively attack and extinguish a fire, hoselines must be removed from the apparatus and advanced to the location of the fire. The techniques used to advance hoselines will depend on how the hose is loaded. Hoselines may be loaded preconnected to a discharge outlet or simply placed in the hose bed and not connected.

Preconnected Hoselines

FLAT LOAD

Advancing the preconnected flat load involves pulling the hose from the compartment and walking toward the fire. This procedure is illustrated in the following steps (Figures 10.102 - 10.103).

Step 1: Remove the nozzle from the hose bed and place it over the shoulder, with one hand on the hoseline, and slip the other arm through the hose loops (Figure 10.102).

Step 2: Face away from the apparatus, pull hose clear of the bed, and walk away from the apparatus (Figure 10.103).

Figure 10.102 Place the nozzle over the shoulder and pull the remaining loops. When both long and short loops are used, the long loop goes over the shoulder and the short loop is grasped with the hand.

Figure 10.103 The flat load can be unloaded by simply walking away from the apparatus.

MINUTEMAN LOAD

The added advantage of the minuteman load is that the entire load can be advanced without dragging any of the hose. This procedure is illustrated in the following steps (Figures 10.104 - 10.106).

Step 1: Grasp the nozzle and pull the load partially out of the hose bed (Figure 10.104).

Step 2: Face away from the apparatus and place the hose load on the shoulder with the nozzle against the stomach (Figure 10.105).

Figure 10.104 Grasp the nozzle and partially pull the hose load out of the bed.

Figure 10.105 Place the hose load on the shoulder with the nozzle on the bottom.

This procedure is described and illustrated in the following steps (Figures 10.107 - 10.109).

Step 1: Place the nozzle and fold of the first tier over the shoulder while facing in the direction of travel (Figure 10.107).

Step 2: Walk away from the apparatus, pulling the hose out of the bed (Figure 10.108).

Step 3: When the hose bed has been cleared, drop the folded end from the shoulder, and advance the nozzle (Figure 10.109).

Figure 10.106 Walk away from the apparatus allowing the hose to feed from the top of the shoulder load.

Step 3: Walk away from the apparatus, pulling the hose out of the bed by the bottom loop (Figure 10.106).

Step 4: When the hose is stretched straight out, allow the load to pay off from the top.

TRIPLE LAYER LOAD

Advancing the triple layer load involves placing the nozzle and the fold of the first tier on the shoulder and walking away from the apparatus.

Figure 10.108 Walk away pulling the entire hose load from the bed. Release the fold from the nozzle bail.

Figure 10.107 Place the nozzle with the first fold locked under the handle (bail) over the shoulder.

Figure 10.109 Drop the fold and continue advancing the nozzle to the fire.

Non-Preconnected Lines

WYED LINES

The reverse horseshoe and other wyed lines are normally used in connection with a reverse layout since the wye connection is fastened to the 2½ or 3 inch (65 or 77 mm) hose. The unloading process involves two operations which can be done consecutively by one person. The following steps for unloading and advancing are illustrated in Figures 10.110 - 10.111.

Step 1: Grasp either nozzle and pull the reverse horseshoe out of the bed until the loop is clear, bring the hose to a vertical position while inserting the arm through the loop. Position the hose on the shoulder with the nozzle to the front of the body (Figure 10.110).

Step 2: Pull the hose clear of the apparatus and proceed toward the fire. The hose will pay off from the outside of the load.

The other side of the reverse horseshoe is pulled by a second firefighter. If available, a third firefighter can take the wye and additional large hose.

The unloading of wyed line loads will depend upon how they are placed in the hose bed. In most cases this involves grasping the nozzle or strap connected to the wyed line load and pulling it clear of the hose bed. Both sides of the load can be pulled to shoulder loads or to the ground and the wye connection placed between them (Figure 10.111). The line can then be advanced by any accepted method. The nozzle can be carried in the hands or over a shoulder.

Figure 10.111 Both sides of the wye finish can be pulled at the same time and moved to the fire area with the gated wye.

SHOULDER LOADS (FROM FLAT OR HORSESHOE LOADS)

Due to the way flat and horseshoe loads are arranged in the hose bed, it becomes necessary to load one section of hose at a time onto the shoulder. The following steps for shoulder loading and advancing are illustrated in Figure 10.112-10.114.

Step 1: Face the hose bed, grasp the nozzle or coupling, and place it against the front of the body about chest high.

Step 2: Pass the hose over either shoulder, keeping the hose flat and form a fold that extends to just behind the knee.

Step 3: Form another fold in front of the body extending to the waist (Figure 10.112).

Figure 10.110 Place the arm through the reverse horseshoe, bring it to a vertical position (if not there) and place it on the shoulder.

Step 4: Continue forming folds at the rear and front of the body until a section of hose has been loaded on the shoulder (Figure 10.113).

Step 5: Additional shoulder loads are formed in the same manner with the hose lead from the previous shoulder load, passing over the shoulder to the back and then folding forward over the shoulder. Couplings should rest in the front of the body.

Step 6: When the desired amount of hose has been loaded, all turn the same direction and proceed, allowing top fold to drop from the shoulder as the line becomes taut (Figure 10.114).

Figure 10.112 Grasp the nozzle in front of the body, form a fold over the shoulder at the back of the knee, and return the hose back to the front forming a fold at the waist.

Figure 10.113 Continue forming folds front and rear until one section of hose is loaded.

Figure 10.114 When sufficient hose is loaded, all firefighters turn and proceed. The hose will pay off the top of the last firefighter's shoulder load.

SHOULDER LOADS (FROM THE ACCORDION LOAD)

Since all of the folds in an accordion load are near the same length, they can be loaded on the shoulder by taking several folds at a time directly from the hose bed. The following steps for shoulder loading and advancing are illustrated in Figures 10.115-10.117.

Step 1: Face the hose bed, grasp the nozzle or coupling and the number of folds needed to make up that portion of the shoulder load with both hands (Figure 10.115).

Step 2: Pull the folds about one-third of the way out of the bed, twist the folds into an upright position, turn and pivot into the folds placing them on top of the shoulder. Make sure the hose is flat on the shoulder with the nozzle or coupling in front of the body (Figure 10.116).

Step 3: Grasp the bundle tightly with both hands and step away from the apparatus, pulling the shoulder load completely out of the bed (Figure 10.117).

Step 4: Additional shoulder loads are removed in the same manner.

Figure 10.116 Pull the folds from the load and turn into them as they are brought flat over the shoulder with the nozzle or coupling against the front of the firefighter.

Figure 10.115 Face the hose bed and grasp an appropriate number of folds from the accordion load.

Figure 10.117 Grasp the shoulder load with both hands and move away from the apparatus removing the remaining hose from the bed.

ADVANCING HOSELINES

The various hose loads and finishes have all been designed to enable firefighters to more easily and effectively unload fire hose and advance it where it can be used. Improvements in loading and unloading fire hose have always been an outgrowth from the efforts of firefighters trying to do a better job. These previously discussed hoseline handling techniques are basically for ground-level operations from any one of the varied hose loads. Ground-level operations bring the firefighter only to the building entrance at which point the firefighter must apply what has been learned to those areas and obstacles inside the building. Hoselines must sometimes be elevated on the outside of buildings to entrances which are higher than ground level by means of ladders, elevating platforms, or ropes. In like manner, hoselines must sometimes be advanced to upper floors through the use of stairways. The application of hoseline advancement techniques to points above ground level is limited to the amount of hose which an average crew of firefighters can carry. Fire department standpipe systems enable fire hose to be advanced beyond these heights.

Advancing Hose Into A Structure

For maximum firefighter safety it is necessary that the firefighter be alert to the ever present danger of backdraft, flashover, and structural collapse. When advancing a hoseline into a burning structure the firefighter should.

- Have the firefighter on the nozzle and the backup firefighters on the same side of the line (Figure 10.118).

- Feel the door using the back of the hand with the glove off.

- Set the nozzle pattern wide for an indirect attack or narrow for a direct attack.

- Bleed the air out of the line before entry into the structure (Figure 10. 119).

- Stay low and out of the doorway (Figure 10.120).

Advancing Hose Up A Stairway

Advancing hose up a stairway presents several conditions that hinder the operation. Hose is

Figure 10.118 Firefighters advancing a line should be on the same side of the hose.

Figure 10.119 Set the nozzle pattern, check the gallonage setting (if necessary), and bleed the air out of the line before entering the building.

Figure 10.120 All firefighters keep low and out of doorways to avoid burns.

difficult to drag in an open space and is exceedingly difficult to drag around the obstructions found in a stairway. The shoulder carry and the underarm carry are adaptable to stairway advancement, since the hose is carried into position and is fed out as it is needed. Hoselines should be advanced before they are charged with water and this technique is particularly important for stairway advancement. During the advancing process, lay the hose on the stairs against the outside wall to avoid sharp bends and kinks. When the line from the outside becomes taut, the hose starts feeding from the last carrier. Advancing hose from the shoulder carry is shown in Figure 10.121. A similar situation is illustrated in Figure 10.122 for the underarm carry.

Figure 10.121 Hose can be advanced up stairs by the accordion shoulder carry. *Courtesy of Rose Fire Company.*

Figure 10.122 Hose can also be advanced up a stairway by the underarm carry. *Courtesy of Rose Fire Company.*

To advance a charged line, take a six to ten foot (2 - 3 m) loop in the line where slack is available. Stand the loop up and roll it toward the nozzle (Figure 10.123).

Figure 10.123 A charged line on the ground can be advanced by forming a loop and rolling the slack toward the nozzle.

Advancing Hose from a Standpipe.

Fighting fires in tall buildings presents the problem of getting hose to upper floors. While hoselines may be pulled from the apparatus and extended to the fire area, it is not considered good practice. It is more practical to have some hose rolled or folded on the apparatus ready for standpipe use.

The manner in which standpipe hose is arranged is a matter of personal preference. It may be in the form of folds or bundles that are easily carried on the shoulder, or in specially designed hose packs complete with nozzles, fitting, and tools. See page 238.

Hose can be brought to the fire floor by taking either the stairway or the elevator. In any case, fire crews should stop one floor below the fire floor and make the connection to the standpipe. This also gives firefighters an idea of the floor layout below the fire so they can possibly relate it to the fire floor.

Upon reaching the standpipe, detach the building hoseline, check the connection for the correct adapters (if needed), and connect the fire department hose to the standpipe. If 1½-inch (38 mm) hose is used, it is a good practice to place a gated wye either on the standpipe or at the end of a short piece of 2½-inch (65 mm) hose connected to the standpipe. The use of 2½-inch (65 mm) hose should also be considered and will depend on the size and nature of the fire. Once the standpipe connection is completed, any extra hose should be taken up the stairs toward the floor above the fire (Figure 10.124). It is much easier to advance the charged hoseline down the stairs than it is to pull it up.

During pickup operations, the water contained in the hoselines should be carefully drained to prevent unnecessary water damage. This can be done by draining the hose out a window, down a stairway, or by some other suitable drain.

Figure 10.124 Once the standpipe connection is completed, extra hose should be taken up the stairs toward the floor above the fire.

Advancing Hose Up a Ladder

Advancing fire hose up a ladder can be best achieved with a line that is not charged. If the hose is already charged with water, it will be safer, quicker, and easier to relieve the pressure and drain the hose before the advancement is made. Two methods by which fire hose can be advanced up a ladder are shown in Figures 10.125 and 10.126. The firefighters should climb about 10 feet (3 m) apart and there should be about 20 to 25 feet

Figure 10.125 Advance a hose up a ladder by looping the hose over the same shoulder of each firefighter.

Figure 10.126 Rope or web hose tools can be used to advance a hose up a ladder.

(6 to 8 m) of hose between each firefighter. How hose can be advanced by looping the hose over a common shoulder of each carrier is illustrated in Figure 10.125. Rope hose tools used for this advancement are shown in Figure 10.126. Whenever possible, it is best to have one firefighter at the base of the ladder to help feed the hose to the carriers and to have one firefighter hold the ladder during the advancement.

Advancing A Charged Line Up Ladder

Firefighters position themselves within reaching distance on the ladder. Each firefighter takes the leg lock on the ladder since both hands will be required to move the charged line. The hose is pushed upward from firefighter to firefighter (Figure 10.127). The firefighter on the nozzle takes the line into the window and the other firefighters continue to hoist additional hose as necessary.

Using a Charged Line from a Ladder

Charged line is passed up ladder as previously stated. The top firefighter holds the nozzle while the second firefighter hoists some slack. The firefighter on the nozzle projects the nozzle through the ladder and secures it with a rope hose tool. Both firefighters use leg lock and when line is anchored into position, the nozzle is opened (Figure 10.128).

Figure 10.128 A nozzle can be operated from a ladder once it is properly secured.

Passing Hose Upward

Hoselines are sometimes advanced by passing them upward alongside a building, fire escape, or ladder. This operation also requires the hose not to be charged with water. If the operation takes place alongside a building, there should be a firefighter at a window on each floor. When passing hose up a ladder, there should be a firefighter for every 10 to

Figure 10.127 A charged hose can be pushed up the ladder from firefighter to firefighter.

12 feet (3 to 4 m) of elevation. Passing a hoseline up a fire escape requires a firefighter on each landing. A rope hose tool, hose strap, or hose chain can be used to fasten a pike pole to the hose. The pike pole provides a means by which the hose can be passed from one firefighter to another. One method of passing hose upward is shown in Figure 10.129.

Hoisting Hose

Hoisting hose with a rope introduces the possibility of damage to the coupling or nozzle as it is being raised. For this reason, it is recommended that the end of the hose and nozzle be folded back upon itself a short distance so that the nozzle will point opposite the direction of travel. A hoseline being hoisted with a rope and hose roller is shown in Figure 10.130. One rope tie that is commonly accepted for hoisting hose is as follows: Tie a clove hitch well down on the hoseline. Place a half hitch over the nozzle and hose where the nozzle is folded back on the hose. Place the second half hitch on the doubled hose about 12 inches (30 mm) from the loop end. With the ties properly placed, the hose will turn on the hose roller, so that the coupling and nozzle will be on top and the hose passes over the roller. An alternate method is to make the tie at the bend with the nozzle facing outward. A running bowline is used so that the rope can be dropped down and used to pull up more hose.

Figure 10.129 A pike pole and hose tool can be used to pass hose upward.

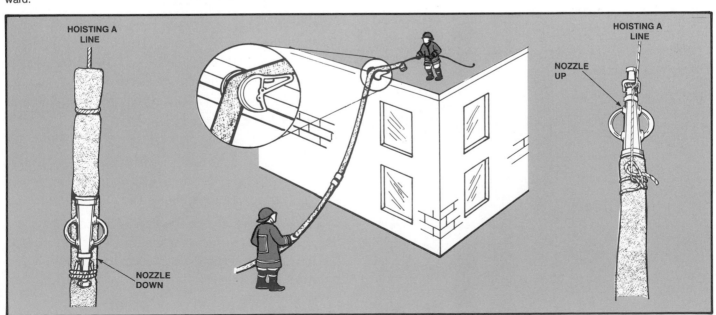

Figure 10.130 Dry and charged hoselines can be hoisted with a rope and hose roller. The rope and hitches should be against the building and the nozzle away from it.

When hoisting a charged hoseline, attach the rope to the hose below the nozzle coupling. This will put all of the strain on the hose and not the coupling. Make sure the nozzle shutoff is securely tied with a half hitch so that it cannot be bumped open while it is being elevated. It is recommended that charged hoselines have the pressure bled off and be drained prior to lowering them. To lower a hoseline using a rope, reverse any of the procedures for hoisting.

Extending a Section of Hose

If a hoseline needs to be extended in length, sufficient extra sections may be added using the following procedure (Figure 10.131).

Step 1: Bring additional sections of hose as needed to the nozzle end of the line.

Step 2: Apply the hose clamp approximately three feet (1 m) behind the nozzle and open the nozzle to release the pump pressure in the line.

Step 3: Remove the nozzle, add the new section of hose and reattach the nozzle.

Step 4: Slowly release the clamp allowing water to flow to the nozzle.

It is also possible to extend a section of hose by using a break-apart nozzle (Figure 10.132).

Figure 10.132 A hose can also be extended by using a break-apart nozzle.

Retrieving a Loose Hoseline

A loose hoseline is one in which water is flowing through a nozzle, an open butt, or a broken line, and which is not under control by firefighters (Figure 10.133). This situation is very dangerous since the loose hoseline may whip back and forth and up and down. Firefighters and bystanders may be seriously injured if they are hit by the uncontrolled whipping end.

Closing a valve to shut off the flow of water is the safest way to control a loose line. These valves may be at the pump or a hydrant, or a hose clamp may be used as a stationary point in the hoseline. If closing a valve is not possible, use the following procedure.

Step 1: Move back down the hoseline as far as necessary for safety and get on hands and knees. Place knees on each side of the hose and place the hands on top of the hose.

Step 2: Move toward the nozzle while keeping a secure grip on the hose with the hands and knees.

Step 3: Continue moving toward the nozzle to accomplish control of the line and to shut the nozzle off.

Figure 10.131 To extend a hoseline place a hose clamp 3-feet (1 m) behind the nozzle, remove the nozzle, add the hose, reattach the nozzle, and release the hose clamp.

Figure 10.133 Four ways in which a hoseline may become loose are open nozzle, slipped coupling, broken line, and slipped clamp.

Replacing a Burst Section of Hose

A burst section of hose under pressure can be dangerous. Since the need for immediate shutoff is urgent and a hose clamp may not be available, the quickest and probably the most efficient method to shut off the flow of water is to kink the hoseline as shown in Figure 10.134. First, obtain sufficient slack in the line, bend the hose over on itself, and apply body weight to the bends in the hose. It is helpful, during the operation, to place one knee directly upon the bend and apply pressure at this point. Of course, if a hose clamp is available, it should be used. The shutoff must be made between the failure and the source of water supply approximately three feet (1 m) behind the coupling of the damaged section (Figure 10.135). Two additional sections of fire hose should be secured from the ap-

paratus for the replacement of one bad section. This practice is necessary because hoselines stretch to longer lengths when under pressure; thus, the couplings in the line are invariably farther apart than the length of the replacement sections.

Figure 10.135 A hose clamp can also be used to control the flow from the pumper while a burst section of hose is replaced.

OPERATING HOSELINES

The following method can be used with medium size attack lines of 1½-, 1¾-, and 2-inch (38 mm, 44 mm, 50 mm) hose.

One-Firefighter Method (Medium Hoselines)

A hoseline and the nozzle must be kept under control at all times. When one firefighter is required to handle a medium size hose and nozzle,

Figure 10.134 The flow of water in a burst hose can be quickly stopped by getting enough slack and placing two bends in the hose.

some means must be provided for bracing and anchoring the hoseline. To accomplish this, the firefighter at the nozzle should hold the nozzle with one hand and with the other hand hold the hose just back of the nozzle. The hoseline should be straightened for at least ten feet (3 m) behind the nozzle, and the firefighter at the nozzle should face the direction in which the stream is to be projected. Permit the hose to cradle against the inside of the closest leg and brace or hold it against the front of the body and hip. Anchor the hose to the ground or floor by placing the foot of the supporting leg upon the hose. If the stream is to be moved or directed at an excessive angle from the center line, close the nozzle, straighten the hose, and resume the operation position. The method is illustrated in Figure 10.136.

Figure 10.136 With the hose straight, one firefighter can hold a small charged line with the foot and body as a brace.

Two-Firefighter Method (Medium Hoselines)

The two-firefighter method of handling a nozzle on a medium size attack line should be used whenever possible, because it provides a greater degree of safety than the one-firefighter method. The two-firefighter method is usually necessary when the nozzle needs to be advanced. The person at the nozzle holds the nozzle with one hand and with the other holds the hose just back of the nozzle. The hoseline is then rested against the waist and across the hip. The back-up firefighter takes a position on the opposite side of the hose about 3 or 4 feet (1 m) back. The firefighter then holds the hose

Figure 10.137 When the medium size charged hoseline must be advanced two firefighters are recommended.

with both hands and rests it against the waist and across the hip. One important function of the back-up firefighter is to keep the hose straight behind the person at the nozzle. The method is illustrated in Figure 10.137.

The following methods can be used with large size attack lines of 2½-, 2¾-, and 3-inch (65 mm, 69 mm, 77 mm) or larger hose.

One-Firefighter Method (Large Hoselines)

Whenever a fog or solid stream nozzle is used connected to a large size attack line fire hose, a minimum of two, and preferably three, firefighters should be used. One firefighter may, however, be found alone with a large charged hoseline that needs to be used. A reasonably safe way by which this task can be performed is illustrated in Figure 10.138. The firefighter secures slack hose from the line, forms a large loop and crosses the loop over the line about two feet (1 m) back of the nozzle. The

Figure 10.138 One firefighter can safely operate a larger hose by forming a loop, passing the nozzle under the line, and sitting on the intersection.

firefighter then sits where the hose crosses and directs the stream. This method does not permit much maneuvering of the nozzle, but the nozzle can be operated from this point until help is available. If the operation is to be of a considerable duration and other equipment or manpower is not available, tie the hose at the cross to permit ease of operation and greater safety as shown in Figure 10.139.

Figure 10.139 The safety of this operation can be increased by tying the hose where it crosses.

Two-Firefighter Method (Large Hoselines)

When only two firefighters are available to handle a nozzle on a large hoseline, some means of anchoring the hose must be provided because of the kickback. One method is for the firefighter at the nozzle to hold the nozzle with one hand and the hose, just back of the nozzle, with the other hand. Then rest the hoseline against the waist and across the hip. The back-up firefighter, in this case, must serve as an anchor, and this position is about 4 feet (1 m) back. Here the back-up firefighter places the closest knee upon the hoseline. In this position the back-up firefighter should be kneeling on one knee with both hands on the hoseline near the other knee. This position prevents the hose from moving back or to either side. Should the hose in front try to come back or up, the back-up firefighter is in a position to push it forward.

Another two-firefighter method utilizes rope hose tools to assist in anchoring the hose. The firefighter at the nozzle loops a rope hose tool around the hose a short distance from the nozzle and places the large loop over the outside shoulder.

The nozzle is then held with one hand and the hose just back of the nozzle with the other hand. The hoseline is rested against the body and leaning slightly toward the nozzle, holds some of the back pressure. The back-up firefighter again serves as an anchor, and this position is about 4 feet (1 m) back. The firefighter also has a rope hose tool around the hose and the outside shoulder and leans forward to absorb some of the back pressure. The method is illustrated in Figures 10.140 and 10.141.

Figure 10.140 A rope hose tool will assist the back-up firefighter in anchoring a larger line.

Figure 10.141 Both the firefighter at the nozzle and the back-up firefighter can use rope or web hose tools to aid in absorbing the nozzle's back pressure.

Three-Firefighter Method (Large Hoselines)

Handling a nozzle on a large size hoseline can be more easily done by three firefighters. There are some differences of opinion as to the location of the back-up firefighter and the anchor firefighter. Some prefer a back-up firefighter (on either side of the hoseline) directly behind the firefighter at the nozzle with the anchor firefighter kneeling behind the back-up firefighter as shown in Figure 10.142. Others prefer both firefighters to serve as anchors kneeling side by side on opposite sides of the hoseline. The firefighter in this position places the

Figure 10.142 The firefighters backing up and anchoring a hoseline may be positioned in several locations and do an effective job.

hoseline in a vise to assure more stability at the anchor point.

HOSE CARRIES

Accordion Shoulder Carry

The accordion shoulder carry has many variable adaptations. Although it is adaptable to the advancement of working hoselines, it is likewise a useful carry for a single section of fire hose. This carry provides a system and method whereby a firefighter can take either a single section of hose or a segment of a working line and carry it in a systematic and usable manner.

Step 1: Stretch a section of fire hose into a straight line. Carry one coupling end toward the other until approximately the middle of the section is reached. Pick up the hose near the center with the same hand and advance to the other coupling (Figure 10.143).

Step 2: Pick up the second coupling and place both couplings on the fold, making sure the ends are even (Figure 10.144).

Figure 10.143 Take one coupling to the middle of a section of hose and pick up the hose with the same hand.

Step 3: Advance back toward the other end (Figure 10.145) and line up all folds and couplings.

Step 4: Go to the center of the hose and squat next to it, placing the folded hose onto the shoulder and lifting with the legs (Figure 10.146).

Step 5: Use caution when laying the bundle down so as not to damage the couplings. This can be accomplished by holding onto the couplings and flipping the hose off the shoulder (Figure 10.147).

Figure 10.144 Hold the fold in the left hand while picking up the second coupling. Then place the couplings on top of the fold.

Figure 10.145 With the ends even, proceed back to the other end of the hose and line up all folds and couplings.

Figure 10.146 Pick up the folded hose section at the center using proper lifting techniques.

Figure 10.147 To unload, hold the couplings to avoid damage and flip the hose off the shoulder.

Single Section Drain and Carry

It frequently becomes necessary to drain excess water from a section of fire hose and to carry it a reasonable distance. Draining the hose is one task and preparing it to be carried is another. The following technique is a way by which both tasks can be performed at the same time for an accordion shoulder carry. The hose does not necessarily need to be in a straight line but sharp bends should be avoided.

Step 1: Pick up either end, allow the water to flow forward, and place the coupling in front of the body near the waist with the hose looped over one shoulder (Figure 10.148).

Step 2: Hold the hose in front of the body with both hands, walk slowly forward, and form a loop in front of the body.

Step 3: Continue to walk slowly down the hoseline, place the gathered hose over the same shoulder, and form loops about knee high in front and behind the body (Figure 10.149).

Step 4: Continue to walk slowly down the entire hose section, guide the hose over the shoulder, and form additional loops in front and behind the body until the section of hose has been drained and loaded on the shoulder (Figure 10.150).

Figure 10.149 Walk forward making additional loops in front and back of the body that extend to the knees.

Figure 10.148 Pick up a coupling, place it in front at the waist, and loop the hose over the shoulder.

Figure 10.150 Continue until the entire section is drained and loaded on the shoulder.

Shoulder Loop Carry

The shoulder loop carry can be used to carry a single section of hose. The steps for picking up the shoulder loop carry are illustrated in Figures 10.151-10.156.

Step 1: Stand facing the outstretched hoseline and place the coupling end (nozzle may be attached) over the shoulder. Permit the hoseline to hang directly in front of the body (Figure 10.151).

Step 2: Step forward sufficiently to form a bight about three feet (1 m) long on the ground. Stoop forward and pick up the hoseline (Figure 10.152).

Step 3: Move forward slightly and stand erect, bring the just formed loop back and over the shoulder to form a loop just short of the ground (Figure 10.153).

Step 4: Continue to load successive shoulder loops in the same manner (Figure 10.154).

Figure 10.153 Stand erect bringing the just formed loop back and over the shoulder. *Courtesy of William Cooper.*

Figure 10.151 Face the stretched out hoseline and place the nozzle or coupling over the shoulder. *Courtesy of William Cooper.*

Figure 10.152 Step forward to form a bight with the hose, and pick up the hose. *Courtesy of William Cooper.*

Figure 10.154 Continue to load successive loops in the same manner. *Courtesy of William Cooper.*

Step 5: When a section of hose has been loaded on the shoulder, grasp the flaked hose at the shoulder.

Step 6: Lift the shoulder loop from the shoulder, (Figure 10.155) turn toward the loop, insert arm through the loop, and place the shoulder loop upon the other shoulder (Figure 10.156).

Figure 10.155 Grasp the hose, lift it from the right shoulder, and turn toward the loop. *Courtesy of William Cooper.*

Figure 10.156 Insert the other arm through the loop placing the hose on the left shoulder. In order proceed away from the apparatus. *Courtesy of William Cooper.*

Working Line Drag

The working line drag is one of the quickest and easiest ways to move fire hose at ground level. Its use is limited by available personnel, but when adapted to certain situations, it proves to be an acceptable method.

Method One: Stand alongside a single hoseline at a coupling and face the direction of travel. Place the hose over a shoulder with a coupling in front and resting on the chest. Hold the coupling in place and pull with the shoulder. Additional firefighters are needed at each coupling and about one-third of the hose section should form a loop on the ground between each firefighter.

Method Two: Two loops, one over each shoulder and the couplings resting on the chest is an adaptation of this drag to advance additional sections of hose. Two lines can be advanced simultaneously by placing loops over both shoulders with the couplings on the chest and pulling (Figure 10.157).

Figure 10.157 Two lines can be advanced simultaneously by placing loops over both shoulders with the couplings on the chest and pulling.

NFPA STANDARD 1001
VENTILATION
Fire Fighter I

3-10 Ventilation

3-10.1   The fire fighter shall define the principles of ventilation, and identify the advantages and effects of ventilation.

3-10.2   The fire fighter shall identify the dangers present, and precautions to be taken in performing ventilation.

3-10.3   The fire fighter shall demonstrate opening various types of windows from inside and outside, with and without the use of fire department tools.

3-10.4   The fire fighter shall demonstrate breaking window or door glass, and removing obstruction.

3-10.5   The fire fighter, using an axe, shall demonstrate the ventilation of a roof and a floor.*

Fire Fighter II

4-10 Ventilation

4-10.1   The fire fighter shall demonstrate the use of different types of power saws and jack hammers.

4-10.2   The fire fighter shall identify the different types of roofs, demonstrate the techniques used to ventilate each type, and identify the necessary precautions.

4-10.3   The fire fighter shall identify the size and location of an opening for ventilation, and the precautions to be taken during ventilation.

4-10.4   The fire fighter shall demonstrate the removal of skylights, scuttle covers, and other covers on roof tops.

4-10.5   The fire fighter shall demonstrate types of equipment used for forced ventilation.

4-10.6   The fire fighter shall demonstrate ventilation using water fog.*

*Reprinted by permission from NFPA Standard No. 1001, *Standard for Fire Fighter Professional Qualifications*. Copyright © 1981, National Fire Protection Association, Boston, MA.

IFSTA's Ventilation Transparencies are designed to complement this chapter.

Chapter 11
Ventilation

Ventilation is the systematic removal of heated air, smoke, and gases from a structure, followed by the replacement of a supply of cooler air, which facilitates other fire fighting priorities.

The importance of ventilation cannot be overlooked. It increases visibility for quicker location of the seat of the fire. It decreases the danger to trapped occupants by channeling away hot, toxic gases and it reduces the chance of backdraft. Unfortunately, ventilation may be misunderstood by the public because it requires doing limited damage to a building; but it results in a much larger reduction in damage.

INCREASED USE OF PLASTICS MAKES VENTILATION MORE IMPORTANT

Modern technology requires new emphasis on ventilation. Consider a single item: plastics. Since the middle of the twentieth century, the use of plastic materials has had phenomenal growth. These materials perform such a vital function that a new industry has been created, taking its place beside such basic industries as those using wood, metal, and textiles. The plastics industry is expected to surpass the steel industry in pounds produced per year early in the 1980s. As a result, the fuel load in all occupancies will be increased and one can also expect increases in the amounts of products of combustion produced during fires. Prompt ventilation for the saving of lives, suppression of fire, and reduction of damage becomes more important every day.

Recent research has indicated that modern energy conservation policies may be creating additional ventilation problems with increased insulation requirements. This is because the insulation will retain heat much better and will raise the temperature of combustibles in the fire area to ignition temperature and may cause flashover to occur much faster. Therefore, the need of ventilation is increased and must be accomplished much sooner than has been practiced in the past. Insulation installed over roof coverings of fire-rated roof construction will effectively retain heat and may reduce the fire rating drastically, causing premature roof failure.

Pre-fire inspection should note the roof construction and where extra insulation has been added to existing roofs and attic areas, so personnel performing ventilation may be aware of possible problems.

When a fire officer determines the need for ventilation, the precautions that may be necessary for the control of the fire and the safety of the firefighters must also be considered. There is a need for protective breathing equipment for respiratory protection, and charged hoselines during the ventilation process. The possibility of fire spreading throughout a building and the danger of exposure fires are always present.

ADVANTAGES OF VENTILATION

The major objectives of a fire fighting force are to reach the scene of the fire as quickly as possible, rescue trapped victims, locate the fire, and apply suitable extinguishing agents with a minimum of fire, water, smoke, and heat damage. Ventilation during fire fighting is definitely an aid to the fulfillment of these objectives. When proper ventilation is accomplished to aid fire control, there are certain advantages that may be obtained from its application.

Aids Rescue Operations

Proper ventilation simplifies and expedites the rescue of victims by removing smoke and gases which endanger occupants who are trapped or unconscious, and by making conditions safer for firefighters.

Speeds Attack and Extinguishment

The removal of smoke, gases, and heat from a building permits firefighters to more rapidly locate the fire and proceed with extinguishment. Proper ventilation not only reduces the danger of asphyxiation, it also reduces the obstacles which hinder firefighters while they perform fire extinguishment, salvage, rescue, and overhaul procedures, by improving vision and removing the discomfort of excessive heat. When an opening is made in the upper portion of a building during ventilation, a chimney effect is created which draws air currents from throughout the building in the direction of the opening. For example, if this opening is made directly over the fire, it will tend to localize the fire. If it is made elsewhere, it may contribute to the spread of the fire (Figure 11.1).

Reduces Property Damage

Rapid extinguishment of a fire reduces water damage. Proper ventilation assists in making this damage reduction possible. One method of ventilation that may prove advantageous is applying water in the heated area in the form of water fog or spray. The gases and smoke may be dissipated, absorbed, or expelled by the rapid expansion of the water when it is converted into steam. In addition to removing gases, smoke, and heat, this method also reduces the amount of water that may be required to extinguish the fire.

Smoke may be removed from burning buildings by controlling heat currents, by dissipating smoke through the expansion of water as it is turned into steam, or by mechanical processes. Mechanical processes include blowers, exhaust

Figure 11.1 Smoke and heat make entrance into a building difficult and dangerous. Proper ventilation will remove the heat and smoke.

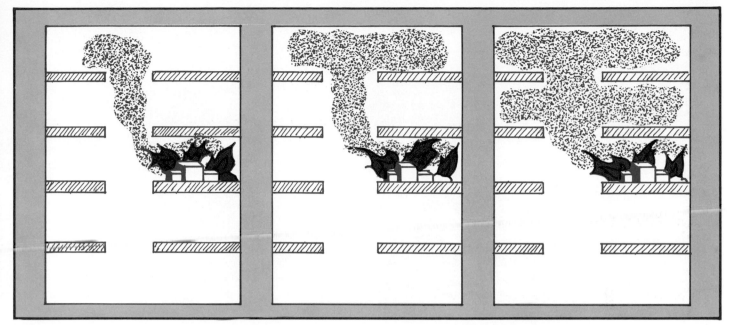

Figure 11.2 When smoke, heat, and gases cannot escape they fill all available space from the top, mushrooming down.

fans, and smoke ejectors. Regardless of the method used, ventilation reduces smoke damage because fuel vapors and carbon particles are removed.

When smoke, gases, and heat are removed from a burning building, the fire can be quickly confined to an area. This will permit effective salvage operations to be initiated even while fire control is being accomplished. Very little merchandise is saved when covers are placed over water-soaked, smoke-filled materials; therefore, it is imperative that salvage operations be started as soon as possible.

Reduces Mushrooming

Heat, smoke, and fire gases will travel upward to the highest point in an area due to convection until they are trapped by a roof or ceiling. As they are trapped and begin to accumulate they bank down and spread laterally to involve other areas of the structure. This phenomenon is generally termed mushrooming (Figure 11.2).

Proper ventilation of a building during a fire reduces the possibility of mushrooming. It tends to draw the fire to a point by providing an escape for the rising heated gases.

Reduces the Danger of Backdraft

When sufficient heat is confined in an area, the temperatures of combustible materials rise to their ignition points. These materials will not ignite, however, unless sufficient oxygen is available to support combustion. In this situation, a very dangerous condition exists because the admittance of an air supply (which provides the necessary oxygen) is all that is needed to change the overheated area into an inferno. This sudden ignition is often referred to as a backdraft. In order to prevent this critical situation from occurring, top ventilation must be provided to release superheated fire gases and smoke.

Firefighters must be aware of this explosion potential and must proceed cautiously in areas where excessive amounts of heat have accumulated. During fire fighting or rescue operations, doors should be opened slightly and carefully so they may be closed quickly, if necessary, to shut off the air supply to extremely hot areas.

Situations which create a backdraft are confinement and intense buildup of heated gases in an atmosphere being depleted of oxygen (Figure 11.3). Proper ventilation is a procedure whereby these heated gases can be released with a minimum of additional damage to the structure.

CONSIDERATIONS AFFECTING DECISION TO VENTILATE

The requirements for a plan of attack must be considered before a fire officer directs or orders to

Figure 11.3 Watch for the indications of a backdraft, and if necessary, perform vertical ventilation.

ventilate. A series of decisions should first be made that pertain to the ventilation needs. These decisions will, by the nature of fire situations, fall into the following order:

- Is there a need for ventilation at this time? The answer to this question must be based upon the heat, smoke, and gas conditions within the structure, and the life hazard.

- Where is ventilation needed? The answer to this question involves construction features of the building, contents, exposures, wind direction, extent of the fire, location of the fire, top or vertical openings, and cross or horizontal openings.

- What type of ventilation should be used? The answer to this question may be derived from a fire officer's knowledge of the following three methods of ventilation: (1) Providing an opening for the passage of air between interior and exterior atmospheres; (2) Using the application of water fog and the expansion of water into steam to displace contaminated atmospheres; and (3) Using forced air ventilation.

Visible Smoke Conditions

Smoke conditions will vary according to how burning has progressed. A free-burning fire must be treated differently than one which is in the smoldering stage. Smoke accompanies most ordinary forms of combustion, and it differs greatly with the nature of the substances of materials being burned. The density and color of the smoke is in a direct ratio to the amount of suspended particles. A fire that is just beginning and is consuming wood, cloth, and other ordinary furnishings will ordinarily give off gray-white or blue-white smoke of no great density. As the burning progresses, the density may increase, and the smoke may become darker because of the presence of large quantities of carbon particles.

Black smoke is usually the result of burning rubber, tar, roofing oil, or other flammable liquids. It has been said that brown smoke may indicate the oxides of nitrogen fumes and that gray-yellow smoke is a danger signal of approaching backdraft. A firefighter should remember that the materials that smoke contains can only be determined by chemical analysis. *Although the color of the smoke may be of some value in determining what is burning, smoke color is not always a reliable indicator.* It is sometimes quite easy for firefighters to classify what is burning by distinctive odors, especially during the early stages of the fire. Smoke from burning rubber, rags, pine wood, feathers, or grass all have a characteristic odor. As the smoke grows more dense and the firefighter is exposed to greater quantities, irritation to the nasal passages soon decreases an ability to recognize odors.

The Building Involved

Knowledge of the building involved is a great asset when making decisions concerning ventilation. Building type and design are the initial factors to consider in determining whether horizontal or vertical ventilation should be accomplished. The number and size of wall openings, the number of stories, staircases, shafts, dumbwaiters, ducts, roof openings, the availability and involvement of exterior fire escapes and exposures are determining factors. Building permits that are issued in one's own area may enable the fire department to know when buildings are altered or subdivided. Checking these permits will often reveal information concerning heating, ventilating, and air-conditioning systems; and avenues of escape for smoke, heat, and fire gases. The extent to which a building is connected to adjoining structures also has a bearing on the decision. In-service company inspection and pre-fire planning may provide more valuable and detailed information.

HIGH-RISE-BUILDINGS

A major consideration in high-rise buildings is the danger to occupants from heat and smoke. High-rise buildings are normally occupied by hospitals, hotels, apartments, and business offices. In any case, a great number of people may be exposed to danger.

Fires and smoke may spread rapidly through pipe shafts, stairways, elevator shafts, air handling systems, and other vertical openings. These openings contribute to a "stack effect," creating an upward draft and interfering with evacuation and ventilation.

The creation of layers of smoke and fire gases on floors below the top floor of sealed multistory buildings is a relatively new phenomenon. Smoke and fire gases produced will accumulate at various levels until the building is ventilated. Pre-fire planning should include tactics and strategy that can cope with the ventilation and life hazard problems inherent in stratified smoke. Smoke and fire gases travel through a building until their temper-

ature is reduced to the temperature of the surrounding air. When this stabilization of temperature occurs, the smoke and fire gases form layers or clouds within the building.

Experience has shown that these dense smoke clouds form at a level below the top floor (Figure 11.4). A classic example of this formation occurred

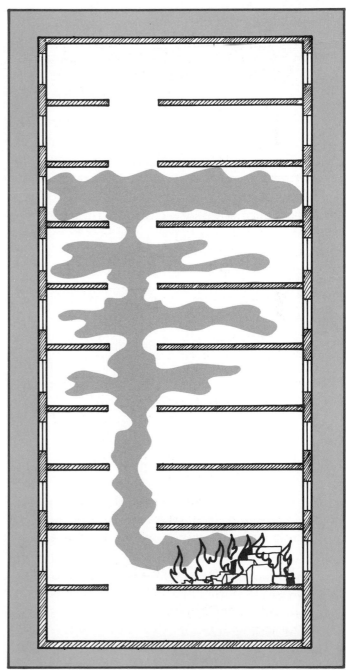

Figure 11.4 In high-rise buildings smoke and gases cool as they rise and stratify or layer before reaching the top. This variant of mushrooming creates a "lid" for other products of combustion. The level above the fire where this occurs varies with a number of factors including fuel, intensity and size of the fire, types of smoke channels, height of the building, and weather.

in a 17-story sealed building. The fire was in the basement, and the dense smoke clouds formed on the tenth, eleventh, and twelfth floors. The fire was extinguished before sufficient heat had built up to move the stratified smoke to the top floor. Ventilating this cooled smoke out of the top of the building was accomplished by creating controlled currents of air up the stairshafts and across the smoke-filled floors.

The mushrooming effect, which is usually expected on top floors, does not occur in tall buildings until sufficient heat is built up to move the stratified smoke and fire gas clouds that have gathered on lower floors in an upward direction.

Ventilation in a high-rise building without specific plans for the effective use of manpower, equipment, and extinguishing agents, should not be attempted. Consideration must be given to the fact that the manpower demand for this type of building is approximately four to six times as great as required for a normal residence fire. In most instances, ventilation must be accomplished horizontally with the use of mechanical ventilation devices. Protective breathing equipment will be in great demand since every firefighter will have to be completely equipped with self-contained breathing apparatus. The problems of communication and coordination between the various attack and ventilating teams become more involved as the number of participants increase.

Top ventilation in modern tall buildings must be considered during pre-fire planning. In many buildings only one stair shaft pierces the roof. This vertical "chimney" must be used to ventilate smoke, heat, and fire gases from various floors. Before the doorways on the fire floors are opened and the stair shaft is ventilated, the door leading to the roof must be blocked open or removed from its hinges. Removal of the door at the top of the shaft insures that it cannot close and allow the shaft to become filled with superheated gases, after ventilation tactics are started.

BASEMENTS AND WINDOWLESS BUILDINGS

Except in private dwellings and where automatic sprinklers are present, outside stairways, windows, and hoistways provide access to fire-

fighters. However, most outside entrances to basements may be blocked or secured by iron gratings, steel shutters, wooden doors, or combinations of these for protection against weather and burglars. All of these features serve to impede attempts at natural ventilation.

Another important factor that should be considered when a basement is involved in fire is its relationship to the rest of the building. Structural features such as stairways, elevator shafts, pipe chases, air handling systems, and other vertical openings contribute to the spread of fire and smoke to upper floors. Ventilation below street level is difficult in that it rarely provides an opportunity to use normal smoke evacuation techniques. Basements will usually require mechanical ventilation.

Many buildings, especially in business areas, have windowless wall areas. While windows may not be the most desirable means for escape from burning buildings, they are an important consideration for ventilation. Windowless building designs create an adverse effect on fire fighting and ventilation operations (Figure 11.5). The ventilation of a windowless building may be delayed for a considerable time, allowing the fire to gain headway or to create backdraft conditions.

Figure 11.5 Windowless structures require vertical or forced ventilation that can be difficult.

Problems inherent in ventilating this type of building are many and varied, depending upon the size, occupancy, configuration, and type of material from which the building is constructed. Windowless buildings usually require mechanical ventilation for the removal of smoke. Air handling systems should be shut down as soon as possible. The means of doing this and the location of the controls should be noted in the pre-fire plan. Most buildings of this type are automatically air-conditioned and heated through ducts. This equipment, in combination with mechanical ventilation equipment, can effectively clear the area of smoke.

It is imperative that firefighters become thoroughly familiar with the design, contents, systems, and access routes in windowless buildings in their area.

Life Hazards

Dealing with the danger to human life is of utmost importance. Certain fire conditions may suggest that ventilation come first to draw away heat and smoke, or that the spreading flames must be attacked immediately; sometimes both must be done simultaneously. All variables cannot be discussed here, but the point is that the first consideration is the safety of occupants. The life hazards are generally reduced in an occupied building involved by fire if the occupants are awake. If, however, the occupants were asleep when the fire developed and are still in the building, either of two situations may be expected: first, they may have been overcome by smoke and gases; second, they might have become lost in the building and are probably panicky. In either case, proper ventilation will be needed in conjunction with rescue operations.

In addition to the hazards that endanger occupants, there are potential hazards to firefighters and rescue workers. The type of structure involved, whether natural openings are adequate, and the need to cut through roofs, walls, or floors (combined with other factors) add more problems to the decision process.

The hazards that can be expected from the accumulation of smoke and gases in a building include:

- Obscurity caused by dense smoke

- Presence of poisonous gases

- Lack of oxygen

- Presence of flammable gases

Another consideration facing firefighters is flammability of the materials and contents within a building. While these combustible materials may be heated above their ignition temperature, they may not burn because of a lack of oxygen. The hazard lies in the fact that preheated combustibles will burst into a free-burning fire when provided with a supply of fresh air.

Location and Extent of the Fire

In most instances ventilation should not be carried out until the location of the fire is established. Opening for ventilation purposes before the fire is located may spread the fire throughout areas of the building that would not otherwise have been affected. Smoke that is coming out of the top floor does not always indicate a fire on that floor since it may be on a lower floor or even, perhaps, in the basement. Likewise, smoke that is gently flowing from an opening is not necessarily close to the seat of the fire. Obviously, extensive roof ventilation may be impractical or extremely dangerous if the location of the fire is such that vertical ventilation will draw the fire into parts of the building which are not involved.

The fire may have traveled some distance throughout a structure by the time fire fighting forces arrive, and consideration must be given to the extent of the fire, as well as to its location. The severity and extent of the fire usually depend upon the kind of fuel, the time it has been burning, installed fire protection devices, and the degree of confinement of the fire. The phase to which the fire has progressed is a primary consideration in determining ventilation procedures. Some of the ways by which vertical extension occurs are as follows:

- Through stairwells, elevators, and shafts by direct flame contact or by convected air currents

- Through partitions and walls and upward between the walls by flame contact and convected air currents

- Through windows or other outside openings where flame extends to other exterior openings and enters upper floors

- Through ceilings and floors by conduction of heat through beams, pipes, or other objects that extend from floor to floor

- Through ceilings and floors by direct flame contact

- Through floor and ceiling openings where sparks and burning material fall through to lower floors

- By the collapse of floors and roofs

Selecting the Place to Ventilate

The ideal situation in selecting a place to ventilate is one in which firefighters have prior knowledge of the building and its contents. There is no rule of thumb in selecting the exact point at which to open a roof except "as directly over the fire as possible." Many factors will have a bearing on where to ventilate. Some of them are:

- The availability of natural openings such as skylights, ventilator shafts, monitors, and hatches

- Location of the fire and the direction which the chief officer wishes it to be drawn

- Type of construction

- Wind direction

- The extent of progress of the fire and the condition of the building and its contents

- Bubbles or melting of roof tar

- Indications of roof sag

The effect that ventilation will have on the fire, the consequences its release will have on exposures, the department's state of readiness, and the ability to protect exposures, prior to actually opening the building, should be considered when determining the place to ventilate.

Before ventilating a building, a fire officer must provide manpower and adequate fire control facilities, because the fire may immediately increase in intensity when the building is opened. These resources should be provided for both the building involved and other exposed buildings. As soon as the building has been opened to permit hot gases and smoke to escape an effort to reach the seat of the fire for extinguishment should be made. Entrance should be made into the building as near the fire as possible, if wind direction will permit. It is at this opening that charged hoselines should be

positioned in case of violent burning or an explosion. Charged lines should also be made available at other points where openings are to protect buildings that are likely to be endangered because of their exposure to the one involved.

VERTICAL VENTILATION

After the fire officer has considered the type of building involved, the location and extent of the fire, moved manpower and tools to the roof, observed safety precautions, and has selected the place to ventilate, the operation is still not complete. Top level ventilation involves all of these factors plus the following precautions and procedures which the officer in charge must consider and practice to be successful.

- Coordinate with ground and attack companies.

- Observe the wind direction with relation to exposures.

- Note the existence of obstructions or weight on the roof.

- Secure a lifeline to the roof as a secondary means of escape.

- Utilize natural roof openings whenever possible.

- Cut a large hole if one is required, rather than several small ones.

- Exercise care in making the opening so that main structural supports are not cut.

- Work with the wind at the back or side to provide protection to the operators while cutting the roof opening.

- Guard the opening to prevent personnel falling into the building.

- Extend a blunt object through the opening to break out the ceiling.

Opening Roofs

In order to properly ventilate a roof, the firefighter must understand the basic types and designs of roofs.

Many designs and shapes of roof styles are used, and their names vary in each locality. Roof

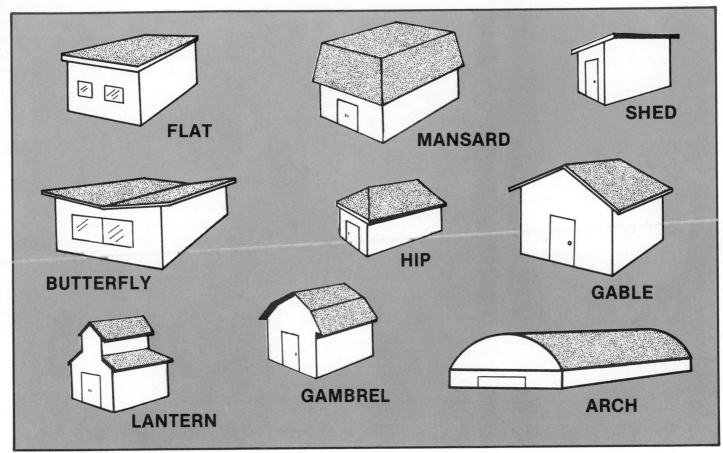

Figure 11.6 Common roof types.

coverings may be wood shingles, composition shingles, composition roofing paper, tile, slate, or a built-up tar and gravel surface. The roof covering is the exposed part of the roof, and its primary purpose is to afford protection against the weather. The selection of a proper roof covering is important from a fire protection standpoint because it may be subjected to sparks and blazing brands. A study of the more common types of roofs and the manner in which their construction affects opening procedures is necessary to develop effective vertical ventilation. The firefighter is concerned with three prevalent types of roof construction: flat, pitched, and arched styles. Buildings may be constructed with a combination of roof designs including these or other types. Some of the more common styles are the flat, gable, gambrel, shed, hip, mansard, dome, lantern, and butterfly, as illustrated in Figure 11.6.

SAFETY PRECAUTIONS

The ranking firefighter aloft should be in constant communication with the chief officer at the scene. Portable radios are most adaptable to this type of communication. Responsibility on the roof includes insuring that only the required openings are made, coordinating the crew's efforts with those of the firefighters who are inside, and insuring the safety of all personnel who are assisting in the opening of the building. Some of the safety precautions that should be practiced include.

- Providing a secondary means of escape.
- Preventing personnel from walking on spongy roofs.

  NOTE: This is usually a sign that structural members have been weakened. (If it is necessary to open a weakened roof, a roof ladder laid on the roof serves to distribute the firefighter's weight over a larger area.)

- Securing a lifeline to any firefighter who is to enter a weakened roof area.
- Protecting personnel from sliding and falling.

Figure 11.7 Scuttle hatches, skylights, bulkheads, and ventilators are possible roof openings. Be sure they are of ample size before using.

- Exercising caution in working around electric wires and guy wires.

- Insuring that the person making the opening is standing to the windward side of the cut and wearing the proper protective equipment.

- Not allowing other persons within range of the axe.

- Cautioning axe users to beware of overhead obstructions within the range of their axe.

- Starting power tools on the ground to insure operation at the site of the cut at upper areas. It is important that the tool be shut off before hoisting or raising the tool to these upper areas.

- Cautioning all cutting equipment operators to make sure the angle of the cut is not toward their bodies.

- Being on the lookout for indications of weakening structures or other hazards.

- Exercising caution, especially when using power tools, not to cut or weaken supporting structural members.

- Keeping firm footing.

NATURAL ROOF OPENINGS

Natural roof openings may exist on various types of roofs in the form of scuttle hatches, skylights, monitors, ventilating shafts, and even stairway doors. Most every type may be expected to be locked or secured in some manner against entry. Scuttle hatches, as shown in Figure 11.7, are normally square and large enough to permit a person to climb to the roof from the cockloft. They may be metal or wood, and generally speaking, they do not provide an adequate opening for ventilation purposes. If skylights contain ordinary shatter type glass, they may be conveniently opened. If they contain wired glass, they are very difficult to shatter and are more easily opened by removing the frame. The sides of a monitor may contain glass (which is easily removed), louvers of wood, or metal. The sides which are hinged are easily forced at the top. If the top of the monitor is not removable, at least two sides should be opened to create the required draft. Stairway doors may be forced open in the same manner as other doors of like type.

FLAT ROOFS

Flat roofs are more common to mercantile buildings, industrial buildings, and multiple dwelling apartments than to individual buildings. This type of roof ordinarily has a slight slope toward the rear of the building and is frequently pierced by chimneys, vent pipes, shafts, scuttles, and skylights. The roof may be surrounded and divided by parapets, and it may support water tanks, air-conditioning equipment, antennas, and other obstructions which may interfere with ventilation operations.

The structural part of the flat roof is generally similar to the construction of a floor which consists of wooden, concrete, or metal joists covered with sheathing. The sheathing is covered with a layer of waterproofing material and an insulating material (Figure 11.8). Instead of joists and sheathing con-

struction, flat roofs are sometimes poured reinforced concrete or precast gypsum or concrete slabs set within metal joists.

The best way for fire departments to determine the material from which roofs are constructed is through inspection surveys where such information can be collected and made available to the firefighters. The materials used in flat roof construction in some ways determine its ability to be cut with a fire axe. When cutting through a roof, the firefighter should make the opening rectangular or square to facilitate repairs being more easily made. One large opening at least 4-foot by 4-foot (1.2 by 1.2 m) is much better than several small ones. A procedure for opening a wood joist or rafter roof with an axe is suggested in the following sequence.

Step 1: Use the following factors to determine the location for the opening to be made:

- Location of intense fires
- Highest point on roof
- Direction of wind
- Existing exposures

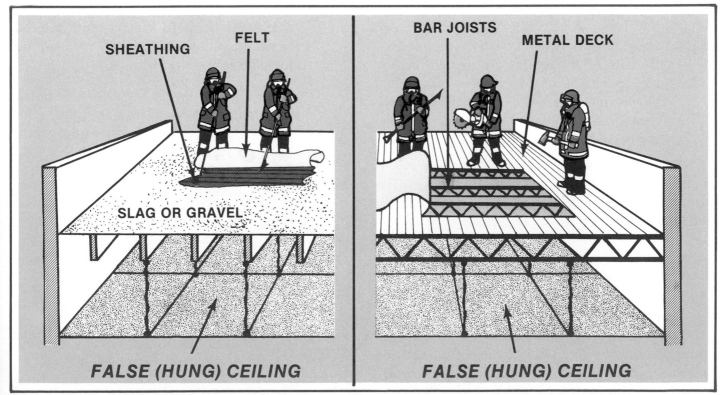

Figure 11.8 Flat roofs are supported generally by wood or metal joists with wood or metal decking, plywood panels, or poured gypsum covered by a watertight roof surface.

- Extent of fire
- Obstructions

Step 2: Locate roof supports by sounding with the axe. Between the rafters or joists it will sound hollow and the axe will bounce. When near or on top of a support it will sound dull and solid.

Step 3: Mark off the location for the opening by scratching a line on the roof surface with the pick head.

Step 4: Remove the built-up roof material or metal by cutting the material and using the pick head to pull the material out of the way.

Step 5: Cut the wood decking diagonally alongside the joist toward the hole.

NOTE: The joist should never be cut. Cutting alongside a joist permits the blows to have a solid base to avoid bouncing.

Step 6: Use short strokes when chopping. If an axe must be swung high to apply more force, check for overhead obstructions and other firefighters. If the blade becomes wedged, care should be taken to avoid breaking the axe handle.

Step 7: Pry up the roof boards with the pick end of the axe. After opening has been cut in the roof, push the blunt end of a pike pole, plaster hook, or some other suitable tool through the roof opening to open the ceiling below.

Power equipment for opening roofs is most useful and often provides a means by which ventilation procedures can be accelerated. This equipment may be driven by an electrical motor or by gasoline engines attached to the saws. Circular power saws can be equipped with a carbide blade. The chain and saber power saws are also useful.

PITCHED ROOFS

The pitched roof is one that is elevated in the center and thus forms a pitch to the edges. Pitched roof construction involves timber rafters or metal trusses that run from the ridge to a wall plate on top of the outer wall at the eaves level. The rafters or trusses which carry the sloping roof can be of various materials. Over these rafters, the sheathing boards are applied either squarely or diagonally across the rafters. These sheathing boards are usually applied solidly over the entire roof. Pitched roofs sometimes have a covering of roofing paper applied before shingles are laid. Shingles may be wood, metal, composition, asbestos, slate, or tile.

Pitched roofs on barns, churches, supermarkets, and industrial buildings may have roll felt applied over the sheathing and then mopped with asphalt roofing tar. Gypsum slabs, approximately two inches (5 cm) thick, may be laid between the metal trusses of a pitched roof instead of wood sheathing. These conditions can only be determined by inspection surveys conducted by fire department personnel.

Pitched roofs have a more pronounced downward incline than those of flat roofs. This incline may be gradual or steep. The procedures for opening pitched roofs are quite similar to those for flat roofs except that additional precautions to prevent slipping must be taken (Figure 11.9). Suggested steps for opening pitched roofs are as follows.

Step 1: Place a roof ladder on the roof and locate the position where the opening is to be made. Bounce the axe on the roof to sound for solid supports or rafters.

Step 2: Move the roof ladder to either side of the selected location and use the ladder for support. The location should usually be at the highest point of the roof.

Step 3: Rip off the shingles or roofing felt sufficiently to permit the initial cut to be made. (In some cases, it is best to first remove all shingles or roofing felt from the entire area where the hole is to be made.)

Step 4: Cut the sheathing along the side of a rafter the distance required for the opening. The opposite side of the opening may then be cut in a like manner. (The opening should be square or rectangular.)

Step 5: Remove sheathing boards with the pick of the axe or some other suitable tool.

Figure 11.9 When opening a pitched roof cut down along the rafters, cut parallel to the ridge, and remove both the roofing materials and the ceiling below.

Step 6: Push the blunt end of a pike pole or other long handled tool through the hole to open the ceiling.

ARCHED ROOFS

Arched roof construction has many desirable qualities for certain types of buildings. One form of arched roof construction uses the bow-string truss for supporting members. The lower chord of the truss may be covered with a ceiling to form an enclosed cockloft or roof space. Such concealed, unvented spaces are definitely forcible entry problems and contribute to the spread of fire.

A trussless arched roof has been developed that is made up of relatively short timbers of uniform length. These timbers are beveled and bored at the ends where they are bolted together at an angle to form a network of structural timbers. This network forms an arch of mutually braced and stiffened timbers. Being an arch rather than a truss, the roof exerts a horizontal reaction in addition to the vertical reaction. The result of these reactions

is the roof thrust, which is distributed to thrust supports. Trussless arch construction enables all parts of the roof to be visible to firefighters. A hole of considerable size may be cut or burned through the network sheathing and roofing at any place without causing collapse of the roof structure, since the loads are then distributed to less damaged timbers around the opening.

Cutting procedures for opening arched roofs are the same as for flat or pitched roofs except that it is doubtful that a roof ladder can always be used on an arched roof. Long straight or extension ground ladders and aerial ladders may sometimes prove satisfactory by placing the ladder as flat on the roof as possible. Regardless of the method used to support the firefighter, the procedure is difficult and dangerous because of the curvature of the roof. Proper safety precautions must be observed.

CONCRETE ROOFS

The use of precast concrete is very popular with certain types of construction. Precast roof slabs are available in many shapes, sizes, and designs. These precast slabs are hauled to the construction site, ready for use. Other builders form and pour the concrete on the job. Roofs of either precast or reinforced concrete are extremely difficult to break through and this method of opening should be avoided whenever possible.

A popular lightweight material made of gypsum plaster and portland cement and mixed with aggregates of perlite, vermiculite, or sand provides a lightweight floor and roof assembly. This material is sometimes referred to as lightweight concrete. Lightweight precast planks are manufactured from this material and the slabs are reinforced with steel mesh or rods. Lightweight concrete roofs are usually finished with roofing felt and a mopping of hot tar to make them watertight.

Lightweight concrete roof decks are also poured in place over permanent formboards, steel roof decking, paper-backed mesh, or metal rib lath. These lightweight concrete slabs are relatively easy to penetrate. Some types of lightweight concrete can be penetrated with a hammer-headed pick, a power saw, a jackhammer, or any other penetrating tool.

METAL ROOFS

Metal roof coverings are made from several different kinds of metal and they are constructed in many styles. Light gauge steel roof decks can either be supported on steel framework or they can span across wider spaces. Other types of corrugated roofing sheets are made from light gauge cold-formed steel, galvanized sheet metal, and aluminum. The light gauge cold-formed steel sheets are used primarily for the roofs of industrial buildings. Corrugated galvanized sheet metal and aluminum are seldom covered with a roof material and the sheets can usually be pried from their supports. Metal cutting tools or power saws with metal cutting blades must be employed to open metal roofs. Metal roofs on industrial buildings are usually provided with adequate roof openings, skylights, or hatches. About the only way firefighters can know when this type of roof is encountered is to have previously discovered its existence through a pre-fire inspection survey.

Precautions Against Upsetting Established Vertical Ventilation

When vertical ventilation is accomplished, the natural convection of the heated gases creates upward currents which draw the fire and heat in the direction of the upper opening. Fire fighting teams take advantage of the improved visibility and less contaminated atmosphere to attack the fire at its lower point. If the "stack effect" is interrupted, the heat, smoke, and steam back up, hampering extinguishment efforts. Stack effect is the vertical natural air movement throughout a building. The magnitude of stack effect is determined by the differences in temperature and the densities between the air inside and outside the building. Some common factors that can destroy the effectiveness of ventilation are as follows: improper use of forced ventilation, breakage of glass, improperly directed fire stream, breakage of skylights, explosions, a burn through, and additional openings between the attack team and the upper opening.

ELEVATED STREAMS PROJECTED DOWNWARD THROUGH VENTILATION OPENINGS WILL HINDER VENTILATION OF STRUCTURE

Elevated streams are frequently used to cut down sparks and flying brands from a burning building or to reduce the thermal column of heat over a building. When elevated or handline streams are projected downward through a ventilation opening or used to reduce the thermal column to a point where ventilation is hindered, they either destroy or upset the orderly movement of fire gases from the building (Figure 11.10). An upset of this nature can materially affect firefighters who may be working at various levels on floors below. Streams that are being operated just above ventilated openings should be projected slightly above the horizontal plane. In this position they will help cool thermal column and extinguish sparks. The movement of the stream may even increase the rate of ventilation.

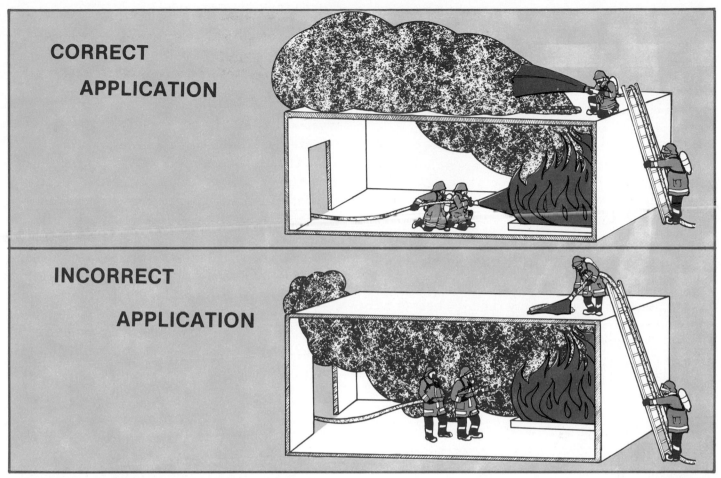

CORRECT APPLICATION

INCORRECT APPLICATION

Figure 11.10 When streams are directed downward through a ventilation opening, they destroy or upset the process. They should be directed slightly above the horizontal plane across the opening.

Vertical ventilation cannot be the solution to all ventilation problems for there may be many instances where its application would be impractical or impossible. Only prompt and accurate size-up, based on a thorough understanding of the many variables which affect, or are affected by, ventilation can provide the answer to the following two questions:

- Is ventilation required?

- What type is most appropriate in this instance?

In addition to selecting the method of ventilation, a fire officer must determine how, when, and where it will serve its most useful purpose. Building type and design are the initial factors which must be considered in determining whether horizontal or vertical ventilation should be accomplished. Type and design features include the number and size of wall openings, the number of stories, and the availability and involvement of ex-

terior fire escapes, and exposures. Directly related to these are protective coverings of the openings, the direction in which the openings face, and the wind direction.

HORIZONTAL VENTILATION

Structures which lend themselves to the application of horizontal ventilation include:

- Residential type buildings in which the fire has not involved the attic area

- Buildings with windows high up the wall near the eaves

- The attics of residential type buildings which have louver vents in the walls

- The involved floors of multi-storied structures

- Buildings with large unsupported open spaces under the roof in which the fire is not contained by fire curtains and in which the structure has been weakened by the effects of burning

Many of the aspects of vertical ventilation also apply to horizontal ventilation. A different procedure must be followed in ventilating a room, a floor, a cockloft, an attic, or basement. The procedure to be followed will be influenced by the location and extent of the fire. Some of the ways by which horizontal extension occurs are as follows:

- Through wall openings by direct flame contact or by convected air

- Through corridors, halls, or passageways by convected air currents, radiation, and flame contact

- Through open space by radiated heat or by convected air currents

- In all directions by explosion or flash burning of fire gases, flammable vapors, or dust

- Through walls and interior partitions by direct flame contact

- Through walls by conduction of heat through beams, pipes, or other objects that extend through walls

Weather Conditions

Weather conditions are always a primary consideration in determining the proper ventilation procedure. Under certain conditions, when there is no wind, cross ventilation is less effective since the force to remove the smoke is absent. In other instances cross ventilation cannot be accomplished due to the danger of wind blowing toward an exposure or feeding oxygen to the fire. The wind plays an important role in ventilation. Its direction may be designated as windward or leeward. The side of the building where the wind is striking is the windward, the opposite side is leeward, as illustrated in Figure 11.11.

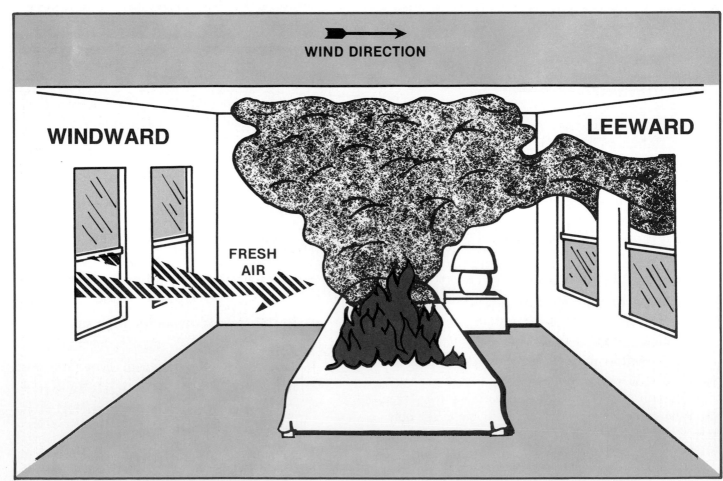

Figure 11.11 Open under windows on the leeward side, then lower windows on the windward side as needed for good cross ventilation.

Exposures

Since horizontal ventilation does not normally release heat and smoke directly above the fire, some routing is necessary. Firefighters should be aware of internal exposures as well as external exposures. The routes by which the smoke and heated gases would travel to the exit may be the same corridors and passageways which occupants will be using for evacuation. Therefore, the practice of horizontal ventilation without first considering occupants and rescue procedures may block the escape of occupants. The theory of horizontal ventilation is basically the same as that of vertical ventilation inasmuch as the release of the smoke and heat is an aid in fighting the fire and reducing damage.

Outside exposures include those previously mentioned plus those which are peculiar to horizontal ventilation. Since horizontal ventilation is accomplished at a point other than at the highest point of a building, there is the constant danger that the rising heated gases will ignite that portion of the structure which they contact when released.

They may ignite eaves of adjacent structures, or be drawn into windows above their liberation point. Unless for the specific purpose of aiding in rescue, a building should not be opened until charged lines are in place at the windward attack entrance point, at the intermediate point where fire might be expected to spread, and in positions to protect other exposures.

Precautions Against Upsetting Horizontal Cross Ventilation

The opening of a door or window on the wrong side of a building may reverse air currents and drive heat and smoke back upon firefighters. Opening doors and windows between the advancing fire fighting crews and the established ventilation exit point will reduce the intake of fresh air from the opening behind the firefighters. Firefighters following established cross ventilation currents are illustrated in Figure 11.12. The smoke and heat intensifying as the established current is interrupted by the blocking of fresh air by the firefighter or other obstruction in the doorway is illustrated in Figure 11.13.

Figure 11.12 Fresh air currents can establish beneficial cross ventilation of a smoke-filled room.

Figure 11.13 If doors or windows opened for fresh air are blocked by a standing firefighter or other obstruction, then smoke and heat will again intensify within the room.

FORCED VENTILATION

Ventilation has thus far been considered from the standpoint of the natural flow of air currents, the currents created by fire, and the effect of fog streams. Forced ventilation is accomplished by blowers, fans, or fog streams. The fact that forced ventilation is effective and can be depended upon for smoke removal when other methods are not adequate proves its value and importance.

It is difficult to classify forced ventilation equipment by any particular type. The principle applied is that of moving large quantities of air and smoke. These portable blowers are all powered by electric motors or gasoline-driven engines. Portable ejectors and several methods of using them to ventilate are shown in Figure 11.14.

Forced air blowers should always be equipped with explosion-proof motors and power cable con-

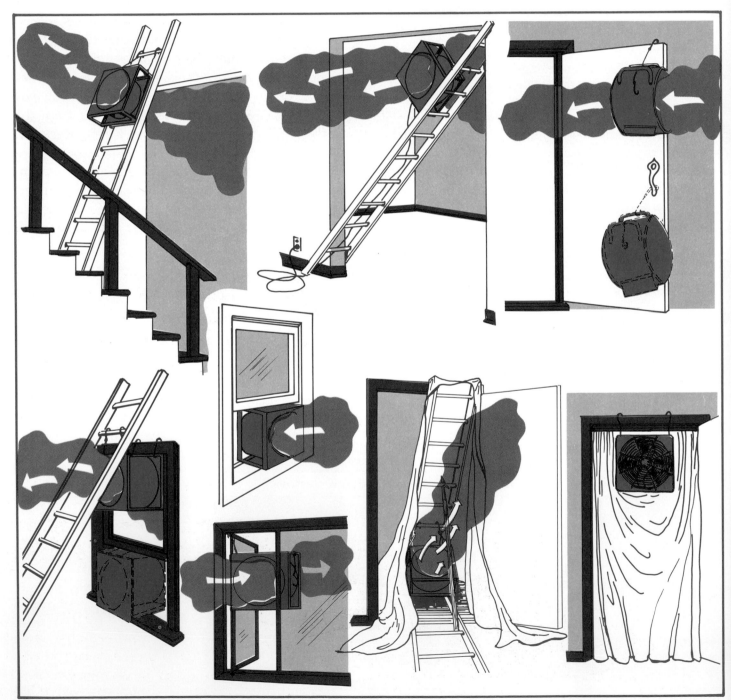

Figure 11.14 Portable smoke ejectors may be placed in any necessary positions with a variety of tools. The higher the placement the better heat will be removed with the smoke.

nections when used in a flammable atmosphere. Forced air blowers should be shut down when they are moved. Before they are started, be sure that there are no persons near the blades and that clothing, curtains, or draperies are not in a position to be drawn into the fan. Blowers should always be moved by the handles which are provided for this purpose. This discharge stream of air should be avoided because of particles that may be picked up and blown by the venting equipment.

Some Advantages of Forced Ventilation

Even though fire may not be a factor, contaminated atmospheres must be rapidly and thoroughly ejected. Forced ventilation, if not the only means of clearing a contaminated atmosphere, is always a welcome addition to normal ventilation. Some of the reasons for employing mechanical or forced ventilation are:

- It insures more positive control.

- It supplements natural ventilation.

- It speeds the removal of contaminants, facilitating more rapid rescue under safer conditions.

- It reduces smoke damage.

- It promotes good public relations.

Disadvantages of Forced Ventilation

If mechanical or forced ventilation is misapplied or uncontrolled, it can cause a great deal of harm. Forced ventilation requires supervision because of the mechanical force that is behind it. Some of the disadvantages of forced ventilation are:

- It can move fire along with the smoke and extend it to lateral areas.

- The introduction of air in such great volumes can cause the fire to spread

- It is dependent upon a power source.

- It requires special equipment.

Forced Ventilation Techniques

- Since the basic goal is to develop artificial circulation and pull smoke out fast, the ejector should be placed to exhaust in the same direction as the natural wind.

- During the early stage of fire, most of the heat, smoke, and congested fumes rise and accumulate near the ceiling. Exhausting smoke ejectors should be placed high for maximum effectiveness in clearing smoke and providing visibility.

- When using both exhausting and blowing ejectors, first clear the area from the windward side in the direction of the exhausting smoke ejector. Then move the blowing smoke ejectors toward the exhausting ejector, keeping the circulation line as straight as possible. The venturi action of the straight line circulation will suck most of the smoke from corners.

- When air is allowed to recirculate around the sides of the smoke ejector and in and out of nearby openings, it causes a churning action that reduces efficiency. If the area surrounding the fan is left open, atmospheric pressure pushes the air through the bottom of the doorway and pulls the smoke back into the room.

  To prevent churning air, cover the area around the unit with salvage covers or other material.

- Establish desired draft path and keep the airflow in as straight a line as possible. Every corner causes turbulence and decreases efficiency. Avoid opening windows or doors near the exhausting smoke ejector unless opening them definitely increases circulation.

- Remove all obstacles to the airflow. Even a window screen will cut effective exhaust by half. Avoid blockage of the intake side of the smoke ejector by debris, curtains, drapes, or anything that can decrease the amount of intake air.

- Have smoke ejectors ready to use when hoselines are in position, and have a charged line at hand when actually placing the ejectors. Since smoke is composed of heated gases, it rapidly condenses and settles on everything in the room or building. Therefore, ejectors should be put into operation as soon as possible after hoselines have been used.

- To speed up the clearing action, place an exhausting smoke ejector on the lee side of the room or building, where it can pull out the remaining smoke and fumes. Then put the blowing ejector in an outside opening on the windward side, where it will admit fresh, clean air. Results are usually most satisfactory when the blowing ejector is located in the lower part of the window or door.

Using Water Fog to Expel Smoke and Gas

The use of water fog in fire extinguishment and in ventilation requires a special technique of operation. The mere fact that firefighters have a good fog nozzle that supplies a protective curtain does not enable them to advance into a heavily charged area and expect to do an effective ventilating and extinguishing job. When water fog is used for ventilating and extinguishing purposes, the degree of effectiveness depends upon how, where, and when the fog stream is applied.

It has been found that a fog stream directed through a window or door opening will draw large quantities of heat and smoke in the direction in which the stream is pointed.

Compared with mechanical smoke ejectors, fog streams have been found to remove two to four times more smoke, depending on the type and size of the nozzle, the angle of the fog pattern, and the location of the nozzle in relation to the opening of the building. A fog nozzle stream directed through the opening with a 60° angle fog pattern covering 85 to 90 percent of the opening has been found to provide the best results for ventilation (Figure 11.15). The nozzle should be about two feet from the opening. Larger openings permit greater airflow, so a door might sometimes be more beneficial than a window. Whatever the size of the opening, wide angle streams should not be used, because when the water path comes close to a right angle to the air path much of the energy that moved the air is lost.

There are three drawbacks to the use of fog streams in forced ventilation. There will be an increase in the amount of water damage within the structure, there will be a drain on the available water supply, and in climates subject to freezing temperatures there will be an increase in the problem of ice in the area surrounding the building.

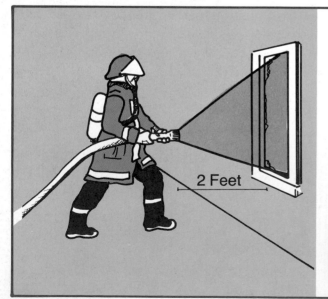

Figure 11.15 When using a fog stream to ventilate, the nozzle set at 60° should be about two feet (.5 m) from the opening and the pattern should cover 85 to 90 percent of the opening.

NFPA STANDARD 1001
SALVAGE AND OVERHAUL
Fire Fighter I

3-6 Salvage

3-6.1   The fire fighter shall identify the purpose of salvage, and its value to the public and the fire department.

3-6.2   The fire fighter, as an individual and as a member of a team, shall demonstrate folds and rolls of salvage covers.

3-6.3   The fire fighter, as an individual and as a member of a team, shall demonstrate salvage cover throws.

3-6.4   The fire fighter shall demonstrate the techniques of inspection, cleaning, and maintaining salvage equipment.

3-17 Overhaul

3-17.1   The fire fighter shall demonstrate searching for hidden fires by sight, touch and smell.

3-17.2   The fire fighter shall demonstrate exposure of hidden fires by opening ceilings, walls, floors and pulling apart burned materials.

3-17.3   The fire fighter shall demonstrate how to separate and remove charred material from unburned material.

3-17.4   The fire fighter shall define duties of fire fighters left at the fire scene for fire and security surveillance.

3-17.5   The fire fighter shall identify the purpose of overhaul.*

Fire Fighter II

4-6 Salvage

4-6.1   The fire fighter, given salvage equipment, operating as an individual and as a member of a team, shall demonstrate the construction and use of a water chute.

4-6.2   The fire fighter given salvage equipment, operating as an individual and as a member of a team, shall demonstrate the construction and use of a water catch-all.

4-6.3   The fire fighter, given salvage equipment except salvage covers, shall demonstrate the removal of debris, and removal and routing of water from a structure.

4-6.4   The fire fighter shall demonstrate the covering or closing of openings made during fire fighting operations.

4-17 Overhaul

4-17.1   The fire fighter shall list the procedures to follow during overhaul.

4-17.2   The fire fighter shall identify the safety precautions necessary during overhaul.*

*Reprinted by permission from NFPA Standard No. 1001, *Standard for Fire Fighter Professional Qualifications*. Copyright © 1981, National Fire Protection Association, Boston, MA.

IFSTA's Salvage and Overhaul Transparencies are designed to complement this chapter.

Chapter 12
Salvage and Overhaul

Salvage work in the fire service consists of those methods and operating procedures allied to fire fighting which aid in reducing fire, water, and smoke damage during and after fires. A portion of these damages can be attributed to the necessary operations of applying water, ventilating a building, and searching for fires throughout a structure. These procedures cannot be entirely eliminated, but improved techniques in fire extinguishment plus prompt and effective use of good salvage procedures result in a more systematic approach to minimize these losses.

Overhaul operations consist of the search for and extinguishment of hidden or remaining fires; placing the building, its contents, and the fire area in a safe condition; determining the cause of the fire; and recognizing and preserving any evidence of arson. Salvage and overhaul may sometimes be performed simultaneously, but overhaul generally follows salvage operations. When effective salvage procedures precede a thorough and systematic overhaul, the result will have a significant effect upon reducing the extent of the loss and facilitate prompt restoration of the property to full productive use.

PLANNING FOR SALVAGE OPERATIONS

Efficient salvage operations are dependent upon planning and training. Those persons responsible should review the department's salvage equipment inventory and make certain that the equipment on hand meets the requirements as specified in the NFPA Standard No. 1901, *Automotive Fire Apparatus*. The officer responsible for training should see that all firefighters are adequately trained in salvage operations and the use of salvage equipment. Chief officers should be instructed to give salvage operations a high priority and not hesitate to call additional help to perform salvage work.

The damage caused by heat, smoke, and water can frequently exceed the direct loss by fire, but these losses may be materially reduced through planned and well-executed salvage and overhaul practices. Also, good salvage and overhaul work is one of the most effective means of building good will. A fire department may often receive complimentary words of appreciation and praise by the news media for good salvage and overhaul operations. This praise gives the firefighters a feeling of accomplishment, particularly when the appreciation comes from people who have had their belongings saved by the firefighters. It is common for fire officials to notice better morale and efficiency among firefighters who have significantly contributed to reducing fire loss by successfully practicing salvage and overhaul.

Arranging Contents to be Covered

The actual arranging of contents to be covered may be limited when large stocks and display features are involved. Display shelves are frequently built to the ceilings and directly against the wall. This construction feature makes it difficult to cover shelving. When water flows down a wall it will naturally come into contact with each shelf and wet the contents. A better construction feature would be to allow adequate clearance between shelves and walls.

One common obstacle to efficient salvage work is the lack of skids under all stock that is susceptible to water damage. Some examples of contents that have perishable characteristics are flour, material in cardboard boxes, feed, paper, and other dry goods. Stock in basements should be placed on skids at least five inches above the floor but stock on upper floors is reasonably safe with skids of lesser heights. If the contents are stacked too close to the ceiling, this will also present a salvage problem. There should be enough space between the stock and ceiling to allow firefighters to easily apply salvage covers. When salvage covers are limited it is good practice to use available covers for water chutes and catchalls even though the water must be routed to the floor and cleaned up afterward.

Arranging household furnishings presents a different type of situation. If a reasonable degree of care is taken, one average-sized cover will usually protect the contents of one room (Figure 12.1). A suggested procedure for arranging furniture in a room is to group the furniture in the center of the room, if possible not under a light fixture, which might leak. If the floor covering is a removable rug, slip the rug from under the furniture as each piece is moved and roll for convenience. A dresser, chest, or high object may be placed at the end of the bed. If there is a roll of rug, place it on top to serve as a ridge pole. Other furniture can be grouped close by and pictures, curtains, lamps, and clothing can be placed upon the bed. It may sometimes be neces-

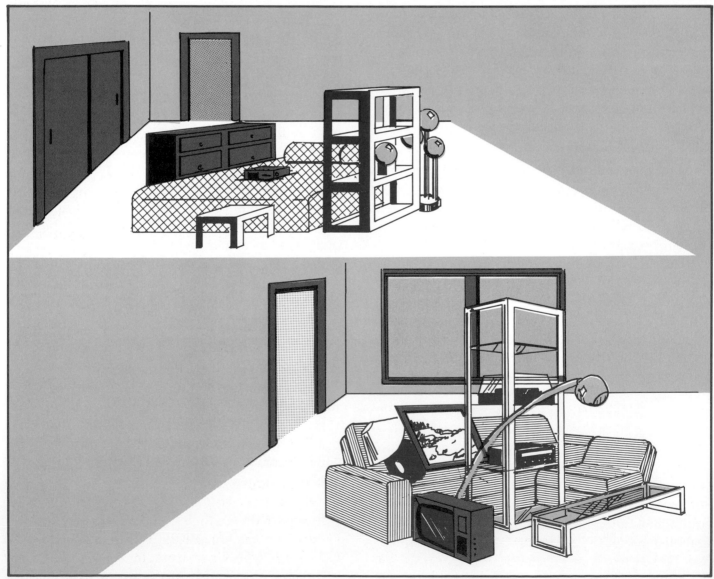

Figure 12.1 When arranging furniture, place those pieces with drawers facing toward the center to reduce the chance of water leaking inside. Make the arrangement slope outward.

sary to place the cover into position before some articles are placed on the bed. In this event, bed and furniture can be protected while other items are placed under the cover.

SALVAGE COVERS

Conventional salvage covers are made from closely woven canvas materials which have been treated for waterproofing. Some covers with a rubber coating have been used but they are not common. Canvas salvage covers have reinforced corners and edge hems into which grommets are placed for hanging or draping the covers. These treated canvas covers can be obtained in various sizes, but the sizes 9 x 12 feet (2.75 x 3.65 m) to 14 x 8 feet (4.26 x 2.43 m) are more common.

A relatively new trend in the use of plastic salvage covers has some promising advantages. These covers, made from 100 percent polyethylene film, are extremely lightweight and are easily handled. A cover made from heavyweight polyethylene and comparable in size to canvas covers weighs about two pounds (1 kg). These covers are quite economical and are practical for indoor and outdoor use. They are chemically inert, available in colors, and are not generally affected by alkalines, oils, acids, caustics, or solvents. They are relatively unaffected by normal temperatures, will remain flexible below zero degrees, and will not mold, mildew, or absorb moisture. Plastic covers have a tendency to slip from highly-piled merchandise and are not well adapted to bagging by rolling the edges.

Salvage Cover Maintenance

Proper cleaning, drying, and repairing of salvage covers will increase their span of service. Ordinarily, the only cleaning that is required for salvage covers is showering with a hose stream and scrubbing with a broom (Figure 12.2). Covers that are extremely dirty and stained may be scrubbed with a detergent solution and then thoroughly rinsed. Permitting salvage covers to dry while in a dirty condition is not good practice because after carbon and ash stains have dried, a chemical reaction takes place which rots covers. When dried, foreign materials are difficult to remove even with a detergent. Canvas salvage covers should be perfectly dry before they are folded and placed in ser-

Figure 12.2 Salvage covers should never be allowed to dry before being cleaned. Wood ash and other products of burned materials may have a chemical effect on the cover.

vice. This practice is essential in order to prevent mildew and rot. There is no particular objection to outdoor drying of salvage covers except that wind tends to blow and whip the covers. After salvage covers are dried, they should be examined for damage. One efficient and quick method is to arrange the cover on the floor so that three or four firefighters can raise it over their heads. Each person watches for holes and torn places as the cover is slowly moved overhead. When a hole is found, a watcher holds the palm of the hand against the cover and another firefighter on the outside marks the hole with chalk. The developments made in filament or plastic adhesive tapes have changed the entire practice of mending covers. It is no longer necessary to hand-sew small holes and tears or patch them with various types of cement or glue. One good adhesive tape, known as Mystic Tape, may be obtained from most any department store or auto supply store. There are also iron-on patches available. If mending tape is applied to both sides of the surface, it will increase the bond for a stronger and more permanent seal.

Salvage Equipment

For conducting salvage work at fires it is suggested that salvage equipment be located in one area on the apparatus. This common storage will eliminate having to hunt for equipment. Tools and equipment should be kept in a tool box or other container in order to make them easier to carry.

EQUIPMENT CARRIED ON APPARATUS

Typical salvage equipment recommended by NFPA Standard No. 1901, *Automotive Fire Apparatus* is listed below. The use of this equipment however is not limited to salvage work.

- Electricians pliers
- Sidecutters
- Chisels
- Aviator tin snips
- Tin roof cutter (can opener)
- Adjustable end wrenches
- Pipe wrenches
- Hammers
- Sledge hammer
- Hacksaw
- Crosscut handsaw
- Heavy-duty stapler and staples
- Linoleum knife
- Wrecking bar
- Padlock and hasp
- Screwdrivers
- Applicable power tools
- Hydraulic jack
- An assortment of nails
- An assortment of screws
- Roofing paper
- Tarpaper or plastic sheeting
- Wood lath
- Mops
- Squeegees
- Water scoops
- Scoop shovels
- Brooms
- Mop wringers with bucket
- Automatic sprinkler kit
- Water vacuum
- Submersible pump
- Sponges
- Chamois
- Assortment of rags
- 100 foot (30 m) length of electrical cable with locking type connectors 14-3 gauge or heavier
- Pigtail ground adapters, 2 wire to 3 wire, 14-3 gauge or heavier with 12 inch (30 cm) minimum length
- Salvage covers
- Floor runners

AUTOMATIC SPRINKLER KIT

The tools that may be supplied in a sprinkler kit are needed when fighting fires in buildings protected by automatic sprinkler systems. A flow of water from an open sprinkler head can do considerable damage to merchandise on lower floors after the fire has been controlled. The following tools are suggested to form a sprinkler kit:

- Sprinkler head wrenches
- Pipe caps and plugs
- Sprinkler tongs
- Sprinkler wedges
- Pipe wrenches
- Assorted sprinkler heads

CARRYALLS

A small waterproof carrier or bag that can be made from an old or damaged salvage cover is illustrated in Figure 12.3. This carrier is usually 6 or 8 feet square (2 to 2.5 m). Handholds are provided by lashing a sash cord through eyelets which permit the carryall to be puckered, thus forming a bag. Such a carryall is useful in carrying debris, catching debris, and providing a water basin for immersing small burning objects.

FLOOR RUNNERS

Costly floor coverings are sometimes damaged by mud and grime which may be tracked in by firefighters when a floor runner is not used. Floor runners are usually 20 or 30 inches (50 to 76 cm) wide. They can be unrolled from an entrance to

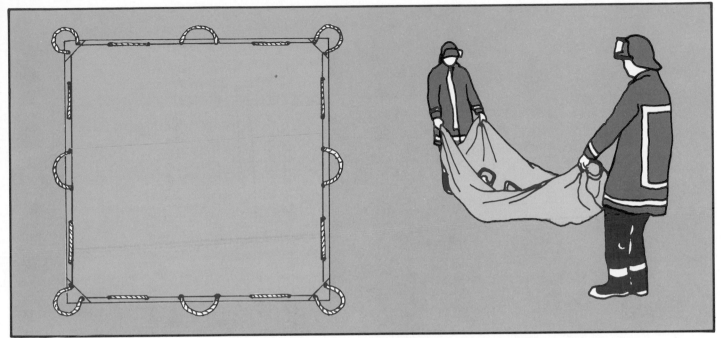

Figure 12.3 A carryall is useful in overhaul work to carry debris or to provide a watertight basin for immersing smoldering materials.

most any part of a building. Commercially prepared vinyl laminated nylon floor runners are lightweight, flexible, tough, heat and water resistant, and easy to maintain (Figure 12.4).

PORTABLE PUMPS

Portable water pumps are used to remove water from basements, elevator shafts, and sumps. Fire department pumpers should never be used for this purpose because fire pumps are intricate and expensive machines and are not intended to pump the dirty, gritty water that is found in such places.

Various kinds of devices known as jets or siphons may be used for the removal of excess water. These devices can be moved to any point where a line of hose can be placed and an outlet for water can be provided.

WATER VACUUM

One of the easiest and fastest ways to remove water, chemicals, acid solutions, and all other non-flammable liquids is the use of a water vacuum device. Dirt and small debris may also be removed from carpet, tile, and other types of floor coverings with this equipment. The water vacuum appliance consists of a tank, worn on the back, and a nozzle. It has a suction powerful enough to extract liquids from deep pile carpeting. The tanks normally have

Figure 12.4 Floor runners will prevent additional damage to floor surfaces or coverings from debris, water, or materials tracked in.

a capacity of 4 to 5 gallons (15 to 19.1 L) and can be emptied by simply pulling a lanyard which empties the water through the nozzle. It is compact, moderately light and very effective for clean-up operations (Figure 12.5).

An industrial type water vacuum may be used instead of the backpack type water vacuum. These industrial type water vacuums range in size from 3 gallon to 20 gallon (11 to 75 L) capacity.

Figure 12.5 The backpack water vacuum is widely used by fire departments. It is efficient, but due to its limited capacity must be emptied frequently.

CARE AND MAINTENANCE OF SALVAGE EQUIPMENT

Proper cleaning, drying, and repairing of salvage equipment is as important as the care of salvage covers. The proper care of salvage equipment will increase their span of service depending upon the intensity of their use. Some procedures for caring for salvage equipment are:

- Water Vacuum
  — Inspect power cords for broken insulation
  — Flush collection tank
  — Clean nozzle
- Mops
  — Clean with soap and water
  — Dry thoroughly
- Tools
  — Dry
  — Lightly oil if needed
- Brooms
  — Clean
  — Sand handles if burned
- Buckets and tubs
  — Clean
  — Without holes

METHODS OF FOLDING AND SPREADING SALVAGE COVERS

Rolling the Salvage Cover for a One-Firefighter Spread

The principle advantage of the one firefighter salvage cover roll is that it can be rolled across the top of an object and then be unfolded by one person. To form this roll, the two firefighters must make initial folds to reduce the width of the cover. The following steps for making the one-firefighter roll are illustrated in Figures 12.6-12.14.

Step 1: Lay the cover out in a flat position on the floor. Place a firefighter at each end to fold one side at a time (Figure 12.6).

Step 2: Grasp the cover with one hand midway between the center and the edge to be folded. Using the other hand as a pivot, pull tightly with the folding hand and bring the fold over toward the center of the cover (Figure 12.7).

Step 3: After the first fold has been placed at the center, grasp the open edge with the outside hand. Using the other hand as a pivot, stretch the cover tight (Figure 12.8).

Step 4: Bring this outside edge over to the center and place it on top of and in line with the previously placed first fold (Figure 12.9).

Figure 12.6 Lay out the cover with the finished side up.

Figure 12.7 Grasp the cover midway between the center and edge being folded and form a pleat.

Figure 12.8 Grasp the open edge and stretch the cover tight.

Figure 12.9 Fold the outside edge to the center on top of the first pleat.

Step 5: Fold the other half of the cover in the same manner by using steps 1, 2, 3, and 4. If the folds are not straight, they should be straightened. The completed fold is shown in Figure 12.10.

Step 6: Fold over about twelve inches (30 cm) at both ends of the cover to prevent the loose ends from flopping and to make a neater roll (Figure 12.11).

Step 7: Start the roll at either end and compress the first few rolls as tightly as possible (Figure 12.12).

Step 8: As the roll progresses, tuck in the slack at the center when wrinkles appear (Figure 12.13).

Step 9: The completed roll may be held tight by inner tube bands or tied with cords (Figure 12.14).

Figure 12.10 Repeat the previous steps to the unfolded side and pull the outside edge to the center on top of the first pleat to complete the fold.

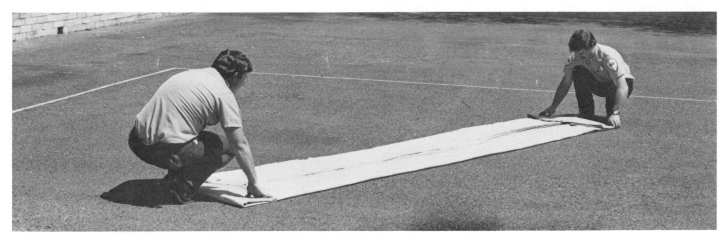

Figure 12.11 Fold cover ends over about 12 inches (30 cm).

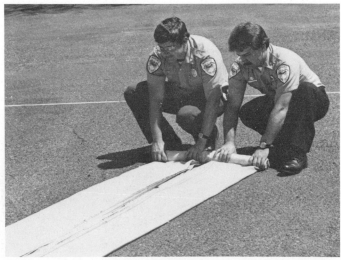

Figure 12.12 Start the roll from either end.

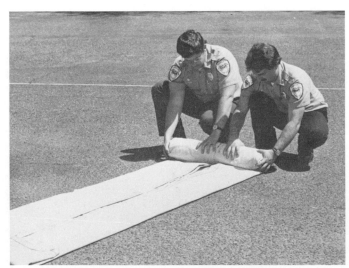

Figure 12.13 Tuck slack in at center when wrinkles appear.

Figure 12.14 Secure the roll with rubber bands or tie with cord.

One-Firefighter Spread With A Rolled Salvage Cover

A salvage cover rolled for a one-firefighter spread may be carried on the shoulder or under the arm. If it is fastened with inner tube bands or cords, its compactness makes it useful in several ways. One use, other than covering objects, is as a floor runner, which can be opened out wide if necessary. Use the following steps when one firefighter spreads a rolled salvage cover (Figures 12.15-12.20).

Step 1: Start at one end of the object to be covered and while holding the roll in the hands, unroll a sufficient amount to cover the end (Figure 12.15).

Step 2: Lay the roll on the object and continue to unroll toward the opposite end (Figure 12.16).

Step 3: Let the rest of the roll either fall into place at the other end or arrange it in a position as shown in Figure 12.17.

Step 4: Stand at one end and grasp the open edges where convenient, one edge in each hand (Figure 12.18).

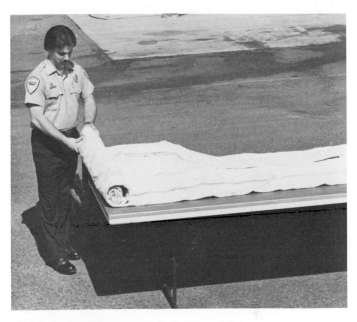

Figure 12.16 Continue to unroll cover over object.

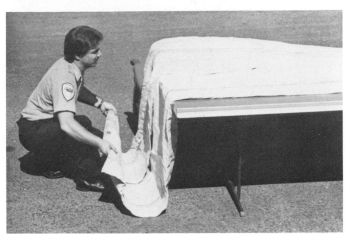

Figure 12.17 Allow roll to fall in place at the opposite end.

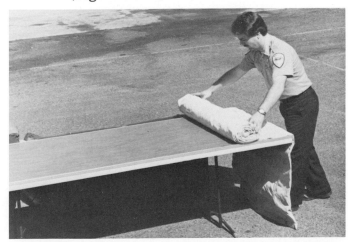

Figure 12.15 Unroll cover to protect the end of the item being covered.

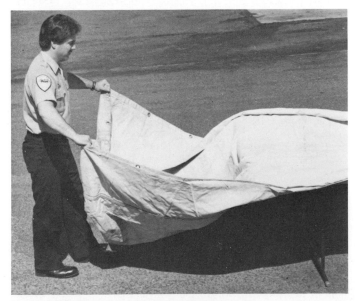

Figure 12.18 Grasp open edges to unfold the cover.

Step 5: Open the sides of the cover over the object by snapping both hands up and out (Figure 12.19).

Step 6: Open the other end of the cover over the object in the same manner and tuck in all loose edges at the bottom (Figure 12.20).

Step 6: With the cover folded to a reduced width, both firefighters grasp one end of the cover and bring it to a point just short of the center. Each firefighter should use the inside foot as a pivot to form the fold (Figure 12.21).

Figure 12.19 Open cover snapping hands up and out.

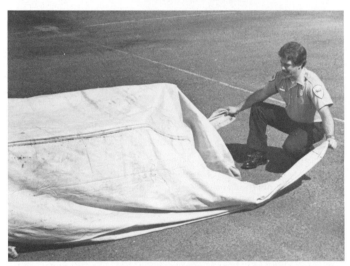

Figure 12.20 To complete coverage tuck in all loose edges at the bottom.

Salvage Cover Fold for A One-Firefighter Spread

As with the one-firefighter roll, initial folds must first be made to reduce the width of the cover. These initial folds are exactly the same as those described in steps 1 through 5 and illustrated in Figures 12.6-12.14, for the one-firefighter roll. Other steps for the one-firefighter salvage cover fold are illustrated in Figures 12.21-12.25.

Steps 1 — 5: Figures 12.6-12.14.

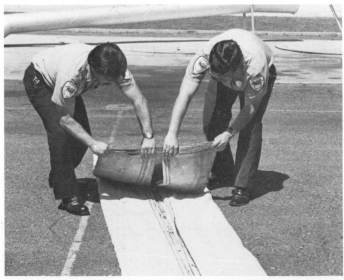

Figure 12.21 Using the feet as pivots, bring the folded cover to a point just short of the center.

Step 7: Each firefighter then uses one hand as a pivot and brings the folded end over and on top of the previous fold (Figure 12.22).

Step 8: Continue, once again, and bring the folded ends over and on top of the previous fold at the middle (Figure 12.23).

Step 9: Fold the other end of the cover toward the center and leave about 4 inches (10 cm) between the two folds (Figure 12.24).

Step 10: The space between the folds now serves as a hinge as the firefighters place one fold on top of the other for the completed fold (Figure 12.25).

Figure 12.23 Make a third fold to the center further reducing the size.

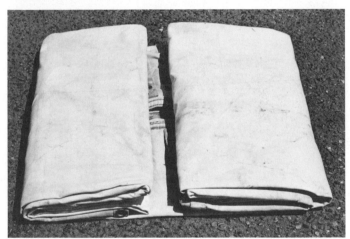

Figure 12.24 Fold the other side in a similar manner with four inches (10 cm) between folds.

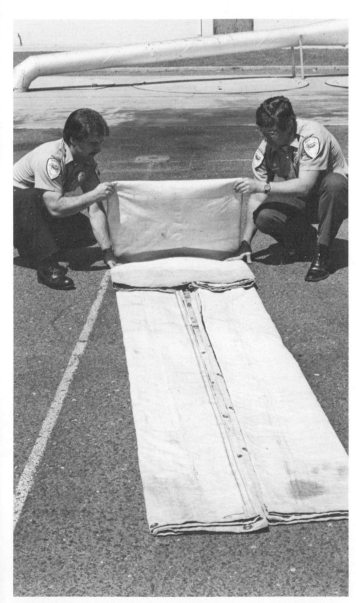

Figure 12.22 Using the hands as pivots, bring the folded end to the center.

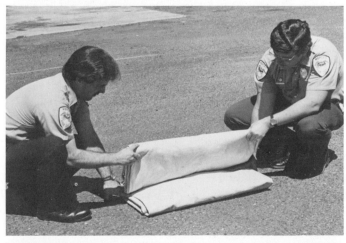

Figure 12.25 Complete the fold by placing one set of folds on top of the other.

One-Firefighter Spread With A Folded Salvage Cover

A salvage cover folded for a one-firefighter spread may be carried in any manner. It is suggested, however, that this fold be carried on the shoulder for convenience. Use the following steps when one firefighter spreads a folded salvage cover (Figure 12.26-12.32).

Step 1: Lay the folded cover on top of and near the center of the object to be covered and separate it at the first fold (Figure 12.26).

Step 2: Select either end and continue to unfold by separating the next fold (Figure 12.27).

Step 3: Continue to unfold this same end toward the end of the object to be covered (Figure 12.28).

Step 4: Grasp the end of the cover near the center with both hands to prevent the corners from falling outward (Figure 12.29).

Step 5: Bring the end of the cover into position over the end of the object being covered (Figure 12.30).

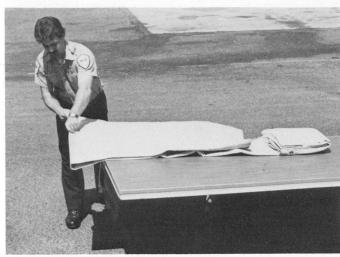
Figure 12.28 Continue to unfold until the end is covered.

Figure 12.26 Place cover at center of object to be covered and separate the sets of folds.

Figure 12.29 Grasp the cover at the center with both hands.

Figure 12.27 Start at either end and unfold the next fold.

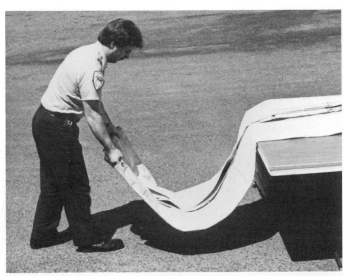
Figure 12.30 Bring the end into position over the end of the object.

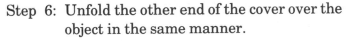

Figure 12.31 Grasp the open edge and prepare to spread the cover.

Step 6: Unfold the other end of the cover over the object in the same manner.

Step 7: Grasp the open edges of the cover at either end and prepare to open the cover as described for the one-firefighter roll (Figure 12.31).

Step 8: Open the sides of the cover over the object by snapping both hands up and out (Figure 12.32).

Step 9: Open the other end of the cover over the object in the same manner.

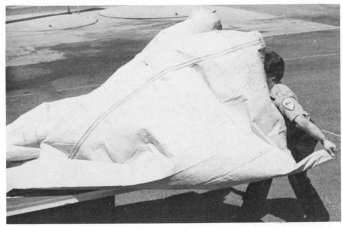

Figure 12.32 Open both ends by snapping the hands up and out to cover the object.

Salvage Cover Fold For A Two-Firefighter Spread

This fold is particularly adapted to a two-firefighter operation and is very versatile with respect to its application. The following steps for making the fold are illustrated in Figures 12.33-12.43.

Step 1: With the cover stretched lengthwise, both firefighters grasp opposite ends of the cover at the center grommet and then pull the cover tightly between them. Raise this center fold high above the floor and shake out the wrinkles to form the first half-fold (Figure 12.33).

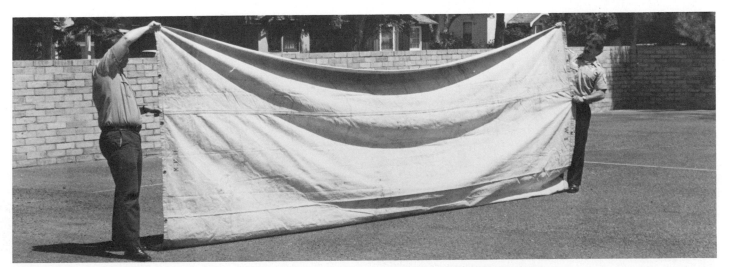

Figure 12.33 Form the first half fold by raising the cover high at the center grommet with finished side to the inside.

Step 2: Spread the half-fold upon the floor and smooth it flat to remove the wrinkles as shown in Figure 12.34.

Step 3: With a firefighter standing at each end of the half-fold and facing the cover, grasp the open-edge corners with the hand nearest to these corners. While in this position, place the corresponding foot at the center of the half-fold and thus make a pivot for the next fold (Figure 12.35).

Step 4: Stretch that part of the cover being folded tightly between the operators. Make the quarter-fold by folding the open edges over the folded edge (Figure 12.36).

Figure 12.34 The cover is spread on the floor and all wrinkles removed.

Figure 12.35 Using the feet as pivots at the center of the half fold, prepare to make the quarter-fold keeping the cover taut.

Figure 12.36 Complete the quarter-fold by folding the open edges over the folded edge.

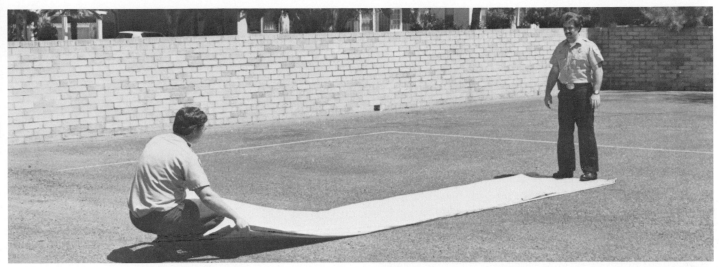

Figure 12.37 Grasp both ends of the quarter-folded cover and shake out the wrinkles.

Step 5: One firefighter should then stand on one end of the quarter-fold while the other person grasps the opposite end and shakes out all wrinkles (Figure 12.37).

Step 6: The firefighter holding the end of the cover then folds the quarter-fold into one hand and carries this end to the partner (Figure 12.38).

Step 7: Both firefighters then crouch at each end of the lengthwise fold. The one at the fold forms a pivot with the hands. The other firefighter places the upper end just short of the other end (Figure 12.39).

Figure 12.38 Fold the quarter-fold into one hand, snap the bottom forward and move to fold it, reducing the length.

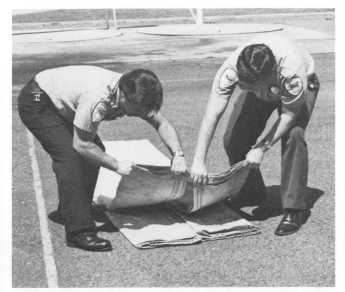

Figure 12.39 As one firefighter forms a pivot with the hands, the other places the upper end just short of the other open end.

Step 8: The two firefighters then grasp the open ends and use their inside foot as a pivot for the next fold (Figure 12.40).

Step 9: Bring these open ends over and place them just short of the folded center fold (Figure 12.41).

Step 10: Continue this folding process by bringing the open ends over and just short of the folded end. During this fold the free hand may be used as a pivot to hold the cover straight (Figure 12.42).

Step 11: Complete the operation by one more fold in the same manner. Bring the open ends over and to the folded end using the free hand as a pivot during the fold (Figure 12.43).

Figure 12.40 The open ends are then folded again using the feet as pivots.

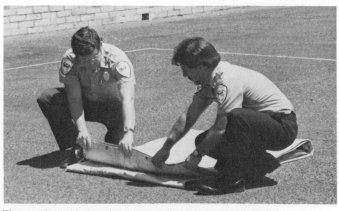

Figure 12.41 Line up the edges and remove any wrinkles.

Figure 12.42 Using the hands as pivots, continue to fold the open edges over the folded edge.

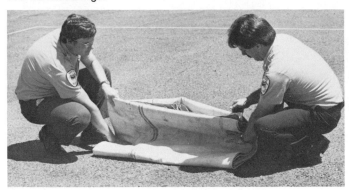

Figure 12.43 One more fold completes the operation with the open edges appearing on top.

Carrying the Salvage Cover Folded For A Two-Firefighter Spread

Probably the most convenient way to carry this fold is on the shoulder with the open edges next to the neck. It makes little difference which end of the folded cover is placed in front of the carrier because two open-end folds will be exposed. Figure 12.44 shows the folded edge next to the neck with

Figure 12.44 The two-person fold is carried on the shoulder so the top two single edges are available to the partner and the bottom two for the carrier.

the fold separated by the carrier and the four corners exposed. The carrier arranges the four corners in pairs with the uppermost pair for the second firefighter and the lower pair for the carrier. After the cover has been stretched lengthwise, the corners may be secured by running the hand down the end hem.

Two-Firefighter Spread from a Folded Salvage Cover

A commonly used method of spreading from the two-person fold is known as the balloon throw. The balloon throw gives better results when sufficient air is pocketed under the cover. This pocketed air gives the cover a parachute effect which tends to float it in place over the article to be covered. The following steps for making the balloon throw are illustrated by Figures 12.45-12.49.

Step 1: Stretch the cover along one side of the object to be covered and separate the last half-fold by grasping each side of the cover near the ends as shown in Figure 12.45.

Step 2: Make several accordion folds in the hand that is to make the throw and place the other hand about midway down the end hem (Figure 12.46).

Figure 12.45 Stretch the cover beside the object to cover and separate the half-fold.

Figure 12.46 Make several accordion folds in the thowing hand.

Figure 12.47 Pull the cover taut and swing down, out, and up in a sweeping movement, trapping air.

Figure 12.48 With the cover at its peak, the accordion folds are carried across the object.

Step 3: Pull the cover tightly between the fire-fighters and prepare to swing the folded part down, up, and out in one sweeping movement so as to pocket as much air as possible (Figure 12.47).

Step 4: When the cover is as high as the fire-fighters can reach, the accordion folds may either be pitched or carried across the object, an action which causes the cover to float over the object (Figure 12.48).

Step 5: As the cover is floated over the object, guide it into position and straighten for better water runoff (Figure 12.49).

Figure 12.49 Guide the cover into position and straighten for water runoff.

Figure 12.50 A salvage cover chute to the outside may be supported by using two pike poles.

IMPROVING WITH SALVAGE COVERS

Removing Water with Chutes

One of the most practical means of removing water that comes through the ceiling from upper floors is by chutes. Water chutes may be constructed on the floor below to drain through windows or doors. Some fire departments carry prepared chutes approximately 10 feet (3 m) long as regular equipment, but it may be more practical to make chutes using one or more covers. Effective water chutes can be made with two pike poles and a salvage cover. This is done by arranging the cover over two poles that reach outside a window or on ladders like the one shown in Figure 12.50. The weight of the water in the trough tends to tighten the rolls. Water chutes may also be used on stairways as shown in Figure 12.51.

Constructing a Catchall

A catchall is constructed from a salvage cover which has been placed on the floor to contain small amounts of water. The catchall may also be used as a temporary means to control large amounts of water until a time when chutes can be constructed to route the water to the outside. Properly constructed catchalls will hold several hundred gal-

Figure 12.51 A stairway can also be covered by rolling the cover edges, flipping the cover over, and forming a chute working from the bottom up the stairs.

Figure 12.52 A catchall is started by rolling each side inward about three feet (1 m) and placing the end of the side rolls perpendicular to the sides to form the corners.

lons of water and often save considerable time during salvage work. The cover should be placed into position as soon as possible even before the sides of the cover are rolled. Two persons are usually required to prepare a catchall in order to make more uniform rolls on all sides. The steps required to make a catchall are as follows and are shown in Figures 12.52-12.54.

Step 1: With the cover spread upon the floor, roll the sides inward approximately three feet (1 m) and lay the ends of the side rolls over at a 90-degree angle to form the corners of the basin (Figure 12.52).

Step 2: Roll one end into a tight roll on top of the side roll and form a projected ear. Using the outside hand to lift the edge roll, tuck the end roll to lock the corners (Figure 12.53).

Step 3: Roll the other end in a like manner and lock the corners. A completed catchall is shown in Figure 12.54.

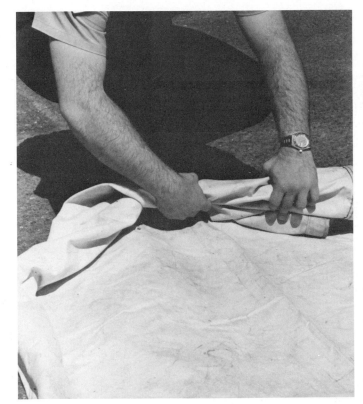

Figure 12.53 Roll one end into a tight roll on top of the side and then tuck the end of the roll under the side wall to lock the corner.

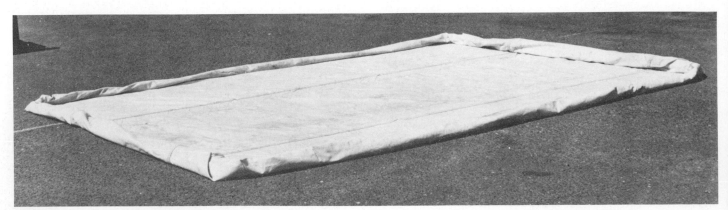

Figure 12.54 A completed catchall ready to use.

USING SPRINKLER STOPS

Considerable time may elapse from the moment of activation of a sprinkler head until authorization to close the main sprinkler valve is received. It is dangerous to prematurely close the main valve before complete extinguishment is certain. Extensive water damage can be prevented by plugging the individual heads which are no longer needed for fire extinguishment.

Each firefighter can conveniently carry a small wedge-shaped sprinkler stop. Most wedges are so designed that they can be driven into position with the heel of the hand. If the wedge is properly made, practically all of the water flow can be controlled by the driven wedge. Wedges can have a band of rubber placed on them to improve their operation. Sprinkler tongs are, to some extent, more positive in their ability to hold back water. Due to the way sprinkler tongs are constructed they are not as conveniently carried as routine equipment in a pocket. If properly applied, the rubber or neo-prene stopper will permit no dripping from a plugged sprinkler head (Figure 12.55).

OVERHAUL

Overhaul is the practice of searching a fire scene to detect hidden fires or sparks which may rekindle and to detect and safeguard signs of arson. Afterwards the building is to be left in as safe and habitable condition as possible. Salvage operations performed during fire fighting will directly affect any overhaul work that may be needed later. Many of the tools and equipment used for overhaul are the same as those used for other firefighter operations. Some of the tools and equipment used for overhaul, along with their uses, may include:

- Pike poles, plaster hooks
  — Opening ceilings to check on fire extension

- Axes
  — Opening walls and floors

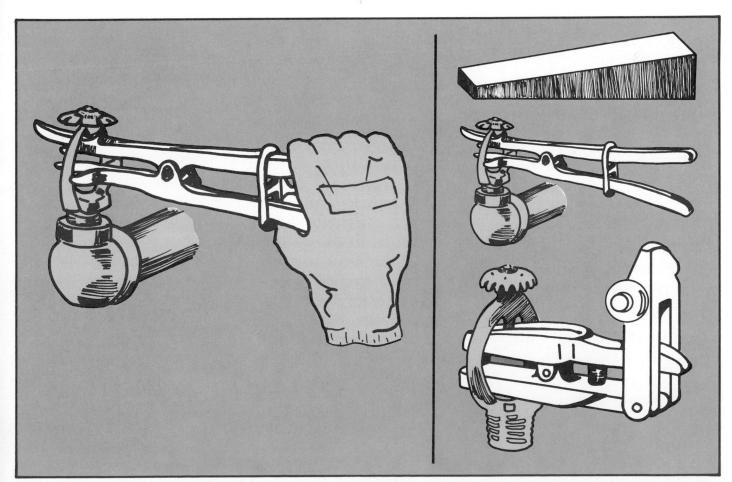

Figure 12.55 Sprinkler stops can be applied to stop the flow of water and enhance salvage operations without reducing the fire protection in other areas.

- Carryall, buckets, tubs
  — To carry debris or to provide a basin for immersing smoldering material
- Shovels, hooks, forks
  — To move baled or loose materials

Searching for Hidden Fires

Since overhaul procedures may not necessarily follow a pattern or plan, one of the first routine operations will probably be searching for hidden fires. It is usually considered to be a mark of inefficiency if it is necessary to return to the scene of a fire to cope with rekindle. A search must be continued at the fire area to be certain the fire has been completely extinguished.

DETERMINE STRUCTURAL CONDITIONS TO ASSURE FIREFIGHTER SAFETY DURING OVERHAUL

Before starting a search for hidden fires it is important to determine the condition of the building in the area to be searched. The intensity of the fire and the amount of water used for its control are two important factors that affect the condition of the building. The first determines the extent to which the structural members have been weakened, and the second determines the additional weight placed on floors and walls due to the absorbent qualities of the building contents. Consideration should be given to these two factors for the protection of personnel during overhaul.

The firefighter should be aware of other dangerous building conditions, such as:

- Weakened floors due to floor joists being burned away
- Spalled concrete from the heat
- Weakened steel roof members (tensile strength is affected at about 500°F [260°C])
- Walls offset due to elongation of steel roof supports
- Mortar in wall joints opened due to excessive heat
- Wall ties holding veneer walls melted from heat

The firefighter can often detect hidden fires by sight, touch, or sound.

- Sight
  — See discoloration of materials
  — Peeling paint
  — Smoke emissions from cracks
  — Cracked plaster
  — Dried wallpaper
- Touch
  — Feel walls and floors with the back of the hand
- Sound
  — Popping or cracking of burning
  — Hissing of steam

An important objective, in searching for hidden fires, should be to make a systematic and careful check to determine whether the fire extended to other areas of the building or other buildings. If it is found that the fire did extend to other areas, it is necessary to determine through what medium it traveled. When floor beams have burned at their ends where they enter a party wall, it is a good policy to overhaul the ends by flushing the voids in the wall with water. The far side of the wall should also be checked to see whether fire or water has come through.

Insulation material, in the form of bats or spray, will often harbor hidden fires for a prolonged period and these bats must be removed in order to locate the hidden fire.

When the fire has burned around windows or doors, there is a possibility that there is fire remaining within the casings. These areas should be opened to assure complete extinguishment. Another point of possible trouble is behind a cornice. When fire has burned around a combustible roof or cornice, it is advisable to open the cornice and inspect for hidden fires.

When concealed spaces below floors, above ceilings, or within walls and partitions must be opened during the search for hidden fires, the furnishings of the room should be moved to locations where they will not be damaged. Only enough wall, ceiling, or floor coverings should be removed to assure complete extinguishment. Weight bearing members should not be disturbed.

The method of opening ceilings from below involves the use of either a pike pole or plaster hook. To open a plaster ceiling, the firefighter must first break the plaster and then pull off the lath. A pike pole is often used for this operation. Metal or composition ceilings may be pulled from the joist in a like manner. When pulling, a firefighter should not stand under the space to be opened. The pull should be down and away to prevent the ceiling from dropping on the head of the firefighter. No firefighter should attempt to pull down a ceiling without wearing full protective clothing (Figure 12.56).

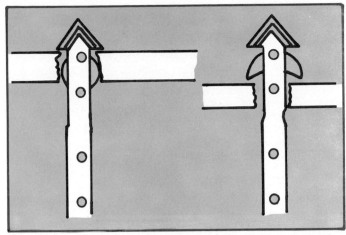

Figure 12.57 The plaster hook also pulls ceilings efficiently. The blades retract to pierce the ceiling, then springs cause them to expand when pulling.

Figure 12.56 When opening a ceiling with a pike pole, wear full protective clothing.

The plaster hook, made of tooled steel with a spear head and two sharp edge blades, is useful in removing metal ceilings, plaster, or other obstructions to gain access to the fire. The blades fold downward when the spear is forced through an object and automatically open after penetration (Figure 12.57). This feature causes ceilings and wall materials to be pulled away when force is applied to the handle.

Extinguishing Hidden Fires

It is essential for firefighters to wear proper protective clothing including positive-pressure self-contained breathing apparatus while performing overhaul and extinguishing hidden fires. It is particularly important for metal insoles to be worn

inside boots, and departments should require gloves. Eye protection should be worn if it is safe to remove breathing apparatus (Figure 12.58). This protection is an aid to the tedious and hazardous task of extinguishing hidden fires.

Charged hoselines should always be available for the extinguishment of hidden fires, although the same caliber of lines as were used to bring the fire under control is not always necessary. Fire department pumpers can often be disconnected from the hydrant but one or more 2½ inch (6.5 cm) or larger hydrant lines should be left for standby. These large lines do not necessarily need to be tugged around by an overhaul crew during their

Figure 12.58 Wear full protective clothing during overhaul. *Courtesy of Steve Taylor.*

search for hidden fires, but they should be kept ready to extinguish any fire outbreak. Smaller hoselines are often extended from these large lines to provide better maneuverability throughout the building. The nozzle should be placed so that if accidentally opened it will not cause additional water damage. Garden hose connected directly to a domestic water supply can also be used to extinguish small fires. When garden hose and other limited extinguishing devices are used during overhaul, provisions should be made to maintain large hoselines within close proximity. Water pumps, cans, and portable fire extinguishers provide limited extinguishing facilities, and are not as reliable as hoselines for overhaul purposes.

Quite frequently small burning objects are uncovered during overhaul. Because of their size and condition, it is better to dunk the entire object in a container of water than to try drenching it with a stream of water. Bath tubs, sinks, lavatories, and wash tubs are all useful for this purpose. The carryall can be used as a vat for this purpose. Larger furnishings, such as mattresses, stuffed furniture, and bed linens should be removed to the outside where they can be easily and thoroughly extinguished. It is important for all firefighters to remember that all scorched or partially burned articles may prove helpful to an investigator in preparing an inventory or determining the cause of the fire. The use of wetting agents is of considerable value when extinguishing hidden fires. The penetrating qualities of wetting agents usually permit complete extinguishment of hidden fires in cotton, upholstery, baled goods, and a large number of other materials. Special care should be taken to eliminate indiscriminate use of and direction of hose streams.

Protecting and Preserving Evidence

Two things should be kept clearly in mind by firefighters regarding the protection and preservation of material evidence.

- Keeping the evidence where it is found, untouched and undisturbed if at all possible
- Properly identifying, removing and safeguarding evidence that cannot be left at the scene of the fire

No changes of any kind should be permitted in the evidence other than what is absolutely necessary in the extinguishment of the fire. Photographs are excellent supporting evidence if they are taken immediately. The photographs should be made before any evidence is washed away. Firefighters should avoid trampling over possible arson evidence and obliterating it so much that it becomes useless (Figure 12.59). The same precaution applied to the excessive use of water may avoid similar unsatisfactory results. Protect human footprints to permit measurements of the prints, comparison of the prints, length of stride, position of feet, and any peculiarities in the gait (walk or run) of a suspect. Boxes placed over prints will prevent dust from blowing over otherwise clear prints and keep them in good condition for either photographs or plaster casts at a later time. Completely or partially burned papers found in a furnace, stove, or fireplace should be protected by immediately closing dampers and other openings upon discovery.

Figure 12.59 During overhaul care must be taken not to destroy arson evidence. *Courtesy of Jim Nichols.*

All evidence collected by firefighters should be properly marked, tagged, identified, and preserved in clean containers. Careful notation should be made of the date, time, and place they are found. Additional identifying marks on cans, bottles, and other articles such as the initials of the person who collected the evidence may also be noted. This procedure may establish unquestionable identity. There must be a record of witnesses and of each person who has had or will have responsibility for care and preservation of the evidence.

When facilities for good paper and ash protection are not available, partially burned paper and ash may be protected between layers of plastic or between pieces of window glass for the investigator and for later transportation to a laboratory. Letters, documents, or bills demanding payment of money should be preserved to assist in establishing a person's financial condition which might indicate a motive for arson.

Place wood suspected of containing paraffin or oil in a clear container and seal until a chemical analysis can be made. All bottles should be labeled with gummed stickers and identified. Objects such as candlewicks and burned matches should be packed in a bottle containing cotton to prevent breaking the evidence by jarring and handling. Samples of materials, such as cotton, wood, rayon, felt, and other fabrics can be stored in clean, large-mouthed bottles and then tightly sealed and properly marked. Volatile liquid, oil samples, oil-soaked rags, waste, and the like should be put in tin cans and sealed. This procedure will preserve

odors as well as the materials themselves and they can then be presented in court in their original condition. Any evidence of this sort should be sealed with wax, so that, when necessary, a laboratory technician may testify in court that the wax seal was unbroken until the proper time.

The firefighter who detects arson and finds the evidence should be able to identify it later. When such material has been tagged, labeled, and properly marked, it is ready to be turned over to the proper authorities. Evidence should be kept under lock and key and as few persons as possible should be permitted to handle it. A record of each person who has handled this evidence should always be kept.

After evidence has been properly preserved, debris may be cleaned up. Charred materials should be removed to prevent the possibility of rekindle and to help reduce the loss from smoke damage. Any unburned materials should be separated from the debris and cleaned. Shoveled debris collected in large containers such as buckets or tubs reduces the number of trips back and forth to the fire area. It is considered to be poor public relations practice to dump debris onto streets and sidewalks or to damage costly shrubbery (Figure 12.60).

Figure 12.60 Overhaul debris should be properly discarded, not merely dumped on the sidewalk or street. *Courtesy of Jim Nichols.*

Covering Windows and Roofs

Salvage and overhaul operations should continue even after the fire is completely extinguished, otherwise merchandise or furnishings previously protected could become damaged by weather. Coverings should be provided for broken windows or opened roofs, whether the fire itself caused them or the fire fighting operation made them necessary (Figure 12.61).

Figure 12.61 Salvage operations should be finalized by covering broken windows, doors, and opened roofs to prevent further damage.

Fire Cause Determination

NFPA STANDARD 1001
FIRE CAUSE DETERMINATION
Fire Fighter II

4-1 General
    4-1.1   The fire fighter shall identify responsibilities of the fire fighter in determining the point of origin, cause, and protection of evidence in fires.*

*Reprinted by permission from NFPA Standard No. 1001, *Standard for Fire Fighter Professional Qualifications*. Copyright © 1981, National Fire Protection Association, Boston, MA.

IFSTA's Prevention and Identification Transparencies
are designed to complement this chapter.

Chapter 13
Fire Cause Determination

Many fire departments have the tendency to misidentify the cause of fires. Sometimes this is because "something has to be written down"; sometimes because of inadvertent inaccuracy. For example, a fire caused by a person using an arc welder near a container of gasoline might be listed as caused by "gasoline" by one department, "electricity" by another, "welding" by another, and "sparks igniting gasoline" by yet another (Figure 13.1). None of these, however, tell the whole story. A fire's cause is a combination of factors: fuel ignited, form of heat of ignition, source of heat of ignition, and, if there is a person involved, the act or omission by the person that helped to bring all these together (Figure 13.2). Inaccurate fire reports cannot be used to develop preventive measures, nor can they be used to indicate fire incidence trends.

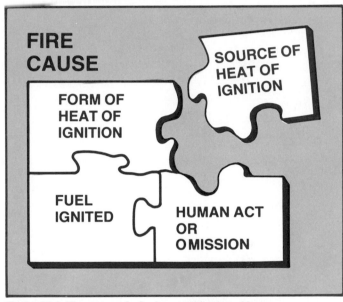

Figure 13.2 The cause of a fire is a combination of factors.

Figure 13.1 If an arc welder ignited the gasoline in this container, would the cause be "gasoline," "electricity," "welding," or something else? *Courtesy of Steve Taylor.*

All too often the fire company rushes to the scene, rushes to extinguish the fire, rushes to clean up the mess — and all evidence with it — and rushes back to the fire station. Seldom does the fire company take time to help determine the fire cause. If a department does not know the causes of fires, whether accidental or intentional, the department cannot work to decrease their number and severity.

THE FIREFIGHTER'S ROLE

Although in most jurisdictions the chief of the fire department has the legal responsibility for fire cause determination, the firefighters at the scene are the ones with the responsibility of being sure the true and specific cause of the fire can be determined. The properly trained firefighter can draw important conclusions by observing the fire and its behavior and other circumstances upon arrival, while entering the structure, and while finding and extinguishing the fire.

The firefighter, more than anyone, is aware of the color and consistency of the smoke, is more apt to recognize unusual odors such as those of flammable liquids and chemicals, and to note the color of the flames. The firefighter is best able to answer important questions: Are the contents of the rooms as they should be, or are they ransacked or unusually bare? Is there evidence of forced entry made before arrival? Are there indications of more than one point of origin, of fire spreading unnaturally fast?

The firefighter's importance cannot be overemphasized. The firefighter's use of a fire stream, for example, often determines whether evidence is saved or destroyed. Judicious, careful overhaul might uncover important evidence (Figure 13.3).

Figure 13.3 Watch for signs of prior forced entry. *Courtesy of C.D. Ray and Associates.*

Investigators are seldom present while the firefighters fight the fire, perform overhaul, and interview occupants and witnesses for report information. Because of this, the firefighters have the important responsibility of noting everything that could point to the cause of the fire. The investigator will seldom be familiar with the history of the case, the circumstances, and the persons involved, so the information furnished by the firefighters who were at the scene is vital.

The items in the following lists have been put under their ideal headings, so the lists are somewhat loose and overlapping. Some of the observations and actions might be noticed or done at different times. For example, evidence of trailers or plants might not be found until overhaul, instead of during fire fighting, and a thorough search for containers and signs of forcible entry will probably be feasible only after the fire is out. The important point is not *when* the firefighters notice something that can point to the cause, but that they do notice it and take the proper steps afterward.

Enroute or in Vicinity

The firefighters' responsibility for gathering information begins as soon as they receive the alarm (Figure 13.4). Items to consider may include:

- Time of day. This will give an indication of the persons and circumstances that should be found at the scene. For example, if the fire is in a dwelling at three in the morning, the occupants would be expected to be wearing night dress, not their best clothes. If the fire is in an office building after working hours, the owner probably should not be at the scene.

Figure 13.4 The firefighters' responsibility in assisting with fire cause determination begins as soon as they receive the alarm. *Courtesy of Robins, Zelle, Larson, and Kaplan.*

- Weather and natural hazards. Is it hot, cold, or stormy? Is there heavy snow, ice, high water, fog? If the outside temperature is high, the furnace at the structure should not be operating. If the temperature is low, the windows should not be wide open. Arsonists often set fires during inclement weather because the fire company's response time is longer.

- Man-made barriers. Are there barricades, felled trees, cables, trash containers, and vehicles blocking access to hydrants, sprinkler and standpipe connections, streets, and driveways?

- People leaving the scene. Most people are intrigued by a fire and will stay to watch. If people are leaving the scene by automobile, make note of the make, model and color, the license number, and a general description of the occupants. If people are leaving the scene on foot, make a note of their dress, general physical appearance, and any peculiarities (Figure 13.5).

Figure 13.5 Make note of people leaving the scene. *Courtesy of C.D. Ray and Associates.*

On Arrival

Additional information that should be gathered after firefighters arrive at the fire scene may include:

- Time of arrival and extent of fire. At a later time, the person who reported the fire can be questioned concerning the extent of the fire at the time it was discovered and reported. If the fire spread unusually fast between the company's receipt of alarm and arrival, accelerants and trailers could have been used.

- Wind direction and velocity. They have a great effect on the natural path of fire spread.

- Location of the fire. Are there separate, seemingly unconnected fires? If so, the fire might have been set in different spots or spread by trailers to several plants.

- Color of smoke. Smoke color gives some indication of what is burning (Table 13.1). If the color of the smoke indicates a fuel that should not ordinarily be in the structure, there may be cause for suspicion and, in some cases, caution.

TABLE 13.1

Color of Smoke Produced by Various Combustibles

Combustible	Smoke Color
Hay/Vegetable compounds	White
Phosphorous	White
Benzine	White to gray
Nitrocellulose	Yellow to brownish yellow
Sulfur	Yellow to brownish yellow
Sulfuric acid, nitric acid, hydrochloric acid	Yellow to brownish yellow
Gunpowder	Yellow to brownish yellow
Chlorine gas	Greenish yellow
Wood	Gray to brown
Paper	Gray to brown
Cloth	Gray to brown
Iodine	Violet
Cooking oil	Brown
Naphtha	Brown to black
Lacquer thinner	Brownish black
Turpentine	Black to brown
Acetone	Black
Kerosene	Black
Gasoline	Black
Lubricating oil	Black
Rubber	Black
Tar	Black
Coal	Black
Foamed plastics	Black

- Color of flame. Flame color can corroborate or enhance inferences drawn from smoke color and is an indication of the fire's intensity.

- Indications of forcible entry (Figure 13.6). Look for signs of forcible entry prior to the arrival of the fire department. The fire could have been set to conceal another crime.

Figure 13.6 Watch for signs of prior forced entry. *Courtesy of Dallas County, Texas, Fire Marshal's Office.*

- Doors or windows locked or unlocked. If the doors should have been locked but are not, and if the investigation leads to a suspicion of intentional setting, key holders will be likely suspects.

- Doors and windows covered. Closed drapery, blankets, and paper covering doors and windows are often used to delay discovery of the fire.

- Discarded containers. Containers found inside or outside the structure could have held flammable liquids.

- Discarded burglary tools.

- Familiar faces. Look for familiar faces in the crowd of bystanders. They might be fire buffs or they might be habitual firesetters.

During Fire Fighting

Firefighters should continue to observe any conditions that may lead to the determination of the fire cause.

- Unusual odors. Gasoline, kerosene, paint thinner, and other common accelerants often can be smelled and identified by the firefighters. Perfumes, deodorants, and ammonia are sometimes used by arsonists in attempts to disguise the smell of accelerants.

- Behavior of fire when water is applied. Flashbacks, reignition, several rekindles in the same area, and an increase in the fire's intensity when ordinary combustibles seem to be the only fuel involved are indications of flammable liquid.

- Obstacles hindering fire fighting. An example is furniture placed in doorways and hallways.

- Incendiary devices, trailers, and plants.

- Alterations to help spread fire: plaster removed to expose wood; holes in ceilings, walls, and floors; fire doors propped open (Figure 13.7).

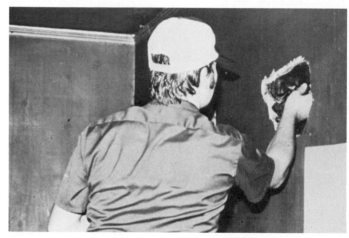

Figure 13.7 Look for alterations purposely done to help spread the fire. *Courtesy of Richard McIntyre.*

- Uneven burning, local heavy charring, or charring in unusual places.

- Heat intensity. High intensity could indicate flammable liquids.

- Speed of spread. Unusually fast spreading could indicate flammable liquids.

- Fire protection systems and devices inoperable because of tampering or intentional damage.

- Burglar alarm tampered with or damaged.

- Fire in unusual places.

- Absence of personal possessions. Absence or shortage of clothing, furnishings, appliances, and similar costly items; absence of personal possessions such as diplomas, financial papers, and toys; items of sentimental value such as photo albums, wedding pictures, and heirlooms; absence of pets that would ordinarily be in the structure. (Remember not to read more into a lack than there really is. Some persons do not have as many material possessions as circumstances might lead one to believe they would have.)

- Absence of equipment or stock. Look for absence of stock, fixtures, display cases, equipment, raw materials, records; business records out of place and endangered by fire.

After the Fire

All facts concerning the fire should be reported to the officer in charge as soon as possible. If the fire is of suspect origin, each firefighter should write a chronological account of important circumstances personally noticed (not hearsay or conjecture). Such an account will be invaluable if the firefighter must testify in court, as cases often come to trial long after the incident, and a person should not rely on memory.

After the fire the officer in charge interviews occupants, owners, and witnesses for information for the fire report, and, if the cause is not already known, conducts the preliminary cause investigation.

Salvage and overhaul are probably the pivotal operations in determining fire cause. Some departments take great pride in their salvage and overhaul work, some boasting that they leave a building neater, cleaner, and more orderly than it was before the fire began. Admirable as this is, it does destroy evidence of how the fire started. Such thorough salvage and overhaul work must be left until the fire cause has been determined.

Salvage and overhaul is necessary, but should be done carefully, the primary consideration being the discovery of the cause of the fire. Debris should not be moved more than is necessary, especially in the area of origin, because the investigation will be hampered. Neither should debris be thrown outside in a pile — evidence is buried this way (Figure 13.8).

Figure 13.8 Evidence of a fire's cause is hidden or destroyed when salvage and overhaul are done indiscriminately. *Courtesy of Jim Nichols.*

When the cause has been proven to be accidental the firefighters can do a thorough cleanup; but if evidence of intentional setting is found, the officer in charge should be notified immediately and all salvage and overhaul operations should be stopped except, of course, those needed to prevent rekindle.

Overhaul is the perfect time for firefighters and the officer in charge to examine the building and premises for some indication of the fire cause.

Conduct and Statements at the Scene

Although the firefighters and officer should obtain all information possible pertaining to the fire, there should be no attempt to cross-examine a potential arson suspect. That is the job of the trained investigator, not of the firefighter or officer. If they are inclined to do so, allow the owners or occupants of the property to talk freely. Some valuable information is often gathered this way. From the moment one suspects a person of arson, an authorized investigator must be called to take over.

Do not make statements of accusation, personal opinion, or probable cause to anyone until the investigator arrives. Then make these statements only to the investigator. Any statement regarding the fire cause should be made only after the investigator and ranking fire officer have agreed to its accuracy and validity and have given permission for it to be made public.

Jesting statements should never be made at the scene. They can easily be overheard by the

Figure 13.9 The premises must be kept secure until the investigation is completed. *Photo: Jim Nichols; Notice: Go-Write, Inc.*

property owner, a news reporter, or other bystanders, all of whom could consider such statements fact.

Most reporters are avid persons, and a microphone could be anywhere. Careless, unauthorized, or premature remarks that are published or broadcast can be very embarrassing to the fire department and many times will impede the efforts of the investigator to prove malicious intent as the fire cause. "The fire is under investigation" is a sufficient reply to any question concerning cause.

SECURING THE FIRE SCENE

The most efficient and complete efforts to determine the cause of a malicious and incendiary fire are completely wasted unless the building and premises are properly secured and guarded until the investigator has finished evaluating the evidence exactly as it appears at the scene.

If an investigator is not immediately available, the premises should be guarded and kept under control of the fire department until all evidence has been collected (Figure 13.9). All evidence should be marked, tagged, and photographed at this time because in many instances a search warrant or written consent to search will be needed for further visits to the premises. This duty might be given to the police if there is a personnel shortage, but whenever possible should be done by fire department personnel.

The fire department has the authority to bar access to any building during fire fighting and as long afterward as is deemed necessary. Be aware, however, of any local laws pertaining to the right of access by owners or occupants. This authority ends as soon as the last firefighter leaves the scene.*

No person should be allowed to enter the premises for any reason unless accompanied by a fire officer or responsible firefighter. A log of any such entry should be kept: the person's name, times of entry and departure, and a description of any items taken from the scene.

The premises can be secured and protected in several ways with the use of few personnel. In fenced areas, gates may be locked, and possibly watched by one person. Areas may be roped off and marked by signs. Goods and materials may be piled around the entrance to a small business or plant to discourage entry. At large manufacturing plants a full-time guard force is often employed, so they could handle the situation. In some extreme instances all doors, windows, or other entrances could be completely closed with plywood or similar material.

LEGAL CONSIDERATIONS

The firefighters at the scene are not through when the investigator has been called. They must keep possession of the premises to prevent unau-

---

*Check local laws and interpretations of *Michigan* vs. *Tyler* in this regard.

thorized entry and to prevent the need for a search warrant for the investigator.

This last point concerning a search warrant is based on *Michigan* vs. *Tyler* (436 U.S. 499, 56 L.Ed. 2d 486 [1978]). The U.S. Supreme Court held in that case that "once in a building [to extinguish a fire], firefighters may seize [without a warrant] evidence of arson that is in plain view. . . . [and] officials need no warrant to remain in a building for a reasonable time to investigate the cause of a blaze after it has been extinguished."

The Court agreed, with modification, with the Michigan State Supreme Court's statement that "[if] there has been a fire, the blaze extinguished and the firefighters have left the premises, a war-rant is required to re-enter and search the premises, unless there is consent. . . ."

The import of these decisions seems to be that if there is evidence of possible arson, the fire department should leave at least one person on the premises until the investigator arrives. To leave the premises and return later without a search warrant and make a search might be enough to make prosecution impossible or for an appellate court to overturn a conviction.

Each department should learn the legal opinions that affect its jurisdiction in this regard. These opinions or interpretations can be obtained from such persons as the district attorney or state attorney general. Write a standard operating procedure around these opinions.

NFPA STANDARD 1001
FIRE SUPPRESSION TECHNIQUES
Fire Fighter I

3-8 Fire Streams
3-8.2 The fire fighter shall manipulate a nozzle so as to attack a Class A fire and a Class B fire.

3-15 Safety
3-15.1 The fire fighter shall identify dangerous building conditions created by fire.

3-15.2 The fire fighter shall demonstrate techniques for action when trapped or disoriented in a fire situation or in a hostile environment.

3-15.3 The fire fighter shall define procedures to be used in electrical emergencies.*

Fire Fighter II

4-1 General
4-1.2 The fire fighter shall identify procedures for shutting off the gas service to a building.

4-1.3 The fire fighter shall identify procedures for shutting off the electrical service to a building.

4-8 Fire Streams
4-8.1 The fire fighter shall define the following methods of water application:
> (a) Direct
> (b) Indirect
> (c) Combination.

4-8.2 The fire fighter, given fire situations, for each situation shall select the proper nozzle and hose size.

4-8.4 The fire fighter shall identify precautions to be followed while advancing hose lines to a fire.*

*Reprinted by permission from NFPA Standard No. 1001, *Standard for Fire Fighter Professional Qualifications*. Copyright © 1981, National Fire Protection Association, Boston, MA.

IFSTA's Ground Cover Fires Transparencies are designed to complement this chapter.

Chapter 14
Fire Suppression Techniques

The success or failure of the fire fighting team often depends on the skill and knowledge of the personnel involved in initial attack operations. Most fires are controlled with the immediate application of water from the first arriving engine companies. Conversely, large-loss fires are sometimes the result of poorly selected or improperly used hoselines. A well-trained team of firefighters with an attack plan and an adequate amount of water, properly applied, will contain most fires.

Firefighters should be aware of the operations to be performed by each team member at the fire scene. A typical engine company may apportion jobs as follows:

- Company Officer — Makes decisions for tactical deployment while assisting and supervising individual members.

- Apparatus Operator — Transports the crew, vehicle, and equipment safely to the scene, and operates apparatus pumping equipment. May make supply line to apparatus, hydrant, or draft hook-ups.

- Firefighters — (Individual assignments will depend upon the number of firefighters available.) Deploys selected hoselines and operates nozzle at direction of the officer. Assists in advancing hoselines and carrying tools. Makes supply line connection to hydrant if part of operation and assists operator or other team members as needed.

Familiarity with the uses of all equipment carried on the apparatus and their limitations is best learned by study and manipulative drills. The quick and efficient use of tools is further enhanced when the companies that frequently work together train together (Figure 14.1).

The need for safe procedures and the wearing of protective clothing cannot be overemphasized.

Figure 14.1 Fireground effectiveness improves when companies that frequently work together train together.

While helmets, gloves, turnout gear, boots, and breathing apparatus protect the firefighter from injury, they also permit the firefighter to apply streams from closer positions. In addition, firefighters should work in pairs while handling attack lines and during search and rescue operations. Firefighters working alone may overexert themselves or be unable to help themselves when trapped.

SUPPRESSING CLASS A FIRES

A fire attack must be coordinated to be successful. Depending on present and anticipated conditions, a fire officer may choose to delay an attack to perform immediate rescue or to place intervening streams between the fire and those persons that may be trapped. Firefighters must perform the evolutions desired by the fire officer at the time that the officer wants them performed. Ventilating a fire, for instance, before attack lines can be brought to readiness may result in the unwanted spread of fire. When properly performed, the ventilation effort will substantially aid the entry and attack of hoseline teams.

Teams advancing hoselines will need to bring some equipment with them needed to enter the structure and perform extinguishment. This equipment would include at least a portable light, rescue strap, and forcible entry tool. The person at the nozzle must bleed the air from the line by opening the nozzle slightly. The operation of the nozzle should also be checked through the range of the stream or to set proper pattern for the attack selector. Any burning fascia, boxed cornices, or other doorway overhangs should be extinguished before entry.

Firefighters should wait at the entrance, staying low and out of the doorway, until the officer gives the order to advance. If the attack is coordinated with ventilation, visibility should improve and a more accurate assessment of fire conditions can be made.

Direct Attack

The most efficient use of water on freeburning fires is made by a direct attack from a close position with a solid stream or penetrating fog pattern (30°

or less) on the base of the fire. The water should be applied directly on the burning fuels in short bursts until the fire "darkens down." Streams should not be applied for too long a time or the thermal balance will be upset. Thermal balance is the movement of heated gases toward the ceiling and after the application of fire streams that includes the spread of expanding steam to all areas of the confined space. If water streams are applied for an excessive length of time, the stream begins to condense causing the smoke to drop rapidly to the floor and move sluggishly thereafter.

Indirect Attack

When firefighters are unable to enter the structure due to intense conditions in confined locations, an indirect attack can be made (Figure 14.2). This attack is not desirable where victims may yet be trapped or where the spread of fire to uninvolved areas cannot be contained. The nozzle setting will range from a penetrating fog (30°) to moderate angle fog (60°) and should be directed at the ceiling and played back and forth in the super heated gases at the ceiling level. Directing the stream into the superheated atmosphere near the ceiling results in the production of large quantities of steam. One cubic foot (.03 m^3) of water (7.4 gallons (27 L) completely vaporized will create 1700 cubic feet (518 m^3) of steam. Once again, the steam should be shut down before disturbing the thermal balance. Once the fire has been darkened down the hoseline can be advanced to extinguish any remaining hot spots with a direct attack.

Combination Attack

The combination method utilizes the steam generating technique of ceiling level attack combined with an attack on materials burning near the floor level. The nozzle may be moved in a "T, Z, or O" pattern starting with a penetrating fog directed into the heated gases at the ceiling level and then dropped down to attack the combustibles burning near the floor level. The "O" pattern of the combination attack is probably the most familiar and most frequently abused method of attack. When performing the "O" pattern the stream should be directed at the ceiling and rotated clockwise with the stream edge reaching the ceil-

Figure 14.2 The indirect attack is used to fight the fire from outside the building.

ing, wall, floor, and opposite wall. Keep in mind that applying water to smoke does not extinguish the fire and only causes unnecessary water damage and disturbance of the thermal balance. Fire-fighters assisting the person at the nozzle should not bunch up behind the nozzle as this makes manipulation of the nozzle difficult. The assisting team members need to advance hose to the person at the nozzle as it is needed. All team members must watch for a number of potential hazardous conditions such as:

- Imminent building collapse

- Fire behind, below, or above the attack team

- Kinks or obstructions to the hoseline

- Holes or fall hazards

- Suspended loads on fire weakened supports

- Hazardous or highly flammable commodities likely to spill

- Backdraft or flashover behavior

- Electrical shock hazards

- Overexertion, confusion, or panic by team members

Stream Selection

As previously mentioned, the technique of water application is only successful if the amount of water applied is sufficient to cool the fuels that are burning. The use of a booster line may not only delay extinguishment, but may be of insufficient volume to protect firefighters from advancing flame fronts. Hoseline selection should be dependent upon fire conditions and other factors such as:

- Volume of water needed for extinguishment

- Reach needed

- Number of persons available to handle hoseline

- Mobility requirements

- Tactical requirements

- Speed of deployment

- Potential fire spread

Obviously, it would be incorrect to choose a 1½-inch (38 mm) hoseline to attack a fire in a large well-involved commercial occupancy. The line would have neither the necessary volume nor reach. It would also be incorrect to attack a fire involving a single room and contents in a family dwelling with a 2½-inch (65 mm) line discharging 250 gpm (940 L). Table 14.1 gives a simple analysis of hose stream characteristics, and is not meant to replace the judgement of firefighters in selecting fire streams.

Once hoselines have been selected and the method of attack determined, firefighters should proceed at the officer's direction. Hoselines are of little value if they are not advanced to the seat of the fire during the initial attack. Attack teams must communicate with their officers if the water being applied is not sufficient to contain the fire or if evidence of fire spread to other areas of the building is found. Likewise, officers need to know when the attack is being successful. This coordination will hasten fire control and reduce damage by permitting salvage and overhaul functions to begin as soon as the fire is contained. Evidence of suspicious fire origin must always be reported and crew members must take care not to further disturb physical evidence of arson.

Once the fire has been contained it may be necessary to relieve the initial attack crew. Breathing apparatus must still be worn during mop-up and overhaul phases of the operation due to the presence of fire gases. Special attention should be directed toward walls, partitions, or overhead loads that may be dislodged by fire fighting activities. Valuables found should be taken immediately to an officer.

SUPPRESSING CLASS B FIRES

Every firefighter is familiar with the phrase, "Never put water on flammable liquid fires," yet necessity and experience have shown that water is

TABLE 14.1
HOSE STREAM CHARACTERISTICS

Size	GPM	Reach (Max.)	No. of Persons on Nozzle	Mobility	Control of Damage	Control of Direction	When Used	Estimate Effective Area
Booster (¾" - 1") (16 to 25 mm)	10 to 30 (38 to 114 L/m)	25' to 50' (8 to 15 m)	1	Excellent	Excellent	Excellent	Very small interior fire. No possible chance of extension. Mop up or overhaul.	Less than one room.
1½" (38 mm)	50 to 120 (190 to 454 L/m)	25' to 50' (8 to 15 m)	1 or 2	Good	Good	Excellent	Developing fire - still small enough or sufficiently confined to be stopped with relatively limited quantity of water. For quick attack. For rapid relocation of streams. When manpower is limited. When ratio of fuel load to area is relatively light. For exposure protection.	One to three rooms.
2½" (65 mm)	150 to 250 (568 to 946 L/m)	50' to 100' (15 to 30 m)	2 to 4	Fair to Poor	Fair	Good	When size and intensity of fire is beyond reach, flow or penetration of 1-½" (38 mm) line. When both water and manpower are ample. When safety of men dictates. When larger volumes of greater reach are required for exposure protection.	One floor or more fully involved.
Master	350 to 2000 (1325 to 7570 L/m)	100' to 200' (30 to 60 m)	1	Poor to None (Aerial master streams can be excellent)	Poor	Good	When size and intensity of fire is beyond reach, flow or penetration of hand lines. When water is ample but manpower is limited. When safety of personnel dictates. When larger volumes or greater reach are required for exposure protection. When sufficient pumping capability is available. When massive runoff water can be tolerated. When interior attack can no longer be maintained.	Large structures fully involved.

Courtesy of Joseph Bachtler, Maryland Fire and Rescue Institute.

highly effective in extinguishing or controlling these fires. Control of flammable liquid fires can be accomplished safely if proper techniques are used. These techniques require a basic understanding of flammable liquid properties and the effects water has on them.

Flammable liquids have some unique properties that affect their behavior during fires and extinguishment. Generally, they:

- Float on water
- Create static electricity when flowing
- Can burn with explosive force
- Create flammable vapors at room temperature
- Propagate flame rapidly over the entire exposed surface
- Will pass through the explosive range as mixtures too rich to burn are ventilated

Along with these properties and the tremendous quantities of liquid that are stored, transported, and used make flammable liquids an ever-present fire danger.

The use of personal protective equipment is required to reduce injuries to firefighters and to permit close approach for extinguishment. Firefighters must be aware that standing in pools of fuel or water runoff containing fuel can result in a "wicking" action of the fuel into clothing. This "wicking" action can lead to contact burns of the skin and flaming clothing if an ignition source is present.

Fires burning around relief valves or piping should not be extinguished unless the leaking product can be shut off. Unburned vapors are usually heavier than air and will form pools or pockets of gas in low spots where they may be ignited. Firefighters must always control all ignition sources in the proximity of flammable gas and liquid leaks. Vehicles, smoking materials, electrical fixtures, and sparks from steel tools can all provide an ignition source sufficient to ignite leaking flammable vapors. An increase in the intensity of sound or fire issuing from a relief valve may indicate that rupture of the vessel is imminent. Firefighters should not assume that relief valves are sufficient to

safely relieve excess pressures under severe fire conditions. Firefighters have been killed by the rupture of flammable liquid vessels both large and small that have been subjected to flame impingement.

In vessels containing flammable liquids, this condition of the sudden release and consequent vaporization of the liquids is called a bleve: boiling liquid expanding vapor explosion. A bleve results in the explosive release of vessel pressure, pieces of tank, and a characteristic fireball with radiant heat. Bleves most commonly occur when flames contact the vapor space of the flammable liquid vessel and insufficient water is applied to keep the tank cooled.

Figure 14.3 Water can be used as a cooling agent to extinguish fires and to protect exposures.

Using Water as a Cooling Agent

Water can be used as a cooling agent to extinguish fires and to protect exposures (Figure 14.3). While water without foam additives is not particularly effective on gasoline or alcohols, fires in the heavier oils can be extinguished by applying water in droplet form in sufficient quantities to absorb the heat produced. Formulas are available to determine the application rates necessary for existing tanks and for making quick fireground calculations. Care must be taken not to allow tanks to overflow. Water will be most useful as a cooling agent for protecting exposures. To be effective, water streams need to be applied so that they form a protective water film on the exposed surfaces. This applies to ordinary combustibles and other materials that might weaken or collapse such as metal tanks or support beams. Water applied to storage tanks should be directed above the level of the contained liquid to achieve the maximum efficient use of the water.

Using Water as a Mechanical Tool

Water from hoselines can be used to move the fuel, whether it is burning or not, to areas where it can safely burn or where ignition sources are more easily controlled. Fuels must never be flushed down drains or sewers. Firefighters should use wide angle to penetrating fog patterns for the protection from radiant heat and to prevent "plunging" the stream into the liquid. Plunging a solid stream into burning flammable liquids causes increased production of flammable vapors and greatly increases fire intensity. The stream should be slowly played from side-to-side and the fuel or fire "swept" to the desired location. Care must be taken to keep the leading edge of the fog pattern in contact with the fuel surface or else the fire may run underneath the stream and flash back around the attack crew. In some cases, the fire may be "cornered" by hoselines against a barrier and then the flame swept off the surface, while the product remains contained. Where small leaks occur, a solid stream may be applied directly to the hole and the escaping liquid forced back. The pressure of the stream must exceed that of the leaking material to perform properly. Care must be taken not to overflow the container.

Water may also be used to dissipate flammable vapors through the use of fog streams. These streams aid in dilution and dispersion, controlling to a small degree, movement of the vapors to a desired location.

Using Water as a Substitute Medium

Water can be used to displace fuel from pipes or tanks that are leaking. Fires that are fed by leaks may be extinguished by pumping water back into the leaking pipe or by filling the tank with water to a point above the level of the leak. This displacement will float the volatile product on top of the water as long as the water application rate equals the leak rate. Due to the large water-to-product ratios, water is seldom used to dilute flammable liquids for fire control. However, this technique may be useful for small fires where the run-off can be contained.

Using Water as a Protective Cover

Hoselines can be used as a protective cover for teams advancing to shut-off liquid or gaseous fuels (Figure 14.4). Coordination and slow, deliber-

Figure 14.4 Hoselines provide a protective cover for teams advancing to shut off liquid and gaseous fuels.

ate, movements provide relative safety from flames and heat. While one hoseline can be used as a protective cover, two lines with a backup line are preferred for fire control. When containers of flammable liquids or gases are exposed to flame-impingement, solid streams should be applied from their maximum effecting reach until the relief valve closes. This can best be achieved by lobbing the stream along the top of the tank so that the water runs down both sides. This film of water will cool the vapor space of the tank. Steel supports under tanks should also be cooled to prevent their collapse. Then hose streams can be advanced under progressively widened protective fog patterns to make temporary repairs or shut-offs. A backup line supplied by a separate pump and water source should be provided to protect firefighters in the event other lines fail or additional tank cooling is needed. Approaches to storage vessels should be made at right angles to the tank, never from the ends. Rupturing vessels frequently split in two pieces and then become projectiles.

Fires Involving Bulk Transport and Passenger Vehicles

The techniques of extinguishment for fires in vehicles transporting flammable fuels are similar in many ways to fires in storage facilities. The major differences are:

- Increased life safety risks to firefighters from traffic
- Increased life safety risk to passing motorists
- Reduced water supply
- Difficulty in determining the products involved
- Difficulty in containing spills and runnoff
- Tanks and piping weakened or damaged by the force of collisions
- Instability of vehicles

While a serious accident may bring traffic to a halt, many incidents will be handled with traffic passing the scene at near-normal speeds. A lane of traffic in addition to the incident lane should be closed from traffic use during initial emergency op-

erations. The use of open flame flares should be avoided due to the possibility of their igniting leaking fuels. Fire apparatus should be positioned to take advantage of topography and weather conditions, uphill and upwind, and to protect firefighters from traffic. Firefighters should exit the apparatus and work as much as possible from the curb side away from traffic. In addition, firefighters should avoid working where the apparatus could be pushed into them if it were struck by another vehicle. Where traffic is passing closely, firefighters should be careful not to allow tool handles to extend into the traffic lane where they may be struck. When law enforcement personnel are unavailable a firefighter should be assigned the role of traffic control officer.

The techniques of approaching and controlling leaks or fires involving vehicles are the same as for storage vessels. Additionally, firefighters should be aware of the failure of vehicle tires that may cause the flammable load to shift suddenly. Crews will need to know the status of their water supply so as not to exceed the limitations of that supply. As is the case in structural fire attack, it may be necessary to protect trapped victims with hoselines until they can be rescued. Firefighters must determine, as soon as possible, the exact nature of the cargo from bills of lading, manifests, placards, or the driver of the transport vehicle (Figure 14.5).

Figure 14.5 Placards aid in identifying the type of cargo carried on transport vehicles. *Courtesy of NAPA (CA.) Fire Department.*

Unfortunately, cases will certainly exist where these items cannot be found, placards are either wrong or obscured, and drivers are unable to identify their cargo. In these cases, contact should be made with the shipper or manufacturer responsible for the vehicle. Pre-incident plans for transportation emergencies should be followed to reduce life loss, property damage, and environmental pollution. Single passenger vehicles usually present less of an extinguishment problem due to the reduced amount of fuel carried. Burning or leaking fuel can be flushed from beneath the vehicle, and then the remaining Class A fire attacked. Firefighters should avoid standing in front of the shock-absorber type bumpers on newer vehicles as they may explode. Large amounts of water will be needed to attack fires that have ignited aluminum or magnesium alloy vehicle components. Firefighters should use extra caution when water is first applied to these burning parts as fire intensity will be greatly increased. Firefighters should not assume that private vehicles or small vans are without extraordinary hazards such as saddle fuel tanks, propane tanks, explosives, or hazardous materials. Vans are often used to transport small amounts of radioactive materials for hospital use. Also, large dollar losses can occur from fires in messenger or courier vehicles. Certainly, firefighters should view any military vehicle as a target hazard.

SUPPRESSING CLASS "C" FIRES

Fires in electrical equipment occur quite frequently, but once de-energized they can be handled with relative ease. The primary danger of electrical fires is the failure of emergency personnel to recognize the hazard. It is the responsibility of the fire officer to insure that main power breakers are opened to control power flow into structures. Similarly, a crew member should be assigned to control power at vehicle fires and other emergencies. Once the power has been shut down these fires may self-extinguish or if they continue to burn they will fall into either class "A" or "B" fires. Usually, carbon dioxide extinguishers will be preferred as they do not require clean-up and they are more economical in comparison to the halogenated agents. However, certain microprocessor equipment can be damaged by the temporary cold resulting from carbon dioxide "snow" residue. Multi-purpose dry chemical agents present a considerable clean-up problem in addition to being chemically active with some electrical components.

The use of water on energized equipment is discouraged unless absolutely necessary because of the inherent shock hazard. If water must be applied it should be from broken streams, and at a distance. When fires occur as a result of transmission lines breaking, an area equal to span between poles should be cleared on either side of the break. Fires in transformers can present a serious health and environmental risk in the form of coolant liquids that contain PCB's (poly-chlorinated biphenyls). These liquids are flammable because of their oil base and are extremely carcinogenic (cancer causing). Transformers at ground level should be extinguished carefully with a dry chemical extinguisher. Transformers above ground should be permitted to burn until a qualified person can extinguish the fire with a dry chemical extinguisher from a power-operated elevating device. Placing a ground ladder against the pole places personnel under risk from both the power source and from the liquid. Applying hose streams to these fires can result in spreading the material onto the ground.

Consultation and cooperation with power company officials is vital in these incidents in order to reduce the risk to life and property. Other unusual electrical hazards can be found in railroad locomotives, telephone relay switching stations, and electrical substations. Procedures for fighting fires in these occupancies should be established in pre-fire plans.

SUPPRESSING CLASS "D" FIRES

Combustible metals present the dual problem of burning at extremely high temperatures and being reactive to water. Water is only effective when it can be applied in large enough quantities to cool the metal below its ignition temperature. The usual method of control is to protect exposures and permit the metal to burn out. Special extinguishing agents may be manually shoveled in quantities large enough to completely cover the burning metal. Directing hose streams at burning metal can result in the violent decomposition of the water and subsequent release of flammble hydrogen gas. Small chips or dust of the metal are more reactive to water than are larger ingots or finished products. These fires can be recognized by a characteristic brilliant white light being given off until an ash layer covers the burning material. Once this layer has formed it may appear that the fire is out. Firefighters should not assume that these fires are extinguished because flames are not visible.

CONTROL OF ELECTRICAL UTILITIES

Firefighters must be able to control the flow of electricity into structures where emergency operations are being performed. In order to avoid injury and to protect electrical equipment the firefighter should be familiar with electrical transmission and its hazards. While high voltage equipment is usually associated with high shock, conventional residential current is sufficiently powerful to deliver fatal shocks. In addition to reducing the risk of injury or fatal shock, controlling electrical flow reduces the danger of igniting combustibles or accidental equipment start-ups.

The consequences of electrical shock include:

- Cardiac arrest
- Ventricular fibrillation
- Respiratory arrest
- Involuntary muscle contractions
- Paralysis
- Surface or internal burns
- Damage to bone joints
- Ultra-violet arc burns to the eyes

Factors most affecting the seriousness of electrical shock are:

- The path of electricity through the body
- The degree of skin resistance - wet (low) or dry (high)
- Length of exposure
- Available current - amperage flow
- Available voltage - electromotive force
- Frequency - AC or DC

Electrical Service Installations

RESIDENTIAL SERVICES

Homes are commonly serviced with a pole-to-structure line called the service connector. The easiest method to de-energize the structure is to throw the main switch at the service box. The lever control may be padlocked or held in place with a lead wire seal. Problems may occur when more than one service enters a common building or downed lines come in contact with telephone and/ or television wires, or metal portions of a building.

CUTTING LIVE WIRES

For maximum safety on the fireground, live wires should not be cut except by experienced power company personnel with proper equipment. Where planning reveals that the response of utility personnel necessitates fire department action, proper training must be provided.

UNDERGROUND LINES

Underground transmission systems consist of cableways and vaults beneath the surface. The most frequent hazard that these systems present is explosions which may blow manhole covers a considerable distance. The cause of this occurrence is the accumulation of gases which are ignited by a spark from fuses blowing or the arcing from a short circuit. This is a danger to the public as well as firefighters. If these situations are suspected, keep the public clear of the area and be certain that apparatus is not parked over a manhole.

Do not enter a manhole except to attempt a rescue as fire fighting can be accomplished from outside. Simply discharge carbon dioxide or dry chemical into the manhole and replace the cover. Place a wet blanket or salvage cover over the manhole to exclude the oxygen and assist in extinguishing the fire. Water is not suggested for extinguishment because of the close proximity of electrical equipment. The runoff of water would also create puddles that could become dangerous conductors of electricity.

When circumstances dictate that a firefighter must enter a manhole it should be done only with approved positive-pressure self-contained breathing apparatus and full protective clothing. A filter-type mask should never be used under any conditions. Any equipment or tools should be nonsparking because the smallest spark might ignite the explosive mixture.

COMMERCIAL HIGH VOLTAGE INSTALLATIONS

Many industries, large buildings, and apartment complexes have electrical equipment that utilizes voltage in excess of 600 volts. The obvious clue to this condition would be a "high voltage" sign on the doors of vaults of fire-resistive rooms housing the high voltage equipment such as transformers or large electric motors (Figure 14.6). Some transformers use flammable oils as coolants which present a hazard in themselves. Water should not be used in this situation, even in the form of fog, because the hazard of shock is greater and extensive damage may occur to electrical equipment not involved in the fire.

Figure 14.6 A sign such as this may be the first clue for the firefighter that high voltage electrical equipment is involved.

Because of toxic chemicals used in plastic insulation and coolants, smoke becomes a hazard. Enter only when rescue operations require it wearing positive-pressure self-contained breathing apparatus and a safety line monitored by someone outside the enclosure. When searching, do so with a clinched fist to prevent reflex actions of grabbing live equipment which may be contacted.

Electrical Power Shutoff

From a safety standpoint, power should remain on as long as possible to provide lighting, ventilation, or to run special pumps. When the building becomes damaged to the point that service is interrupted and an electrical hazard exists, power should be shut off by a power company employee if possible. When the fire department must do this, only trained personnel who are aware of the effects should be assigned the task. When a fire only involves one area it would be pointless to shut down the entire building.

Power can be shut off by opening a switch or removing a fuse. Care must be taken to do so without contacting bare skin. In residential fires a common practice is to "pull the meter." This is easily done, but then the empty meter socket will have electrical contacts exposed (Figure 14.7). If contacts are exposed, a protector should be inserted and the meter tagged to give warning and prevent its use until qualified personnel have examined the system. With some commercial building meters their removal does not discontinue the flow of electricity (Figure 14.8). Also, firefighters should be

Figure 14.7 The back of a residential meter and the empty socket showing the exposed electrical contacts.

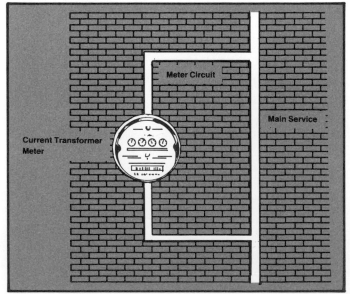

Figure 14.8 When a current transformer meter is used, the meter is connected in parallel with the main service conductor. Therefore, removing the meter will not shut off power to the building.

alert for installations with emergency power capabilities, such as emergency generators. In such cases, pulling the meter or master switch does not shut off the power entirely.

Guidelines for Electrical Emergencies

Listed below are some tips to help deal with electrical emergencies. The list is not totally inclusive but gives some principles which should be considered to maintain a safe working environment for personnel.

- When downed wires are encountered, a danger zone of one span in either direction should be considered for safety. This is because other wires may have also been weakened by the short and fall at a later time.

- A firefighter must guard not only against electrical shock and burns but also eye injuries from electrical arcs.

- Treat all wires as "hot" and being of high voltage.

- From a safety standpoint, firefighters should not cut any wires but wait and let trained utility workers do any necessary cutting. Only in the most extreme circumstances should this rule be excepted.

- When an electrical hazard exists, always wear full protective clothing and use only insulated tools.

- Care must be exercised in raising or lowering ladders, hoselines, or equipment near overhead lines (Figure 14.9).

- Proceed carefully in an area where wires are down and heed any tingling sensation felt in the feet. Because of the carbon in the boots a slight charge is transmitted indicating that the ground may be charged.

- Do not touch any vehicle or apparatus that is in contact with electrical wires since body contact will complete the circuit to ground resulting in electrical shock.

Figure 14.9 When hoisting hose and tools near overhead electrical lines, use extreme care so that the tools do not contact the lines.

- When more than one electrical wire is down, consider all wires equally dangerous when one is arcing and the other is not.

- Straight streams are discouraged when energized electrical equipment is encountered. Fog patterns should be used with at least 100 psi (700 kPa) nozzle pressure and applicators should not be used since they may act as a conductor.

- Special considerations for fences should be given since once an energized electrical line contacts a fence or metal guardrail the entire length becomes charged as long as it is continuous. This can present a difficult hazard to protect people from because of the length of the fence.

CONTROL OF GAS UTILITIES

A working knowledge of the hazards and correct procedures in handling incidents related to natural gas and Liquefied Petroleum Gas (LPG) is important to every firefighter. Almost every house, mobile home, and business uses natural gas or LPG for cooking, heating, or industrial processes. The firefighter familiar with gas distribution and usage will be able to prevent or reduce damage caused by incidents involving these gases.

Natural gas is mostly methane with small quantities of ethane, propane, butane, and pentane added. The gas is lighter than air so it will tend to rise and diffuse in the open. Natural gas is nontoxic but it is classified as an asphyxiant because it may displace normal breathing air and lead to asphyxiation. The gas has no odor of its own but a very distinctive odor is added by the utility. It is distributed from gas wells to its point of usage by a nationwide network of above and below ground pipes. The pressure in these pipes ranges from ¼ to 1,000 pounds per square inch (2 to 6900 kPa³). However, the pressure is usually below 50 psi (350 kPa) at the local distribution level. Natural gas is explosive in concentrations of between four and fourteen percent.

The local utility should be contacted when any emergency involving natural gas occurs in their service area. They will provide an emergency response crew equipped with special tools, maps of the distribution system, and the training and experience needed to help control the flow of gas. The response time of these crews is usually less than an hour but the time may be extended in rural areas or in times of great demand. Good relations between the fire department and the utility are encouraged.

Liquefied petroleum gas (LPG), or bottled gas as it is sometimes known, is a fuel gas stored in a liquid state under pressure. It is used primarily as a fuel gas in campers, mobile homes, agriculture, and rural homes. There is an increased use of LPG as a fuel for motor vehicles. The gas is composed mainly of propane with small quantities of butane, ethane, ethylene, propylene, iso-butane, or butylene added. LPG has no natural odor of its own but a very distinctive odor is added. The gas is nontoxic but it is classified as an asphyxiant because it may displace normal breathing air and lead to asphyxiation. LPG is about one and a half times as heavy as air so it will generally seek the lowest point possible. The gas is explosive in concentrations of between 1.5 and 10 percent. The gas is shipped from its distribution point to its point of usage in cylinders and in tanks on cargo trucks. It is stored near its point of usage in cylinders and tanks, and then connected to the appliances they serve by underground piping and copper tubing. All LPG containers are subject to BLEVE's (Boiling Liquids Expanding Vapor Explosion) when exposed to intense heat or open flame.

Incidents involving the distribution system are most often caused by excavation around underground pipes. If gas is escaping but not burning, the utility should be contacted immediately. Apparatus should not be parked close to the scene because of the possibility of ignition. Firefighters should be prepared for the event of an explosion and any accompanying fire. The firefighter's first concern should be the evacuation of the area around the break and elimination of ignition sources in that area. The broken main may have damaged service connections near the break, so surrounding buildings should be checked for gas buildup. Firefighters should not attempt to operate main valves because incorrect action may worsen the situation or cause unnecessary loss of service to areas unaffected by the break. If the gas is burning, the flame **should not** be extinguished. Exposures can be protected by hose streams if necessary.

The most common situation firefighters will be faced with involves the service meter. The meter is usually located outside of the building and normally visible from the street. The flow of gas into the building may be stopped by turning the cutoff valve to the closed position, which will be at a right angle to the pipe (Figure 14.10).

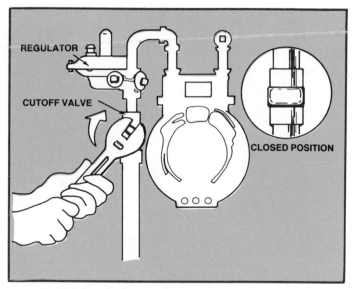

Figure 14.10 The flow of gas into a building may be stopped by turning the cutoff valve to the closed position.

Firefighters involved in this operation should advance a hoseline on a fog pattern in order to protect themselves. It must be stressed that if the gas is burning, the fire should not be extinguished. If for some reason the meter cock is inoperable, the pipe may be partially closed by pinching the line with a hydraulic rescue tool. This action may not stop the flow of gas but it will reduce the flow.

LPG or bottled gas is most often stored in one or more outdoor tanks or cylinders. The supply of gas into a structure may be stopped by shutting a valve on the pipe leading to the building. If the valve is inoperable, the gas may be stopped by pinching the copper line leading into the structure with a pair of pliers or, in the case or larger lines, a rescue tool. Unburned gas may be dissipated by a fog stream of at least 100 gpm (378.5 L). If there are any problems with a cylinder or tank, the company responsible for it should be contacted.

FIRE COMPANY TACTICS

Some fire departments have a predetermined plan or written policy for nearly every type of emergency that they can conceive of occurring. These procedures are usually carried out by the first fire companies reaching the scene and have become known as standard operating procedures or SOP's (Figure 14.11). Standard operating procedures may vary considerably in different localities, but the principle is usually the same. The proce-

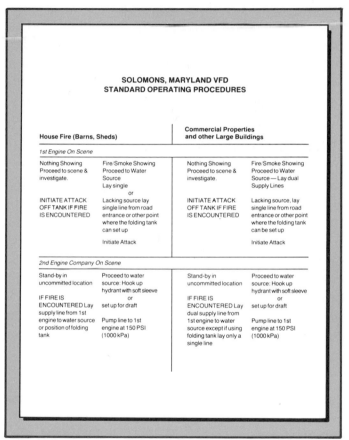

Figure 14.11 Standard operating procedures provide direction for companies arriving at the fire scene.

dure is primarily a means to get the fire attack started, and its use does not replace size up, and decisions based on a common sense evaluation, or command. In addition, there may be several SOP's from which to choose depending on the fire severity, location, and the ability of first-in units to achieve control. In some instances, a standard operating procedure may be all that is necessary to confine and extinguish a fire. At more complex incidents, the SOP may only be used to provide adequate fire defense until additional forces can be brought to bear on the fire. SOP's are generally used to address general activities at any emergency scene, but often can be included to make operations at preplanned target hazards more efficient. This inclusion of SOP's into special hazard plans reduces confusion on the fireground and the need for extensive memorization. Efficiency is increased at unusual incidents when engine companies have regularly used the SOP in their "ordinary" fire attacks. Also, during the process of putting SOP's into effect, the fire officer will have an opportunity to apply other specific pre-fire plans and to develop a fire offense tailored to the needs of the incident. Following are some examples of fire department standard operation procedures for fires in various locations. The scenarios described used standard response of two engine companies, one ladder company, and one chief officer.

Fire Tactics for Single Family Dwellings

The first engine company to arrive at the scene will initiate the fire attack taking into consideration the present and expected behavior of fire. The first engine company makes a radio report to the dispatch center regarding the exact location, exposures, and conditions found at the incident. The company officer, after assessing the situation, may lay a line or order the next incoming engine to do so. It is the job of the second due engine companies to assist the first engine company as ordered.

Once the location of the fire is known the officer of the first company will position the initial attack line to cover the following priorities:

- Placement of lines to intervene between trapped occupants and the fire or to protect rescuers
- Protection of primary means of egress
- Interior exposure protection (other occupancy)
- Exterior exposure protection (other occupancy)
- Initiates extinguishment
- Master stream operations

The second engine company must first establish the fireground water supply and then proceed according to the following priorities:

- Back up the initial attack line.
- Protect secondary means of egress.
- Prevent fire extension (confinement).
- Protect the most severe exposure.
- Assist in extinguishment.

When positive-pressure SCBA is not donned enroute, it should be done upon arrival and before proceeding with the interior attack on the fire (Figure 14.12). In extreme emergencies it may be necessary for the first arriving company to begin rescue operations and for the second company to stretch the first fire attack line. Usually, however, the rescue function will be performed by the ladder company under the cover of hose streams from the engine companies.

Figure 14.12 Second arriving companies donned in SCBA can be used to relieve initial attack crews and continue the interior attack on the fire. *Courtesy of Sheldon Weaver.*

The ladder company most often arrives after the first engine company and is responsible for search and rescue and aiding the fire attack by ventilating. Initially, the ladder company will observe the outside of the building for signs of victims needing immediate rescue. The ladder company can then begin to search for victims either using interior or exterior routes of entry. The ladder company will also ventilate as it advances being on the alert for signs of fire spread above the fire floor. The ladder company must always be prepared to enter the building upon arrival, this means the donning of positive-pressure SCBA should take place enroute to the incident. Generally, it is desirable to have the ladder company personnel assigned as interior or exterior team members. The officer and one or two members with tools enter to begin the interior search before moving up. Simultaneously, the driver with one or two firefighters raises the necessary ladders to enter or ventilate the building from the outside. For instance, a ground ladder may be used to effect a second floor window rescue while the elevating device is raised to the roof for the purpose of ventilation.

Both interior and exterior teams should search first in areas that are most likely to be inhabited according to the situation. As previously described, searches must be conducted systematically to avoid missing areas.

In addition to search and ventilation procedures, it is often necessary for the ladder company to assist the engine companies in making the attack. This can be done by the placement of ground ladders as requested by the officer in charge. The ladder company can frequently be used to knock down large fires above the first floor with a pre-plumbed aerial master stream device. This action must be coordinated with other operations both to prevent injury to personnel inside and to prevent spreading the fire to uninvolved parts of the building. Interior attack teams have frequently been injured or driven out by poorly directed outside master streams. Elevating devices, ladders, platforms, and ladder towers can also be used as a substitute standpipe from which engine company fire fighting teams can advance hoselines.

Whether fire departments have the resources described above or not, the primary concern should be that each necessary function is assigned to someone whether that be an individual or an engine or ladder company. The rescue, exposure protection, ventilation, confinement, and extinguishment functions must be performed for the operation to be successful. It follows that, if these assignments are made, understood, and practiced by the responsible parties, they will be much more efficient and safe.

Fire Tactics at Protected Properties

Certain buildings are provided with automatic sprinkler protection and standpipe systems.

This is due to the potential for high life loss, process hazard, or construction types. Fire department operating procedures at these occupancies must take into account the necessity of supporting these systems. Particularly, the fire department must establish as a high priority the support of fixed fire suppression systems when:

- Initial attack companies must be immediately committed to victim search or evacuation.
- Streams will be unable to reach areas in large undivided or high-rise occupancies.
- Water is flowing due to sprinkler heads operating.
- Systems are dependent on fire department support for their ability to function.
- Tactics require the use of standpipes that are not tied in to water supplies.
- Standpipes are supplied by private water sources and pressure is provided by private pumps.

Standard operating procedures used at these occupancies are most likely to be incorporated as a part of a pre-fire plan written specifically for the occupancy. These plans include a detailed account of the construction features, contents, protection systems, and surrounding properties. This plan will also outline the procedures to be used by each company according to the conditions that they find. A building map showing water supplies, protection system connections, and company placement should be an integral part of the plan and must be updated to reflect changes affecting fire department operations.

Fires and Emergencies in Confined Enclosures

Fire fighting and rescue operations must often be carried out in locations that are below ground or otherwise cut off from natural or forced ventilation. Basements, caves, sewers, storage tanks, and trenches are just a few examples of these types of areas. The single most important factor in safely operating at these emergencies is recognition of the inherent hazards of confined enclosures. The atmospheric conditions that may be expected include:

- Oxygen deficiencies
- Flammable gases and vapors
- Toxic gases
- Elevated temperatures

In addition, physical hazards may also be present such as:

- Limited means of entrance and egress
- Cave-ins or unstable support members
- Water or other liquids in depth
- Utility hazards - electricity, gas

The wearing of personal protective clothing, particularly positive-pressure SCBA, cannot be overemphasized. Canister-filter masks are useless in oxygen deficient atmospheres and must not be used. When firefighters must enter confined spaces separately from the bottle and harness of SCBA, extreme caution must be exercised not to pull the mask off of the rescuer. Air supply masks are also available that have long supply hoses without the need for bulky tanks.

A lifeline should be tied to each rescuer before entry. This line must be constantly monitored and a standby crew, properly outfitted, equal in number to the rescuers working inside must be available. A system of communication between inside and outside team members should be prearranged as portable radios may prove unreliable. One method of signaling is called the O-A-T-H method. Each letter stands for a signal tug from either end of the lifeline.

TABLE 14.2
LIFELINE SIGNALS

	Exterior	Interior
O - 1 pull	OK?	OK!
A - 2 pulls	Advance Line	Advance Line
T - 3 pulls	Take Up Line	Take Up Line
H - 4 pulls	Help Is On The Way	HELP!

Any signal given, such as a sharp pull on the line, should be acknowledged by the other party on the line. Another method of safe communication is the use of sound-powered phones which do not require a power source. Also, the rescuer must be able to use the selected communication system without removing the SCBA mask.

Communications with supervisors or other knowledgeable persons at the scene is also important as they may be able to give valuable information on hazards that are present, the number of victims and their probable location. Likewise, preincident plans of existing enclosed spaces in the fire department's jurisdiction will reduce guesswork and should be referred to during operations in these locations. The firefighter should be ready to implement prearranged methods of extinguishment or rescue without delay. These plans should include provisions for victim and rescuer protection, control of utilities and other physical hazards, communications, ventilators, and lighting. Power equipment that is used during nonfire rescue operations should be rated for use in explosive atmospheres; this includes flashlights, smoke ejectors, and radios.

As the entrances to these incidents are generally restricted, the establishment of a command post and staging area will be vital to a successful operation. The staging area should be near to but not obstructing the entrance and be supplied with the manpower and equipment to be used. Firefighters should not enter these enclosures until the incident commander has decided upon a course of action and issued specific orders. A safety officer should be stationed at the entrance to keep track of personnel and equipment entering and leaving the enclosure. This officer should check and record the mission, tank pressure, name, and estimated safe working time of each entering member (Figure 14.13). This procedure allows for the accounting of all team members and reduces the possibility of a member being unaccounted for after the safe working time limit has passed.

Once firefighters enter the enclosure, searches or other operations should be performed systematically and with minimum effort. If entrance must be made down stairs, the firefighter should proceed as

Figure 14.13 Entry control sheets provide for the accounting of all members working in hazardous areas.

quickly as it is safe to do so to the bottom of the stairs. Under fire conditions the heat and smoke will be concentrated at the top of the stairs and conditions should improve as the firefighter descends. The fire fighting team can then proceed to search for victims and the seat of the fire with the right-hand or left-hand turn method. Advancing teams should take advantage of opportunities to ventilate as they proceed. Team members must also maintain communication with the command post to report progress, difficulties, or to receive further instructions. Firefighters should not hesitate to evacuate themselves if interior conditions or external reports indicate that structural collapse is imminent. It will be desirable to have other companies locate potential impact load hazards above the fire fighting teams. Fire weakened supports such as unprotected steel beams and columns can be "frozen" if a hoseline is played on them. Firefighters should be aware that unprotected steel supports yield quickly when exposed to temperatures in excess of 600°F (315°C). The longer supports have been subjected to fire the more likely they are to fail, regardless of their composition. In addition, piled stock will become more unstable as fire damage progresses and water is absorbed by the stock. The application of hose streams must be performed with prudent care because of the difficulty of ventilating generated steam. Fires in confined areas may also be attacked indirectly with piercing nozzles, cellar nozzles, distributors, or high expansion foam. While the fire is being attacked special attention should be paid to vertical means of fire spread. Due to the confinement of heat, firefighters may find that they tire more quickly and use their positive-pressure SCBA air supply faster. Firefighters should call for relief before they are exhausted and pay rigid attention to air conservation techniques and pressure gauge valves. Firefighters must not advance into confined spaces farther than their air supplies will allow them safe margins for retreating.

Ground Cover Fire Techniques

Ground cover fires include fires in weeds, grass, field crops, brush, and similar vegetation, excepting forests. Ground cover fires have characteristics of their own that are not comparable to other forms of fire fighting. Local topography, fuel type, and climate present different problems. The local experiences of fire suppression forces and their decided mission will determine the methods and techniques used to control ground cover fires.

Once a ground cover fire starts, burning is generally rapid and continuous. There are many factors that affect ground cover fire behavior, but the three most important are fuel, weather, and topography. Any one of these three may be dominant in influencing what an individual fire will do, but usually the combined strength of all three dictates a fire's behavior.

FUEL

Fuels are generally classified by grouping together fuels with similar burning characteristics. This method classifies ground cover fuels as ground, surface, and crown fuels (Figure 14.14).

- Ground Fuels (duff) - small twigs, leaves, and needles that are decomposing on the ground.

- Surface Fuels - grass, brush, and other low vegetation. Nonliving surface vegetation includes downed logs, heavy limbs, and the like.

- Crown Fuels - suspended and upright fuels physically separated from the ground fuels to the extent that air can circulate freely around the fuels causing them to burn more readily.

Several factors affect the burning characteristics of fuels, such as

- Fuel size - small or light fuels burn faster.

- Compactness - tightly compacted fuels such as the ground or surface type burn slower than the crown type.

- Continuity - when fuels are close together, the fire will spread faster because of the effects of heat transfer. Patchy fuels may spread irregularly or not at all.

- Volume - the amount of fuel present in a given area. The volume will determine the fire's intensity and the amount of water needed to perform extinguishment.

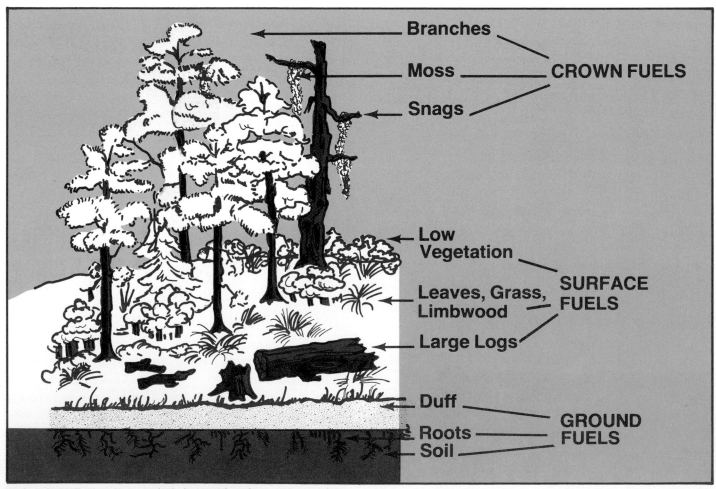

Figure 14.14 Fuel components include ground, surface, and crown fuels.

WEATHER

All aspects of the weather have some effect upon the behavior of a ground cover fire. Some weather factors that influence ground cover fire behavior are:

- Wind - that fans the flames into greater intensity and supplies fresh air that speeds combustion. Medium and large-sized fires may create their own winds.

- Temperature - has effects on wind, is closely related to relative humidity, and primarily affects the fuels as a result of long-term drying.

- Relative Humidity - the greatest impact is on dead fuels that no longer have any moisture content of their own.

- Precipitation - largely determines the moisture content of live fuels. Although dead flash fuels may dry quickly, large dead fuels will retain this moisture longer and burn slower.

TOPOGRAPHY

Topography refers to the lay of the land and has a decided effect upon fire behavior. The steepness of the slope affects both the rate and direction of the spread. Fires will usually move faster uphill than downhill, and the steeper the slope, the faster the fire will move. Other topographical factors influencing ground cover fire behavior are:

- Slope Aspect - is the direction the the slope faces. Full southern exposures (north of the equator) receive more of the sun's direct rays and therefore receive more heat. Ground cover fires typically burn faster on southern exposures.

- Local Terrain Features - directly affect air movements. Obstructions, such as ridges, trees, and even large rock outcroppings may alter air flow and cause turbulence or eddies resulting in erratic fire behavior.

- Canyons - or other wind flow restrictions result in increased wind velocity. Wind movement can be critical in chutes of steep "V" drainages. These terrain features create turbulent updrafts causing a chimney effect. Fires in these chutes or drainages can spread at an extremely fast rate and are very dangerous.

PARTS OF A GROUND COVER FIRE

The typical parts of a ground cover fire are shown in Figure 14.15.

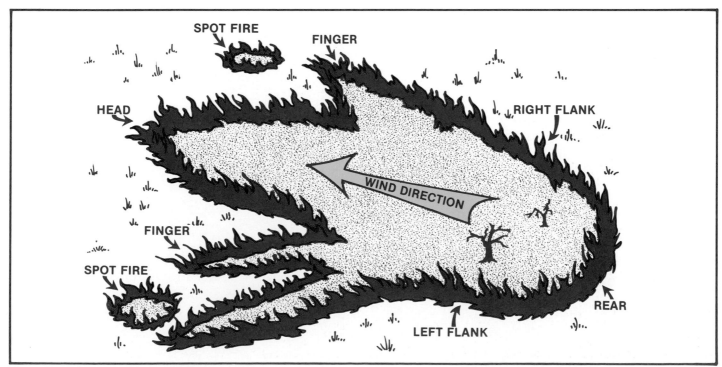

Figure 14.15 Parts of a ground cover fire.

Head. The head is the part of a ground cover fire that travels or spreads most rapidly. The head is usually found on the side of the fire opposite the direction from which the wind is blowing. The head burns intensely and usually does the most damage. To control the head and prevent the formation of a new head is usually the key to the control of the fire.

Finger. Long narrow strips extending out from the main fire are called fingers. They will usually occur when the fire hits an area that has both light and heavy fuels in patches. The light fuel burns faster than the heavy fuel which gives the fingered effect. When not controlled, these fingers form new heads.

Rear. The rear, or heel of ground cover fire is the side opposite the head. The rear usually burns slowly and quietly and is easier to contol. In most cases the rear will be found burning downhill or against the wind.

Flanks. The flanks are the sides of a ground cover fire. The right and left flank separate the head from the rear. It is from these flanks that fingers are formed, which accounts for the importance of controlling them. A shift in the wind direction can change a flank into a head.

Perimeter. The perimeter of a ground cover fire is the boundary of the fire. It is the total length of the outside edge of the burning or burned area. Obviously, the perimeter is ever changing until the fire is suppressed.

Spot Fire. A fire caused by flying sparks or embers landing outside the main fire. Spot fires present a hazard in that personnel and equipment working on the main fire could become trapped between the two fires. These fires must be extinguished quickly or they will form a new head and continue to grow in size.

ATTACKING THE FIRE

The method used to attack ground cover fires revolves around perimeter control. The control line may be established at the burning edge of the fire, next to it, or at a considerable distance away. The objective is to establish fire breaks that completely encircle the fire with all the fuel inside the breaks rendered harmless.

The direct and indirect approaches are the two basic attack methods for attacking ground cover fires (Figure 14.16). The direct method is action taken directly against the flames. The indirect method consists of control techniques applied at varying distances from the advancing fire to halt its progress, and is generally used against fires that are either "too hot," "too fast," or "too big." Since a ground cover fire is constantly changing, it is quite possible to begin with one attack method and end with another. Size up must be continued during the fire so these adjustments can be made when required. The essential rules that firefighters should follow are contained in U.S. Forest Service "Ten Standards of Firefighting Orders."

"Ten Standard Firefighting Orders"

1. Keep informed of fire weather conditions and forecasts.
2. Know what your fire is doing at all times; observe personally and use scouts.
3. Base all actions on current and expected behavior of the fire.
4. Have escape routes for everyone and make them known.
5. Post a lookout when there is possible danger.
6. Be alert, keep calm, think clearly, act decisively.
7. Maintain prompt communication with your personnel, your boss, and adjoining forces.
8. Give clear instructions and be sure they are understood.
9. Maintain control of personnel at all times.
10. Fight fire aggressively, but provide for safety first.

Fighting ground cover fires is a very dangerous occupation. Many firefighters have lost their lives or have been seriously injured while trying to control these fires. Thoroughly think out the situation, then do what will most likely be correct for that situation. Remember: The safety of personnel and equipment always comes first.

DIRECT ATTACK

DIRECTION OF ATTACK

INDIRECT ATTACK

FIRE IS
INDIRECTLY
CONTROLLED

BY BURNING
OFF
INTERVENING
FUEL

FROM A
PLANNED
CONTROL
LINE

Figure 14.16 The direct and indirect approaches are the two basic methods for attacking ground cover fires.

NFPA STANDARD 1001
COMMUNICATIONS
Fire Fighter I

3-14 Fire Alarm and Communications

3-14.1 The fire fighter shall define the procedure for a citizen to report a fire or other emergency.

3-14.2 The fire fighter shall demonstrate receiving an alarm or a report of an emergency, and initiate action.

3-14.3 The fire fighter shall define the purpose and function of all alarm-receiving instruments and personnel-alerting equipment provided in the fire station.

3-14.4 The fire fighter shall identify traffic control devices installed in the fire station to facilitate the response of apparatus.

3-14.5 The fire fighter shall identify procedures required for receipt and processing of business and personal calls.

3-14.6 The fire fighter shall identify prescribed fire department radio procedures.

3-14.7 The fire fighter shall define policy and procedures concerning the ordering and transmitting of multiple alarms of fire and calls for special assistance from the emergency scene.

3-14.8 The fire fighter shall define all fire alarm signals, including multiple alarm and special signals, governing the movements of fire apparatus, and the action to be taken upon the receipt of each signal.*

Fire Fighter II

4-15 Fire Alarm and Communications

4-15.1 The fire fighter shall identify areas assigned for first-alarm response.

4-15.2 The fire fighter shall demonstrate both mobile and portable radio equipment.

4-15.3 The fire fighter shall identify fire department radio procedures.

4-15.4 The fire fighter shall identify supervisory alarm equipment provided in the fire station and the prescribed action to be taken upon receipt of designated signals.

4-15.5 The fire fighter shall identify fire location indicators provided to direct fire fighters to specific locations in protected public or private properties.*

IFSTA's Communications Transparencies are designed to complement this chapter.

Chapter 15
Communications

Fire department communications include all the methods by which the public can notify the fire department communication center of an emergency and all the methods by which the center can notify the proper fire fighting forces and then relay information between all personnel involved at the scene.

THE COMMUNICATIONS CENTER

The communications center houses the equipment and personnel to receive alarms and dispatch apparatus. It may be located in a fire station or in a separate building (Figure 15.1). According to the NFPA standards, a variety of equipment and methods may be used to meet a department's communications needs:

- Automatic receipt and recording of box alarms
- Visual and audible indication of alarm as to location
- Automatic recording of time and location of alarm

The building or portion of the building used as the communications center should not be near high-risk fire hazards. Combustible materials should not be permitted in its construction except floor covering laid directly upon noncombustible base. All lights should not be on a single branch line fuse. Adequate emergency power is essential, including a reliable secondary source independent of other sources which can apply its power to the line within 10 seconds.

Figure 15.1 Communications centers in smaller cities are usually in fire stations while larger cities may have separate facilities.

Figure 15.2 Larger communications centers require a minimum of two operators on duty continuously.

For municipalities receiving more than 2,500 alarms per year, NFPA standards require at least two fully trained operators on duty at all times (Figure 15.2). Actually, more may be needed. In smaller areas where only one operator is required, a major incident may overload the abilities of a single operator. This same possibility also exists in large municipalities with two or more operators.

One method of efficiently handling the calls that may be received by any number of dispatchers is to use a recording and retrieval system (Figure 15.3). Equipment is available to log both phone and radio transmissions protecting both citizens and the department by:

- Providing an instant recheck of a hurriedly given name or location to allow the proper dispatch of personnel and equipment

- Providing a sequentially-timed record of call receipt, dispatch instructions, and apparatus response as proof of proper disposition of an emergency situation

- Providing, with one certain unit, a source for voice print analysis

REPORTING A FIRE

Many properties have been destroyed — and lives lost — through delays in calling the fire department immediately after a fire has been discovered. At other times when there is no intended time delay, traveling motorists have spotted fires and stopped to phone in the alarm only to find that they cannot accurately describe the location. Every department has heard from the caller who

Figure 15.3 Recording and retrieval equipment is available for both telephone and radio communications to assure no loss of messages.

quickly said, "Come quick, my house is on fire!" and hung up without giving an address.

Every fire department should include, in its fire safety education program, information on how to report a fire correctly. The following information should be included:

- Telephone
  - Dial Number
    Fire Department Direct
    911
    "O"perator
  - Report Type of Incident
  - Address
  - Cross Street
  - Your Name and Location
  - Callback Number

- Fire Alarm Box
  - Send Signal
  - Stay at Location

- Fire Alarm Station
  - Send Signal
  - Notify Fire Department

METHODS OF RECEIVING ALARMS FROM THE PUBLIC

Fire alarms may be received in several different ways. This may include municipal alarm systems, telephone, radio, and persons walking into a fire station to report an emergency. Alarms may also be received by private alarm systems.

Municipal Alarm Systems

WIRED TELEGRAPH CIRCUIT BOX

This alarm system is operated by pressing a lever in the alarm box which starts a wound spring mechanism. The rotating mechanism transmits a code by opening and closing the circuit. Each box transmits a different code to specify its location. Wiring may be overhead or underground. The system is limited in that no other information except location is transmitted, and malicious false alarms are a problem (Figure 15.4).

TELEPHONE FIRE ALARM BOX

A telephone is installed in the fire alarm box for direct voice contact, allowing for exchange of more information on the type of response needed.

Figure 15.4 The telegraph fire alarm box has a coded wheel that transmits a signal when activated.

This system is generally leased from the telephone company and each box is connected to the communication center by a separate circuit of telephone company wire. In one type when the handset is lifted from its cradle, a buzzer sounds and a light appears on the communication center's alarm display panel identifying the location of the box visually (Figure 15.5).

Figure 15.5 Telephone type fire alarm boxes are used for direct voice transmission to the communications center. Lifting the handset indicates the box location.

If the number of these boxes is unusually large, causing an increased panel size, some departments have used the combination telegraph and telephone type circuits. This would give the best of both systems where the basic pull-down hook is used to send the coded signal and a telephone is included for additional use.

RADIO FIRE ALARM BOX

A radio alarm box contains an independent radio transmitter with battery power supply (Figure 15.6). Solar recharging is available for some systems. Others feature a spring wound alternator to provide power when the operating handle is pulled down. There are different types of radio boxes. Activating the alarm in radio boxes alerts the fire department dispatcher by an audible signal, visual light indicator, and a permanent record indicating the location. Some models have, besides a red alarm light indicator, a different light color indicator that signals a test or tamper signal. By using a time clock within the box, it can test itself every 24 hours. If the box pole is struck or tampered with, the tamper light comes on and gives the box location. Some boxes are numbered and this number also appears on the display panel, informing the dispatcher of the box involved and its location. Some printing systems when activated by the incoming radio signal, print the day of the year, time of day in 24 hour time, the message sent by the box, the box number, and a coded signal that indicates the strength of the battery within the box. Some radio alarm boxes are so designed that a person can select the need of fire, police, or ambulance service.

Telephone

The public telephone system performs a very valuable function in reporting fires. NFPA Standard No. 1221, *Standard for the Installation, Maintenance and Use of Public Fire Service Communications,* says that the alarm communication center should have at least one telephone line and in larger municipalities additional lines depending upon the traffic handled.

Due to availability today public telephones are the most common form of turning in a fire alarm. Many business calls will also come in on the

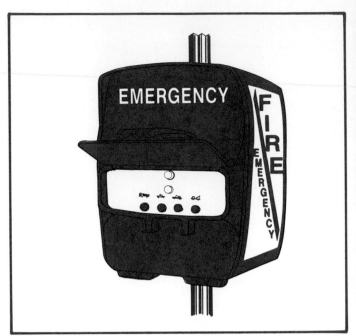

Figure 15.6 Radio alarm boxes transmit their location by an independent transmitter and are not restricted by wired circuits.

public telephones. For this reason it is important to know the correct procedure for processing business calls. The following list indicates some of these procedures.

- Answer calls promptly.
- Identify department, company, and self to help avoid wrong numbers.
- Be prepared to take messages.
- Take accurate messages by including date, time, name of caller, caller's number, message, and your name.
- Never leave the line open or someone on hold.
- Post the message or deliver the message to the person it's intended for promptly.
- Terminate calls courteously.

Some fire alarms have been delayed by confusion when several fire departments were in the same telephone area. Now telephone authorities supply one easily remembered number, "911," for emergencies when requested by local governments. More and more cities are also using the number, but the system has its disadvantages. Police dispatchers commonly receive those calls, and delays in relaying the calls are still reported. One advantage is that the simple number can be

Figure 15.7 Stickers can be placed on the telephone to assist callers when under stress.

dialed in the dark. Emergency telephone number stickers that may be placed directly on the telephone help reduce time delays when calling the fire department (Figure 15.7).

Radio and Walk-In

Fire alarms may be received over the radio from fire department or police department person-nel out on the street who see a fire. People may also walk in to a fire station to report an emergency (Figure 15.8). When incidents are reported in this manner be sure to collect and record all of the information that is needed and notify the dispatcher before alerting the company to respond.

Private Fire Alarm Signaling Systems

Besides municipal alarm systems, private protective signaling systems are used to detect and transmit alarms to a fire department communications center. These systems can be used to:

- Notify occupants to evacuate the premises.
- Summon the fire department or other organized assistance.
- Supervise extinguishing systems to assure their operability when needed.
- Supervise industrial processes to warn of abnormalities that may contribute to a fire hazard.
- Supervise personnel to assure performance of assigned duties.
- Actuate fire control equipment.

Walk-In Report of an Emergency

- **Record Information**
- **Notify the Dispatcher**
- **Alert the Company**
- **Respond**

Figure 15.8 When a fire is reported directly to the fire station record all the data necessary to respond the first due companies.

Types of private signaling systems are:

- Central station protective signaling systems
- Local protective signaling systems
- Auxiliary protective signaling systems
- Remote station protective signaling systems
- Proprietary protective signaling systems
- Household fire warning systems

CENTRAL STATION PROTECTIVE SIGNALING SYSTEMS

A central station system is a commercial fire alerting system or group of systems that signal the alarm to a central station. This agency has trained and experienced personnel continually on duty to receive signals, retransmit fire alarms to the fire department, and take whatever action the supervisory signals indicate is necessary.

LOCAL PROTECTIVE SIGNALING SYSTEMS

An alarm or supervisory signal operates in the protected premises and is primarily for the notification of occupants and supervisory personnel. The fire department should be notified anytime a system is activated, even if the occupancy has a fire brigade and even if the fire is small.

AUXILIARY PROTECTIVE SIGNALING SYSTEMS

An auxiliary alarm system is, of itself, insufficient for notifying the fire department, but in combination with a municipal system does summon a response. The alarm devices are usually owned and maintained by the property owner, but the connecting facilities between the protected property and the fire department are part of the municipal fire alarm system. Areas protected are usually public buildings such as schools and hospitals through a master box.

Visual indicators are provided at a location acceptable to the fire department to further direct firefighters to the precise location of an alarm. This is especially necessary in a large complex of warehouses where at times no personnel are on duty to direct them. Indicators connected to all the alarm signal devices must show:

- Particular building of the complex
- Which floor of the building
- Which section of that floor

REMOTE STATION PROTECTIVE SIGNALING SYSTEMS

Remote station protective signaling systems are installed to protect private premises. Generally, these systems are installed and maintained by a commercial central station company. However, instead of the alarm going to the commercial central station, it goes to a panel at a public fire alarm office or fire station by a direct circuit where operators are available to act on the alarm (Figure 15.9).

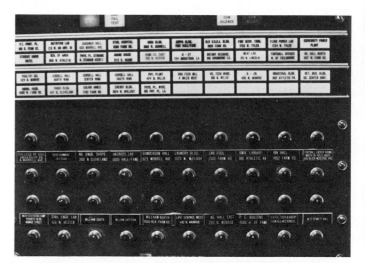

Figure 15.9 Remote station signaling systems send alarms from private premises directly to the public fire alarm communications center.

PROPRIETARY PROTECTIVE SIGNALING SYSTEMS

These are systems protecting large plants and installations which maintian their own proprietary central supervising station at the property protected (Figure 15.10). From this supervising station alarms are retransmitted manually or automatically to the fire department alarm office or fire station. Personnel in the supervising station can call the department after investigation to give information about gate number, type of fire, or building number.

METHODS OF ALERTING FIRE DEPARTMENT PERSONNEL

There are a variety of methods for alerting firefighters of an emergency. The particular way that fire stations and personnel are alerted will depend on whether or not the station is manned (Figure 5.11).

Alerting Manned Stations

Technological advances have brought about new, modern alerting systems to accompany the more traditional types. These types of systems include:

- Computerized lineprinter
- Vocal alarm
- Teletype
- House bell or gong
- House light
- Telephone from dispatcher
- Telegraph register
- Radio
- Walk-in
- Private alarm system

Alerting Unmanned Stations

Devices to alert firefighters are quite common

Figure 15.10 Proprietary signaling systems are privately owned and maintained, and retransmit signals to the public alarm center.

in their home or their possession. Some departments also use outside alerting systems. These systems include:

- Pagers
- Telephones
- Sirens
- Whistles or air horns
- Electro writer

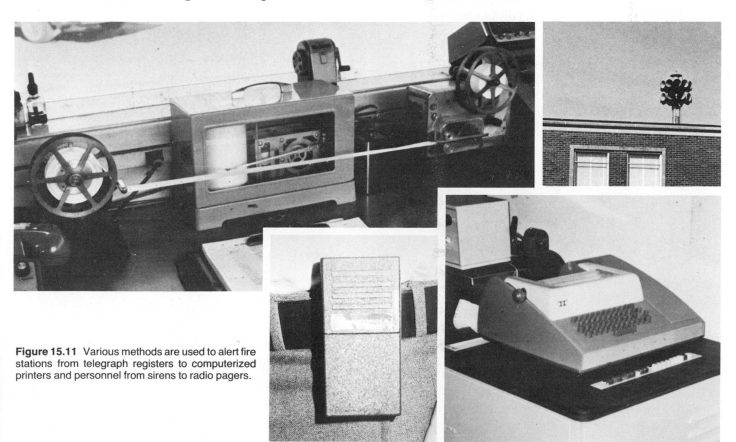

Figure 15.11 Various methods are used to alert fire stations from telegraph registers to computerized printers and personnel from sirens to radio pagers.

TRAFFIC CONTROL DEVICES

Traffic control devices are used to interrupt or stop traffic until the apparatus has cleared the station or an intersection. These traffic control devices include manually, light, and siren activated devices.

Manually Activated Devices

Traffic control devices of this type operate by a switch that is manually held until the apparatus has cleared the area. The switch may be set up on an automatic timer that interrupts traffic flow until after the apparatus responds, then returns to the normal cycle.

Light Activated Device

This type of traffic control device is activated by an apparatus mounted, pulsating, high intensity light (Strobe) that sends a signal to a detector located at each protected intersection (Figure 15.12). The detector then activates the traffic light selector in the control box which holds the traffic light green or speeds up the normal cycle to green in the desired direction of travel. An indicating light, located next to the light detector, assures the driver the traffic signal is actually under the control of the system.

Figure 15.12 Traffic lights can be modified for control from responding apparatus equipped with a high intensity light signaling device.

Siren Activated Device

The siren of the apparatus activates this traffic control device. A sound pick-up unit is located at each protected intersection. This unit filters out all other noise except the siren and sends a signal to the traffic light selector in the control box. The traffic light selector holds the amber light on for a few seconds and then switches to red which flashes at double the normal rate.

RADIO PROCEDURES

All radio communication in the United States is under authorization from the Federal Communications Commission (FCC). Over 7,000 fire departments and fire service organizations have been issued fire radio licenses from the FCC. The actual number of departments is higher because some licenses to some areas cover several individual departments using one communication center.

There should be a certain amount of instruction on the local level in the correct use of the radio. This would include becoming familiar with the equipment and how it operates. Correct voice procedures, which would include tone and speed of transmission, are needed. Department operating procedures such as department codes, test procedures, and time limits on radios need to be established. Local department rules should specify who is authorized to transmit on the radio. It is a federal offense to send personal or nonfire service messages over the radio.

Other important considerations that should be remembered include:

- Avoid unnecessary transmissions. Be brief, accurate, and to the point.

- Do not transmit until determining if the air is clear.

- Any unit working at a fire or rescue has priority over any other transmission.

- Do not use profane or obscene language on the air.

- Hold the microphone 1 inch to 2 inches (2.5 to 5 cm) from the mouth. Do not shout. Pitch the voice high rather than low.

- Speak calmly, clearly, and distinctly in a natural conversational rhythm at medium speed.

- Avoid laying the microphone on the seat of the vehicle where the switch may be pressed to cause interference.

- Do not touch the antenna when transmitting. Radio frequency burns might result.

ARRIVAL AND PROGRESS REPORTS

Whether or not codes or verbal descriptions are used, first arriving companies should use the radio to provide a description of the conditions found at the scene. This establishes a time of arrival and allows other responding units to anticipate what actions might be taken upon their arrival.

A typical arrival report might sound like this:

"Engine 25 to fire control. Light smoke showing from three-story frame store and dwelling at 101 Balsam Avenue."

or like this:

"Engine 15 to fire control. Heavy fire showing at 101 Balsam Avenue. Engine 14 laying two lines at the rear. Have Engine 20 bring in two more lines to the front. This is a working fire."

When giving a report of conditions upon arrival, the following information should be included:

- Report arrival and address if other than the one reported.

- Building and occupancy description.

- Nature and extent of fire, attack mode selected.

- Rescue and exposure problems.

- Instructions to other responding units.

- Location of incident command position.

Once fire fighting operations have begun it is important that the communications center be kept advised of the actions taken at the emergency scene. Such progress reports should include:

- Indicate command officer.

- Indicate change in command location.

- Progress (or lack of it) in situation control.

- Direction of fire spread.

- Exposures by direction, height, occupancy, and distance.

- Indicate problems.

- Describe needs.

- Anticipated actions - holding, doubtful.

CALLS FOR ADDITIONAL RESPONSE

At some fires it may be necessary for additional units to be called. Normally, only the ranking officer at the fire may order multiple alarms or additional response. This may be varied by local conditions. For example, when a single company sent to investigate an unknown fire discovers a working fire, the communication center might go ahead and start additional help, depending on the arrival report.

All firefighters need to know the local procedure for requesting additional alarms. They must also be familiar with the alarm signals, such as for a multiple or special alarm and what to do when they are received. Personnel should know the number and types of units that will respond to these alarms.

When multiple alarms are given for a single fire, maintaining communications with each unit becomes more difficult as the radio traffic increases. To reduce the load on the communication center, a mobile, radio-equipped, command vehicle can be used at large fires (Figure 15.13).

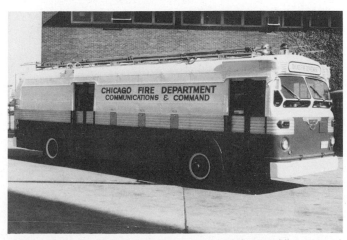

Figure 15.13 For large emergency scene operations mobile communications/command vehicles are used. *Courtesy of Chicago Fire Dept.*

NFPA STANDARD 1001
AUTOMATIC SPRINKLER SYSTEMS
Fire Fighter I

3-13 Sprinklers

3-13.1 The fire fighter shall identify a fire department sprinkler connection and water motor alarm.

3-13.2 The fire fighter shall connect hose line(s) to a fire department connection of a sprinkler or standpipe system.

3-13.3 The fire fighter shall define how the automatic sprinkler heads open and release water.

3-13.4 The fire fighter shall temporarily stop the flow of water from a sprinkler head.*

Fire Fighter II

4-14 Sprinklers

4-14.1 The fire fighter shall identify the *Main Drain* valve on an automatic sprinkler system.

4-14.2 The fire fighter shall open and close a *Main Drain* valve on an automatic sprinkler system.

4-14.3 The fire fighter shall identify the *Main Control* valve on an automatic sprinkler system.

4-14.4 The fire fighter shall operate a *Main Control* valve on an automatic sprinkler system from "open" to "closed" and then back to "open."

4-14.5 The fire fighter shall define the value of automatic sprinklers in providing safety to life of occupants in a structure.

4-14.6 The fire fighter shall identify and define the dangers of premature closure of sprinkler *Main Control* valve, and of using hydrants to supply hose streams when the same water system is supplying the automatic sprinkler system.

4-14.7 The fire fighter shall identify the difference between an automatic sprinkler system that affords complete coverage and a partial sprinkler system.

4-14.8 The fire fighter shall identify at least three sources of water for supply to an automatic sprinkler system.

4-14.9 The fire fighter shall identify the following:
   (a) Wet sprinkler system
   (b) Dry sprinkler system
   (c) Deluge sprinkler system.

4-14.10 The fire fighter shall demonstrate removing one head from a sprinkler system and replacing it with a head of the same type.*

*Reprinted by permission from NFPA Standard No. 1001, *Standard for Fire Fighter Professional Qualifications*. Copyright © 1981, National Fire Protection Association, Boston, MA.

IFSTA's Automatic Sprinkler Systems Transparencies are designed to complement this chapter.

Chapter 16
Automatic Sprinkler Systems

Early types of sprinkler systems were rather crude and unreliable, but present-day systems have been perfected to the point that they are extremely reliable when properly maintained. The reduction of insurance rates for property that is equipped with sprinkler protection has been a very influential factor in growth of the number of installations in properties.

The automatic sprinkler heads and all component parts of the system should be listed by a nationally recognized testing laboratory, such as Underwriters Laboratories or Factory Mutual. Automatic sprinkler systems are now recognized as the most reliable of all fire protection devices and an understanding of the system of pipes and valves and their operation is essential to the firefighter.

Automatic sprinkler protection consists of a series of devices so arranged that the system will automatically distribute sufficient quantities of water to either extinguish a fire or to hold it in check until firefighters arrive. Water is supplied to the sprinkler heads through a system of piping. The sprinkler heads can either extend from exposed pipes or protrude through the ceiling or walls from hidden pipes.

The process of spacing sprinkler heads in a building must conform to well-established and tested standards, such as NFPA Standard No. 13, *Installation of Sprinkler Systems*. Standards are also set for the size of pipe to be used, the proper method of hanging the pipe, and all other details concerning the installation of a sprinkler system. The design of automatic sprinkler systems is based upon the assumption that only a portion of the sprinkler heads will be opened during a fire. Most

ordinary public waterworks systems could not be expected to adequately supply 500 or 1,000 operating sprinklers. This statement emphasizes the fact that a sprinkler system must be properly designed and an adequate source of water must be provided. Water volume and pressure must be adequate for the number of sprinkler heads that are calculated to operate at one time in any given building.

In general, reports reveal that only in rare instances do automatic sprinkler systems fail to operate. When failures are reported, the reason for failure is usually due to a lack of water or from the water supply being shut off instead of failure of the actual sprinklers.

VALUE OF SPRINKLER SYSTEMS FOR LIFE SAFETY

The life safety of building occupants is enhanced by the presence of a sprinkler system because it discharges water directly on the fire while it is relatively small. Since the fire is extinguished or controlled in an early stage, the combustion products are limited. Most fire fatalities in a sprinklered building are caused by asphyxiation either because of a small fire which does not generate sufficient heat to fuse a sprinkler, or because the victim had suffered fatal injuries by the time the sprinkler operated. Other fatal instances include sleeping, intoxicated, or handicapped persons and ignition of clothing or bedding which causes fatal burns; however, the sprinkler system protected the lives of persons in other parts of the building.

An individual's life safety in a sprinklered high-rise building is increased many times. The probability of occupant rescue in an unsprinklered building is lower because of manual fire suppression.

A sprinkler system piping layout, as shown in Figure 16.1, consists of different size pipe. The system starts with a feeder main into the sprinkler valve. A riser is a vertical pipe supplying the sprinkler system, generally a one-way check valve is installed. System piping decreases in size from the riser outward. The pipes connecting the riser to the cross-mains are known as the feed mains. The cross-main directly services a number of branch lines on which the sprinklers are installed. Cross-mains extend past the last branch lines and are capped to facilitate flushing. The entire system is supported by hangers and clamps. All pipes in dry systems are pitched to help drain the system back toward the main drain.

SPRINKLER HEADS

Automatic sprinklers, often called "sprinkler heads" or just "heads," discharge water after the release of a cap or plug which is activated by some heat-responsive element. This head may be thought of as a fixed-spray nozzle that is operated individually by a thermal detector. There are numerous types and designs of sprinkler heads.

Installed heads in wet- and dry-pipe systems are kept in a closed position by various devices. Four of the most commonly used release mechanisms are fusible links, glass bulbs, chemical pellets, quick response, all of which fuse or open in response to heat. (Figure 16.2).

Fusible-Link Head. The design of the fusible-link head involves a frame which is screwed into the sprinkler piping. Two levers press against the frame and against a cap over the orifice in the frame holding the water back. The fusible link holds the levers together until the link is melted during a fire, after which the water pushes the levers and cap out of the way and strikes the deflector on the end

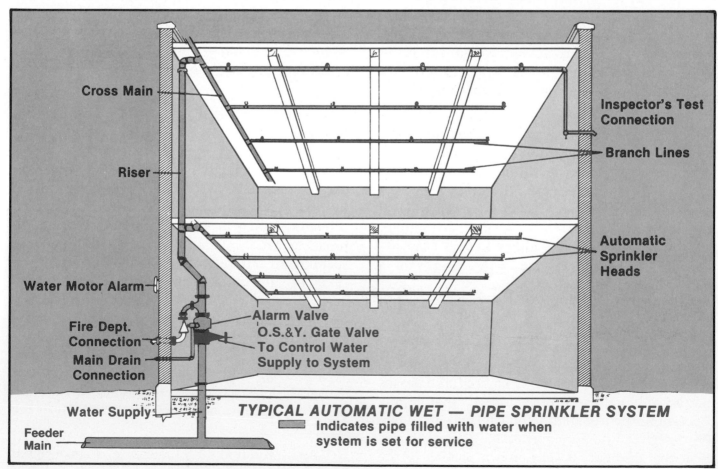

Figure 16.1 These are the components of a typical wet-pipe automatic sprinkler system.

of the frame. The deflector converts the standard half-inch (33 mm) stream into water spray for more efficient extinguishment (Figure 16.3).

Pellet-Type. A pellet of solder under compression within a small cylinder melts at a predetermined temperature, allowing a plunger to move down and release the valve cap parts.

Bulb-Type. A small bulb partially filled with liquid supplants the levers and fusible link of that type. Heat expands the liquid until the bulb shatters at the proper temperature depending on the amount of liquid in the bulb. When the bulb shatters, the valve cap is released. The quantity of liquid in the bulb determines when it will shatter.

Quick-Response Type. The quick-response head was developed for life safety purposes. The fusible link offers increased surface area to collect the heat.

Standard sprinkler heads are designed to discharge a spray of water downward in a hemispherical pattern. The upright-type standard sprinkler heads cannot be inverted for use in the hanging or pendant position because in the inverted position the spray of water would be directed toward the ceiling due to the design of the deflector. Pendant-

Figure 16.2 Automatic sprinklers have different mechanisms that release under fire conditions. These include fusible links, quick response fusible links, glass bulbs, and chemical pellets.

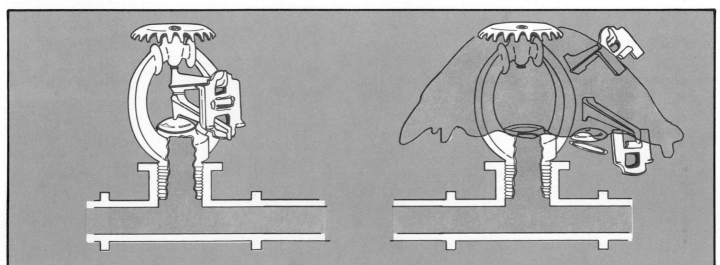

Figure 16.3 As the fusible link melts, the levers force it apart and the water pushes the gasket and cap aside.

Figure 16.4 Standard sprinklers are designed to discharge the water spray downward and are either the upright type or the pendant type that hangs from the supply pipe.

type sprinklers should be used in locations where it is impractical to use sprinklers in an upright position (Figure 16.4).

Special-type sprinkler heads with corrosive-resistant coatings are available and should be installed in areas where chemical, moisture, or other corrosive vapors exist. Without this special coating, the operational parts of the head will corrode and may become inoperative in a very short time.

A storage cabinet should be installed in the area protected by the sprinkler system to house extra heads and a sprinkler head wrench. Normally these cabinets hold a minimum of six heads for small systems. It is necessary to use a sprinkler wrench and to be careful when changing heads to prevent damage (Figure 16.5).

SPRINKLER TEMPERATURE RATINGS

The sprinkler head used for a given application should be based on the maximum temperature expected at the level of the sprinkler under normal conditions, and anticipated rate of heat release that would be produced by a fire in the particular area. The temperature rating shall be indicated by color coding the frame arms of the head (Figure 16.6) except for coated sprinklers and decorative heads (Table 16.1). Coated sprinklers have colored frame arms, coating material, or use a colored dot on the top of the deflector. Decorative sprinklers such as plated or ceiling sprinklers are not required to be color coded; however, some manufacturers use a dot on the top of the deflector.

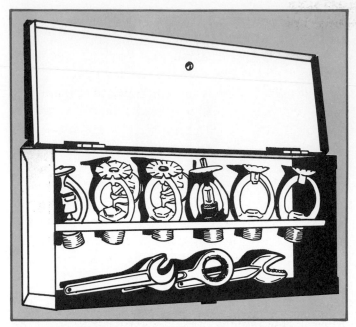

Figure 16.5 Replacement sprinklers and a sprinkler wrench should be available in a storage cabinet.

Figure 16.6 Sprinkler frames are color coded to indicate their operating temperature range.

TABLE 16.1 SPRINKLER COLOR CODES		
TEMPERATURE RATINGS°F.	**TEMPERATURES CLASSIFICATION**	**FRAME COLOR**
135 to 170 (57.2 to 76.7°C)	Ordinary	Unpainted (or Partly black or chrome)*
175 to 225 (79.4 to 107.2°C)	Intermediate	White*
250 to 300 (121 to 148.9°C)	High	Blue
325 to 375 (162.8 to 190.6°C)	Extra High	Red
400 to 475 (204.4 to 246.1°C)	Very Extra High	Green
500 to 575 (260 to 301.7°C)	Ultra High	Orange

*The 135° sprinklers of some manufacturers are half black and half painted.
The 175° sprinklers of these same manufacturers are yellow.

CONTROL AND OPERATING VALVES

Every sprinkler system is equipped with a main water control valve and various test and drain valves. Control valves are used to cut off the water supply to the system when heads must be replaced, when maintenance is performed, or when operation must be interrupted. The main control valve should always be returned to the open position after maintenance is completed. These valves are located between the source of water supply and the sprinkler system (Figure 16.7). These control valves are an "indicating" type and are manually operated. An indicating control valve is one that shows at a glance whether it is open or closed.

These valves are usually located immediately under the sprinkler alarm valve, the dry-pipe or deluge valve, or outside of the building near the sprinkler system that it controls. Separate control valves are required for each system. Each valve should be secured in the open position or supervised.

There are three common types of indicator control valves used in sprinkler systems. One of these valves is an outside screw and yoke valve, usually called an OS&Y valve. This valve has a yoke on the outside with a threaded stem which controls the opening and closing of the gate. The threaded portion of the stem is out of the yoke when

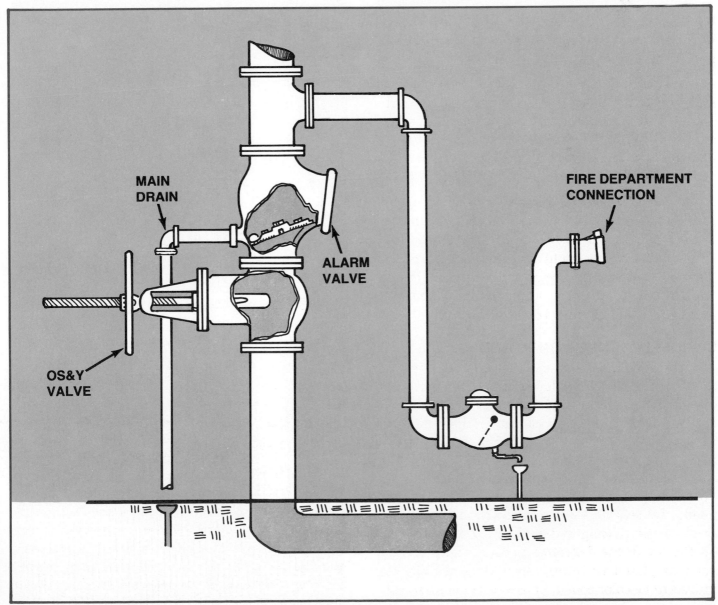

MAIN DRAIN

FIRE DEPARTMENT CONNECTION

ALARM VALVE

OS&Y VALVE

Figure 16.7 The main control valve is between the water supply and the sprinkler system.

Figure 16.8 The outside screw and yoke valve is an indicating control valve, stem out is open, stem in closed.

Figure 16.9 A post indicator valve is a gate type control valve that has a target indicating "OPEN" and "SHUT."

the valve is open and inside the yoke when the valve is closed (Figure 16.8).

Other types of control valves used are the post indicator valve (PIV), the wall post indicator valve (WPIV), and the post indicator valve assembly (PIVA). The PIV is a hollow metal post that is attached to the valve housing. The valve stem is inside of this post; on the stem is a movable target on which the words "OPEN" and "SHUT" appear at the opening. A PIV with the operating handle in both the stored and locked position is shown in Figure 16.9. WPIV is similar to a PIV except that it extends through the wall with the target and valve operating nut on the outside of the building. A PIVA is similar to the PIV except that the valve used is a butterfly type (Figure 16.10) while the PIV and the WPIV use a gate valve.

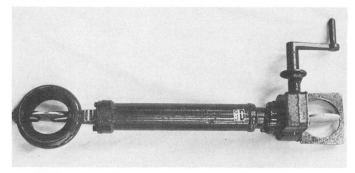

Figure 16.10 A post indicator valve assembly with a butterfly valve rather than a gate valve.

In addition to the main water control valves, sprinkler systems will employ various operating valves such as globe valves, stop or cock valves, check valves and automatic drain valves. The alarm test valve is located on a pipe which connects the supply side of the alarm check valve to the retard chamber. This valve is provided to stimulate actuation of the system by allowing water to flow into the retard chamber and operate the waterflow alarm devices.

WATERFLOW ALARMS

Activation of fire alarms is accomplished by the operation of the alarm check valve, dry-pipe valve, or deluge valve. Sprinkler waterflow alarms are normally operated either hydraulically or electrically to warn of a waterflow. The hydraulic alarm is a local alarm used to alert the personnel in a sprinklered building or a passerby that water is flowing in the system (Figure 16.11). This type of alarm uses the water movement in the system to branch off to a water motor, which drives a local alarm gong. The electric waterflow alarm is also employed to alert building occupants and, in addition, it can be arranged to notify the fire department. With this type of alarm the water movement presses against a diaphragm which in turn causes a switch to operate the alarm.

Figure 16.11 A waterflow alarm sounds when water is moving through the system. This is a water motor operated local alarm.

WATER SUPPLY

Every sprinkler system should have an automatic water supply of adequate volume, pressure, and reliability. In some instances, a second independent water supply is not only desirable but required. A minimum water supply must be able to deliver the required volume of water to the highest sprinkler head in a building at a residual pressure of 15 psi (100 kPa). The minimum flow is established by the hazard to be protected and is dependent upon the occupancy and fire loading conditions. A connection to a public water system that has adequate volume, pressure, and reliability is a good source of water for automatic sprinklers. This type of connection is often the only water supply. A gravity tank of the proper size also makes a reliable primary water supply. In order to give the minimum required pressure, the bottom of the tank should be at least 35 feet (11 m) above the highest sprinkler head in the building. Pressure tanks are another source of water supply and they are used in connection with a secondary supply.

Pressure tanks are normally located on the top floor or on the roof of buildings. This type of tank is filled two-thirds full with water and it carries an air pressure of at least 75 psi (520 kPa). Adequate fire pumps that take suction from large reservoirs are used as a secondary source of water supply. When properly powered and supervised, these pumps may be used as a primary source of water supply.

One or more fire department connections through which the fire department can pump water into the sprinkler system may serve as an auxiliary water supply source (Figure 16.12). Under most conditions, especially when a large number of sprinklers are open or other water

Figure 16.12 Fire department connections are used to supplement the water supply to the sprinkler system.

sources are weak, it is very important for firefighters to connect pumpers to these connections in order to boost the water volume and pressure. On a single-riser system, the fire department connection is attached on the sprinkler-system side of the main water supply valve. On multiple-riser systems, the fire department connection enters the supply piping between the main supply shutoff and the individual riser shutoffs. This design is so that in case the main water supply valve is inadvertently or otherwise closed, water can still be pumped into the system. On multiple systems, however, if sectional or floor valves are closed, the support of the fire department connection will be negated in that area. Fire department support through these connections is important enough to require the first or second pumper responding to hook up to these connections.

In addition, a check valve is installed in the feeding water main to prevent fire department pumpers from pumping back into the supply main instead of into the system. Clapper valves in the fire department connection prevent water from flowing out of the fire department connection when only one hoseline is connected. The check valve and ball drip in the fire department connection line are installed to keep the fire department connection dry to prevent freezing.

Most sprinkler systems are designed on the premise that a fire can be controlled by the operation of only a few sprinkler heads. If a large number of sprinklers are opened by the fire, the regular water supply may be inadequate and the flow from each sprinkler head may be so small that it is ineffective. Fire department pumpers operating from hydrants in the area may also further reduce the pressure in mains and rob the system of water. Whenever possible, fire department pump-

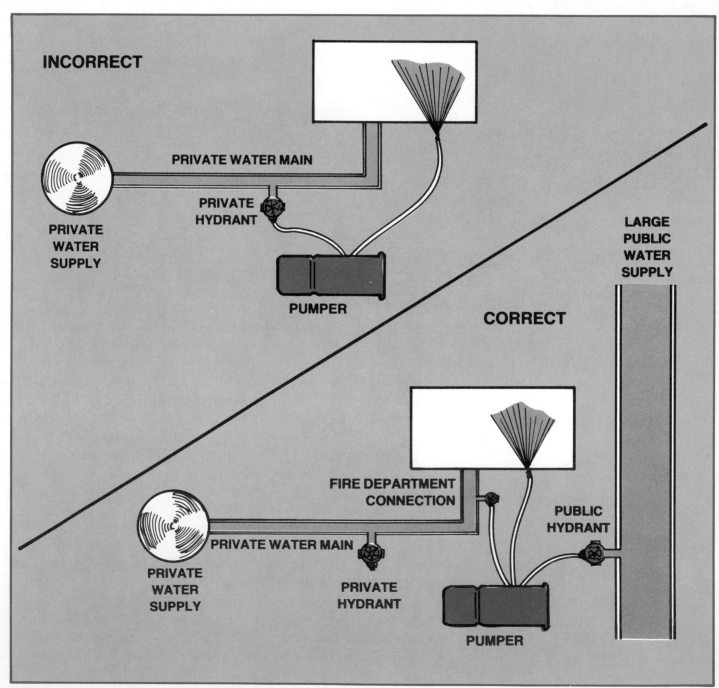

Figure 16.13 Fire departments pumpers must be connected to water supplies that do not reduce the flow to the sprinkler system.

ers other than those supporting the system should operate from mains other than the primary water supply main for the system (Figure 16.13).

Fire department support of standpipes or automatic sprinkler systems should include a standard plan of operation. Develop 150 psi (1000 kPa) at the pumper and maintain this pressure if possible. Circumstances, such as those which exist in high-rise buildings or with deluge systems, may warrant a different pump discharge pressure. Such a plan cannot be established until fire department personnel become familiar with sprinklered properties under their jurisdiction. This should cover which buildings, the type of occupancy, type of system, and the extent of the system. Therefore, an inspection survey is a prerequisite. A thorough knowledge of the water system is important, including volume and pressure.

Important factors to be considered during fire fighting operations at sprinklered property are:

- In addition to normal fire fighting operations, an early arriving pumper should connect to the sprinkler system siamese in accordance with the pre-fire plan. The capacity of the pipe system and the discharge from sprinkler heads can be improved by increasing the pressure on the system. It is recommended that pressure at the pumpers should be in the vicinity of 150 psi (1000 kPa) unless hoselines serving the sprinkler siamese are in excess of 100 feet (30 m) in length. In this event, somewhat higher pressure will be needed. A maximum effort should be made to supply adequate water through a sprinkler system and the water supply should be conserved for this purpose by sparing the use of direct hoselines.

- Check control valves to see that they are open. Observe the discharge of sprinklers in the area of the fire and maintain pressure at the pumper to adequately serve the needs of the sprinkler system.

- Sprinkler control valves should not be closed until it is determined that the fire has been extinguished, or when the fire officers are convinced that further opera-

tions will simply waste water or tend to produce heavy water damage. When a sprinkler control valve is closed, a firefighter should be stationed at the valve in case it needs to be reopened should the fire rekindle. Pumpers should not be disconnected until after extinguishment has been determined by a thorough inspection.

- Sprinkler equipment should be restored to service before leaving the premises, if the weather and the extent of the damage permit. Fused heads should be replaced with new heads of proper temperature rating and the control valve should then be opened.

TYPES OF SPRINKLER SYSTEMS
The Wet-Pipe System

Wet-pipe systems contain water under pressure at all times and are connected to the water supply so that a fused sprinkler head will immediately discharge a water spray in that area and actuate an alarm. This type of system is usually equipped with an alarm check valve that is installed in the main riser adjacent to where the feed main enters the building. This valve actuates an alarm when water flows through the system. Incorporated in this valve is a clapper, which is simply an automatic one-way check valve with some additional features. Clappers are normally in the closed position. When a sprinkler head operates, this clapper opens and permits water to flow to the sprinkler and through the auxiliary valve to a retarding chamber to initiate an alarm signal.

RETARDING DEVICES

The retarding chamber is installed between the alarm check valve and alarm-signaling equipment and is employed since water is subjected to variable pressures. This chamber is a time-delay device that retards the flow of water to the alarm equipment from the alarm check valve since it must be filled before the water can continue to the alarm equipment. Water drains through the small opening in the bottom of the chamber. If it were not for the retarding chamber, surges or increases in water pressure would cause the clapper in the alarm valve to rise momentarily and permit water

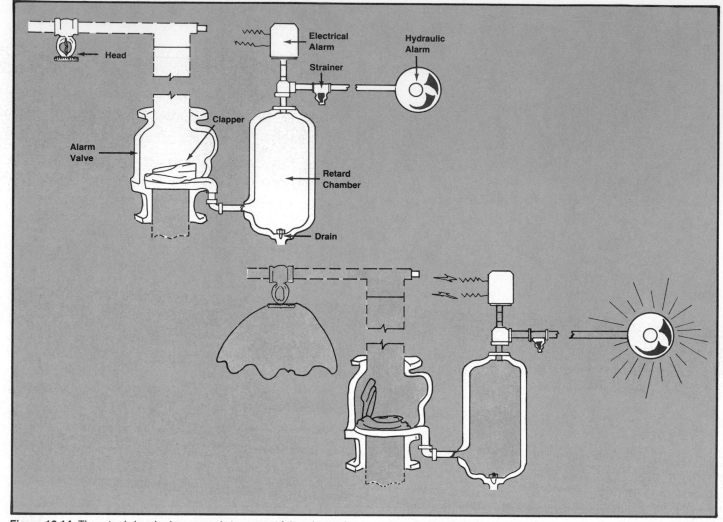

Figure 16.14 The retard chamber's purpose is to prevent false alarms due to pressure surges in the water supply.

to flow directly to the alarm equipment and thus transmit false alarms. A typical wet-pipe alarm check valve and retarding chamber in both the standby and fire positions are illustrated in Figure 16.14. This illustration will give a better idea of how the alarm check valve and retarding chamber operate.

OPERATION OF THE SYSTEM

A diagram of a system is shown in Figure 16.15. When one or more sprinkler heads open, the flow of water lifts the main clapper off its seat and opens the auxiliary valve. The water pressure in the main riser pushes the clapper to full open position and continues to supply the open sprinklers. Water also enters the auxiliary valve alarm line and continues to fill the retarding chamber. When the retarding chamber is full, the water activates the pressure switch, sending an electrical alarm

signal and a flow to activate the water-alarm gong. To shut down the system, turn off the main water control valve and open the main drain.

SPRINKLER HEAD REPLACEMENT

When changing a sprinkler head, replace it with a head of the same type. Normally, buildings with a sprinkler system will have a box (in close proximity to the control valve) containing spare heads and a wrench. Heads are removed counterclockwise. Check piping for obstruction and condition of threads, then replace head.

Steps to follow when replacing a sprinkler head on wet-pipe systems:

1. Close main control valve.

2. Open main drain.

3. Remove head.

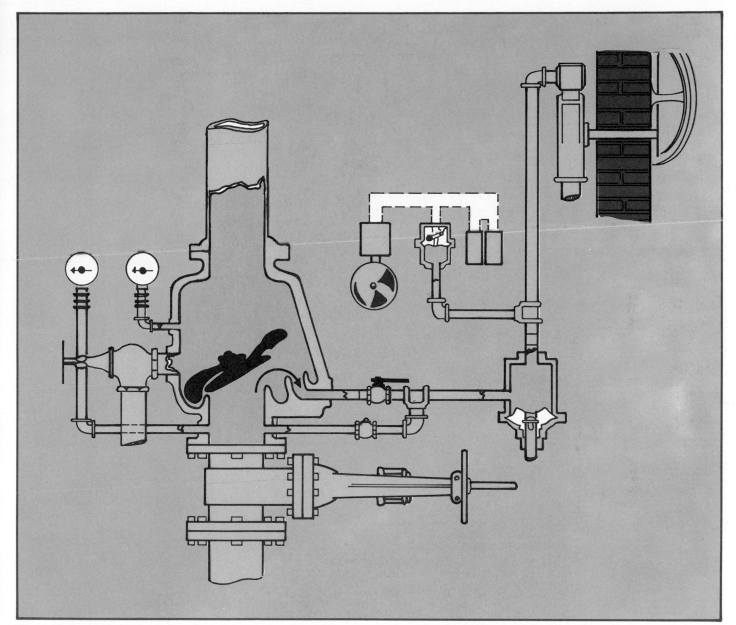

Figure 16.15 This is the flow of water in an alarm check valve and related parts during a fire condition.

4. Replace head.

5. Open inspector's test valve (normally at furthermost and highest point away from the control valve) to relieve system of trapped air.

6. Close main drain.

7. Open main control valve.

8. If water starts to leak around replaced head, tighten head.

NOTE: Care should be exercised when replacing heads as components are under stress.

Dry-Pipe Sprinkler System

A dry-pipe sprinkler system is one where air under pressure replaces water in the sprinkler piping above the dry-pipe valve. A dry-pipe valve is a device that keeps water out of the sprinkler piping until a fire actuates a head or heads. Dry systems should be used in buildings where insufficient heat is maintained to keep water from freezing. When a sprinkler head fuses, permitting the air pressure to escape, the dry-pipe valve automatically opens to permit water to replace the air pressure in the lines. Dry-pipe valves are designed so that a small amount of air pressure above the dry-pipe valve in the sprinkler piping will hold back a much greater water pressure on the water supply side of the dry-pipe valve. Dry systems are equipped with either electric or hydraulic alarm-signaling equipment. Figure 16.16 illustrates the dry-pipe valve in both the standby and fire positions.

The required air pressure for dry systems usually ranges between 15 and 50 psi (100 and 350 kPa). Air pressure that is needed to service a dry system may be derived from two different sources. These sources are either from plant air service or from unit air pressure that is supplied by a compresser and tank used exclusively for the sprinkler system.

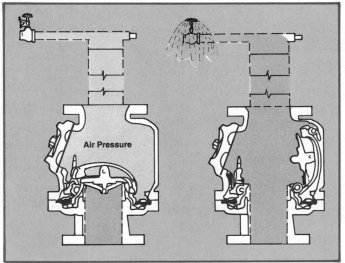

Figure 16.16 The dry-pipe alarm valve is shown in the standby position and in the fire position. Note the prime water used to make a seal between the clapper and seat when the valve is in the standby position.

ACCELERATOR AND EXHAUSTERS

In a large dry system several minutes could be lost while the air is being expelled from the system. Rules have been established which normally require a quick-opening device to be installed in systems that have a water capacity of over 500 gallons (2273 L). There are two types of quick-opening devices: accelerators and exhausters. The accelerator unbalances the differential in the dry-pipe valve, causing it to trip more quickly, whereas the exhauster quickly expels the air from the system. When a sprinkler head is fused and air pressure in the accelerator-type system drops a few psi (usually one or two pounds), a diaphragm in the accelerator becomes unbalanced. This unbalanced condition causes a valve to open, which permits the

air pressure in the system to enter the intermediate chamber of the dry-pipe valve. As soon as air is equalized on both sides of the air clapper (normally 10 to 15 seconds), the valve is automatically tripped by water pressure. In the exhauster type the fusing of a sprinkler head causes a diaphragm to open a large valve. This action permits air pressure to quickly escape to the outside and the dry-pipe valve to trip. Both of these devices are complicated mechanisms, and they require proper care and maintenance. They should be tested in the spring and fall by a competent individual. Although it will take longer, the dry-pipe valve will operate, even if the quick-opening devices do not operate.

OPERATION OF THE SYSTEM

When one or more sprinklers open, the air pressure is vented through the open heads from the system, thus upsetting the differential within the dry-pipe valve. An accelerator or exhauster will be an aid in the speeding-up operation of the system. As the differential is upset, the water pressure in the riser raises the clapper assembly into the wide-open locked position. As the water fills the upper chamber of the dry-pipe valve, it also enters the intermediate chamber where it forces the drip check valve closed. It then flows into the alarm line to activate the alarm equipment. The water continues to the open sprinklers.

Pre-Action System

A pre-action system is a dry system which employs a deluge-type valve, fire detection devices, and closed sprinkler heads. This type of system is used when it is especially important that water damage be prevented, even if pipes should be broken. The system will not discharge water into the sprinkler piping except in response to the detection system. A system that contains over 20 heads must be supervised so that if the detection system fails the system would still operate automatically.

Fire detection and operation of the system introduces water into the distribution piping prior to the opening of any sprinklers. In this system, fire detection devices operate a release located in the system-actuation unit. This release opens the deluge valve and permits water to enter the distribu-

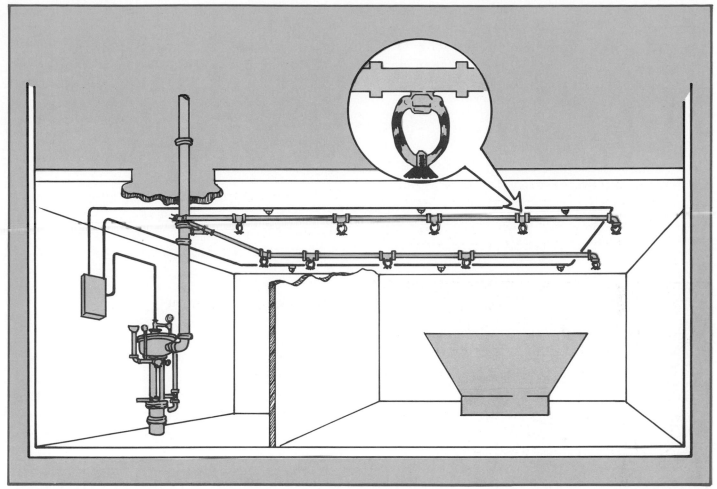

Figure 16.17 A deluge sprinkler system has open heads, is activated by a detection system and is used in areas where there is high hazard or danger of a fast-spreading fire.

tion system so that water is ready when the sprinkler heads fuse. When water enters the system, an alarm sounds to give a warning prior to the opening of the sprinkler head.

Inspecting and testing the system will be essentially the same as that for a deluge system.

Deluge Sprinkler System

This system is ordinarily equipped with open sprinkler heads and a deluge valve. Fire detection devices are installed in the same area as the sprinker heads (Figure 16.17). Upon fire detection, the deluge valve is opened, which permits water to flow into the system and out of all the sprinkler heads. The purpose of a deluge system is to wet down the area in which a fire originates by discharging water from all open heads in the system. This system is normally used to protect extra-hazardous occupancies. Many modern aircraft hangars are equipped with an automatic deluge system which may be combined with an automatic sprinkler system. A system using partly open and partly closed heads is considered a variation of the deluge system.

OPERATION OF THE SYSTEM

Activation of the deluge system may be controlled by fire, heat, or smoke detecting devices, and a manual device. Since the deluge system is designed to operate automatically and the sprinkler heads do not have heat-responsive elements, it is necessary to provide a separate detection system. This detection system is connected to a tripping device which is responsible for activating the system. As there are several different modes of detection there are also many different methods of operating the deluge valve. Deluge valves may be operated electrically, pneumatically, or hydraulically.

NFPA STANDARD 1001
INSPECTION
Fire Fighter I

3-11 Inspection

3-11.1   The fire fighter shall identify the common causes of fires and their prevention.

3-11.2   The fire fighter shall identify the fire inspection procedures.

3-11.3   The fire fighter shall define the importance of public relations relative to the inspection programs.

3-11.4   The fire fighter shall define dwelling inspection procedures.*

Fire Fighter II

4-11 Inspection

4-11.1   The fire fighter shall prepare diagrams or sketches of buildings to record the locations of items of concern during pre-fire planning operations.

4-11.2   The fire fighter shall collect and record in writing information required for the purpose of preparing a report on a building inspection or survey.

4-11.3   The fire fighter shall identify school exit drill procedures.

4-11.4   The fire fighter shall identify life safety programs for the home.

4-11.5   The fire fighter shall identify common fire hazards and make recommendations for their correction.*

*Reprinted by permission from NFPA Standard No. 1001, *Standard for Fire Fighter Professional Qualifications*. Copyright © 1981, National Fire Protection Association, Boston, MA.

IFSTA's Prevention and Identification Transparencies
are designed to complement this chapter.

Chapter 17
Fire Inspection

Inspection practices are usually considered to be the most important nonfire fighting activity performed by firefighters. A carefully planned inspection program carried out by conscientious well-trained personnel can prevent many serious fires. Through its use, many hazardous conditions are discovered and effective control measures are established before fires occur (Figure 17.1). It provides the property owner with a valuable consult-

Figure 17.1 Since fire inspections prevent fires, they are the most beneficial fire department activity for the community.

ing service and is a means by which firefighters can more effectively carry out their responsibility of protecting lives and property.

Inspections by fire department personnel should be complete and thorough. A sketchy, walk-through inspection is not sufficient. As firefighters become acquainted and more familiar with buildings through inspections, the knowledge they acquire is an aid to fire control.

In practically all municipalities, fire prevention work is the responsibility of the head of the fire department. This work is in addition to fire fighting operations and is usually delegated to various fire personnel. The manner by which this work is assigned and performed may vary with each jurisdiction, organization, and available personnel.

PERSONAL REQUIREMENTS FOR INSPECTORS

A uniform cap and a badge do not by themselves make a firefighter a fire inspector. To the public, however, the cap and the badge indicate that the firefighter who wears them is qualified to discuss fire protection matters and that reliable advice can be given as to how fire hazards can be corrected. No firefighter should be permitted to make a fire inspection without special training in this area. The use of the term "fire inspector" in this manual has reference to either a fire prevention or a fire company inspector.

A fire prevention inspector is an individual, directly associated with a fire department, whose assigned duty is to inspect, or to cause to be inspected, all buildings used by the public and to enforce, or to cause to be enforced, all laws and ordinances of a community in accordance with existing laws.

A fire company inspector is any firefighter who, with other fire company members, makes inspections and submits written reports to the fire chief or the fire marshal. Firefighters are part of an inspection team that collects valuable information and reports hazardous conditions.

A prerequisite to any personal requirement necessary to make an inspection is "confidence in one's ability." A fire department inspector must have confidence in the ability to meet the public, to make a favorable impression, and to judge conditions. Ability to judge can be developed only by training and using reference material that is available. An inspector's confidence in the ability to transpose visual information into written reports and sketches can be accomplished through practice. Confidence in the officer in charge of the fire prevention section and the ability to assist when unfamiliar hazards and conditions are encountered is extremely important. If inspectors have confidence in themselves, the public will have confidence in them.

Figure 17.2 Certain equipment must be available to the inspector for making inspections.

The Inspector's Equipment

The equipment needed by an inspector to do a good job of inspecting may be divided into equipment needed at the fire station and equipment used at the place of inspection (Figure 17.2). A recommended list of equipment that may be needed includes:

AT THE INSPECTION

- Neat coveralls, for crawling into attics and concealed spaces
- Clipboard, inspection forms, and standard plan symbols
- Materials for preparing sketches
- 50-foot tape
- Flashlight, because most concealed spaces are not lighted
- Camera, equipped with flash attachment
- Pitot tube and gauges, when water flow tests are required
- Reference books

AT THE FIRE STATION

- Inspection forms
- Inspection reports
- Inspection manuals
- Adequate records
- Quiet place to work
- Sanborn map
- Inspection file
- Drawing materials
- Drawing board

FIRE COMPANY INSPECTIONS

Fire department administrators must consider the number of competing demands for the time of the inspectors. Firefighters cannot choose the time to do fire fighting, but they have the prerogative of selecting the hour of the day, the day of the week, and even the season of the year to do inspection services. Because of this choice, the fire department administration should set up a schedule to carry out the organization plan. The officer in charge should inform each occupant of the purpose of fire prevention inspection, and find out whether there are certain days that such regular inspection would be impractical. This procedure enables fire inspections to be scheduled at a time that will not inflict a hardship on either the occupant or the inspector.

The Approach

The occupant should be informed of the purpose of the inspection and asked whether there are certain days when such inspections would be impractical. This action is not intended to establish a date for the inspection but is to avoid any hardship that might be imposed on either the occupant or fire personnel. Inspections are usually made during normal business hours, but night inspections are sometimes necessary. Occupancies such as theaters, night clubs, and halls may require a part of the inspection to be done in the daytime and completed at night while the building is occupied.

Certain observations should be made before the inspector enters the property. These observations should include the location of fire hydrants, fire alarm boxes, and exposures (Figure 17.3). The condition of the streets and the general housekeeping of the areas are extremely important from the standpoint of fire apparatus response. The inspector should record the type of buildings, occupancy, and general appearance of the neighborhood. All this should be recorded in the inspector's preliminary notes before entering the property.

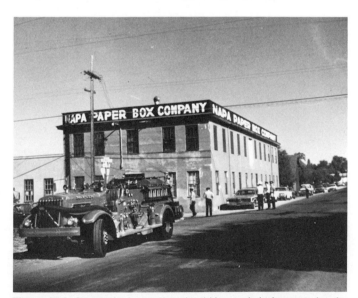

Figure 17.3 An exterior inspection should be made before entering the building and pertinent data recorded.

An earnest effort by fire inspectors to create a favorable impression upon the owner will help to establish a courteous and cooperative relationship. The inspector should enter the premises at the main entrance and obtain permission to make the inspection. An inspection should never be attempted without the proper permission. If necessary inspectors may have to wait to see the proper authority, for this person may be busy with other important matters. The company officer should introduce the crew and state their business. If the owner has been informed of the purpose of the inspection beforehand, this introduction will be much easier. An opportunity should always be given to the owner to have a representative present during the inspection. Such a guide will help obtain ready access to all parts of the building and provide answers to necessary questions. The inspection crew and the fire department, in general, will benefit from the time spent with the owner to make sure that the purpose of the inspection is understood.

Making the Inspection

In order to make a thorough inspection, inspectors should take sufficient time to make notes and sketches of all important features. Additional time to discuss fire protection problems with the owner will pay dividends. A complete set of notes and a well-prepared sketch of the building provides dependable information from which a complete report can be written.

From an inspection point of view, it does not matter whether the inspection is started on the roof and worked downward or started in the basement and worked upward. From a practical standpoint, however, many inspectors find it helpful and less confusing to start on the roof. In either procedure, the route should be planned so that the inspector can systematically inspect each floor. In large or complicated buildings, it may be necessary to make more than one visit to complete the inspection.

After permission to inspect the property has been obtained, the inspector should return to the outside of the building and first conduct an inspection of the exterior. This procedure makes the inspection of the interior easier and provides the necessary information for drawing the exterior walls on the floor plan sketch. When the survey of the exterior is completed, the inspection team should go directly to the roof or basement and proceed with a systematic inspection. Each floor should be inspected in succession and all areas on each floor should be inspected. The inspector should ask for the key to all locked rooms or closets and tactfully explain why it is necessary to see these areas. For example if the guide says, "There is nothing in this locked room," the inspector may say, "Yes, we understand, but it is not particularly the material in the room that we must see. The size, the shape, and the construction of the room are important features." If admission to an area or room is refused because of scientific research or secret process, the inspector should suggest that a cover or a screen be set up to permit the inspection. Secret areas from which the inspection crew is barred should be reported to the fire chief for appropriate action.

If the property includes several buildings, each should be inspected separately. It is a good idea to start on the roof of the highest building from which the inspector can get a general view. A sketch of each floor should be completed before proceeding to the next floor. If a floor plan that was made on a previous inspection is used the inspection can proceed more rapidly. Make sure, however, to record any changes that have been made and change the floor plan sketch accordingly.

Final Interview

To leave the premises without consulting with the person who gave the inspector permission to inspect the property might give the impression that the inspection did not take long and that it was an unimportant venture. Reporting to the person with authority, after one or two days of intensive inspection, can do much to maintain a cooperative attitude of the owner towards matters that need attention. During this interview a fire inspector should first comment favorably on the good conditions that were found and let it be known that these good points will certainly be contained in the inspection report. Unfavorable conditions may be discussed in general, but the inspector should avoid technicalities and direct conclusions at the time. Explain that such conditions will be studied more fully and that recommended solutions will be submitted in a written report. Too much talk concerning unfavorable conditions will give the impression of petty fault finding and may lead to an argumentative discussion. A final interview also gives the inspector an opportunity to express

thanks for the courtesies extended to the fire department and opens the way to explain how firefighters will study these reports from the standpoint of fire fighting procedures.

Follow-Up

A follow-up system is one of the best ways to impress the importance of inspections upon occupants or owners. The follow-up should not necessarily be a surprise visit. The occupant should be told, during the final interview of the inspection, that a representative will return. This follow-up should be made within a few weeks after the occupant or the owner has received the written report and should be conducted much in the same procedure as the inspection. The follow-up representative should first go to the person to whom the report was written to discuss the inspection. Entering the property without this contact may give the impression that the inspector is snooping. Being fair and honest in all dealings will prove more beneficial in the end.

Map and Sketch Making

Maps that convey information in legible form relative to construction, fire protection, occupancy, special hazards, and other details of buildings are available for fire protection and fire insurance agencies. Maps of this nature are usually prepared by the map companies and standard map symbols are used almost universally. A key to standard plan symbols is shown in Figure 17.4.

Inspections made by fire department personnel should include some sort of sketch to show the general arrangement of the property with respect to streets, other buildings, and important features that will help determine fire fighting procedures. This sketch is commonly called a plot plan of the area. Sketches of a property that have been made during an inspection do not need to be drawn again during a reinspection except to make changes or additions. An inspector's sketch of an area frequently constitutes the most informative part of an inspection report and should be made with neatness and accuracy.

A sketch that is made during inspections may be done freehand with the aid of a clipboard and 6-

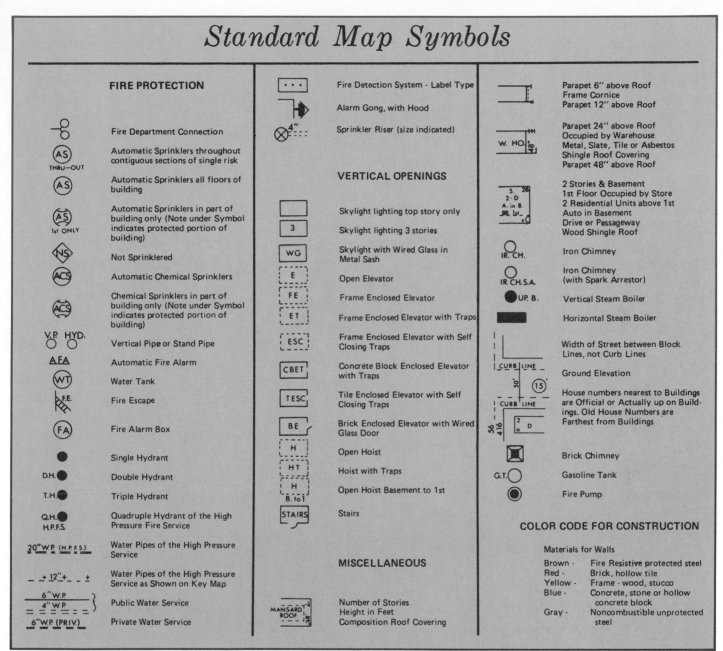

Standard Map Symbols

FIRE PROTECTION

Fire Department Connection	
(AS) THRU-OUT	Automatic Sprinklers throughout contiguous sections of single risk
(AS)	Automatic Sprinklers all floors of building
(AS) 1st ONLY	Automatic Sprinklers in part of building only (Note under Symbol indicates protected portion of building)
(NS)	Not Sprinklered
(ACS)	Automatic Chemical Sprinklers
(ACS)	Chemical Sprinklers in part of building only (Note under Symbol indicates protected portion of building)
V.P HYD.	Vertical Pipe or Stand Pipe
AFA	Automatic Fire Alarm
(WT)	Water Tank
F.E.	Fire Escape
(FA)	Fire Alarm Box
●	Single Hydrant
D.H.●	Double Hydrant
T.H.●	Triple Hydrant
Q.H.● H.P.F.S.	Quadruple Hydrant of the High Pressure Fire Service
20"W.P (H.P.F.S.)	Water Pipes of the High Pressure Service
_ _ +12"+_ _ ±	Water Pipes of the High Pressure Service as Shown on Key Map
6"W.P. 4"W.P.	Public Water Service
6"W.P. (PRIV.)	Private Water Service

⊡ ...	Fire Detection System - Label Type
	Alarm Gong, with Hood
⊗ 4"	Sprinkler Riser (size indicated)

VERTICAL OPENINGS

▭	Skylight lighting top story only
3	Skylight lighting 3 stories
WG	Skylight with Wired Glass in Metal Sash
E	Open Elevator
FE	Frame Enclosed Elevator
ET	Frame Enclosed Elevator with Traps
ESC	Frame Enclosed Elevator with Self Closing Traps
CBET	Concrete Block Enclosed Elevator with Traps
TESC	Tile Enclosed Elevator with Self Closing Traps
BE	Brick Enclosed Elevator with Wired Glass Door
H	Open Hoist
HT	Hoist with Traps
H B. to 1	Open Hoist Basement to 1st
STAIRS	Stairs

MISCELLANEOUS

MANSARD ROOF 4 48	Number of Stories Height in Feet Composition Roof Covering

	Parapet 6" above Roof Frame Cornice Parapet 12" above Roof
W. HO.	Parapet 24" above Roof Occupied by Warehouse Metal, Slate, Tile or Asbestos Shingle Roof Covering Parapet 48" above Roof
S. 28 2-D A. in B. BR. 1st x	2 Stories & Basement 1st Floor Occupied by Store 2 Residential Units above 1st Auto in Basement Drive or Passageway Wood Shingle Roof
IR. CH.	Iron Chimney
IR. CH.S.A.	Iron Chimney (with Spark Arrestor)
UP. B.	Vertical Steam Boiler
▬	Horizontal Steam Boiler
	Width of Street between Block Lines, not Curb Lines
CURB LINE 50' (15)	Ground Elevation
CURB LINE 56 416 2 D	House numbers nearest to Buildings are Official or Actually up on Buildings. Old House Numbers are Farthest from Buildings
▣	Brick Chimney
G.T. ○	Gasoline Tank
◉	Fire Pump

COLOR CODE FOR CONSTRUCTION

Materials for Walls

Brown -	Fire Resistive protected steel
Red -	Brick, hollow tile
Yellow -	Frame - wood, stucco
Blue -	Concrete, stone or hollow concrete block
Gray -	Noncombustible unprotected steel

Figure 17.4 A well-drawn map using standard symbols can communicate important information at a glance.

inch rule (160 mm). Data should be recorded by using standard plan symbols as much as possible.

The floor plan consists of an outline of each floor of the building. By using standard symbols, the inspector can show the type of construction, thickness of walls, partitions, openings, roof types, parapets, and other important features. In addition to this, protection devices, water mains, valves, and other miscellaneous items can be included (Figure 17.5).

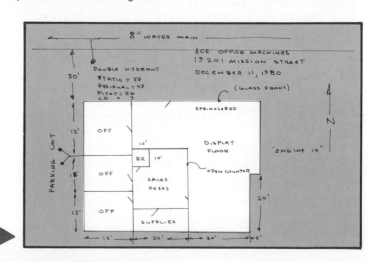

Figure 17.5 The field sketch is a rough drawing showing the general information on the property. Details can be recorded on the survey notes. ▶

A sectional elevation sketch of a structure consisting of a cross section or cutaway view of a particular portion of a building along a selected imaginary line may be needed to show elevation changes, mezzanines, balconies, or other structural features. The easiest section view to portray is to establish the imaginary line along an exterior wall. This location theoretically removes the exterior wall and exposes such features as roof construction, floor construction, parapets, basements, attics, and other items that are difficult to show on a floor plan. Establishing the imaginary line along an exterior wall may not always show the section of the building that is desired. In this case, it may be better to divide the building near the center or along a line where a separate wing is attached to the main structure (Figure 17.6).

From the inspector's sketch and notes a permanent drawing can be made to be filed for future reference and classroom study (Figure 17.7).

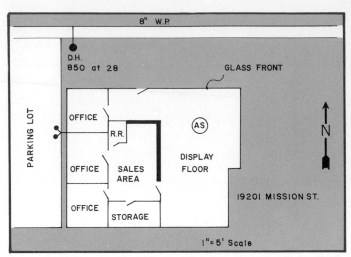

Figure 17.7 The report drawing should be drawn to scale with the details shown by using accepted map symbols.

FIRE HAZARDS

A "fire hazard" may be defined as a condition that will encourage a fire to start or will increase the extent or severity of the fire (Figure 17.8). In order to prevent a situation from being hazardous, the fuel supply, heat source, and oxygen supply hazard must be considered. If any one of these hazards can be eliminated, a fire cannot occur. The oxygen supply hazard is normally present, and very little control can be maintained over the oxygen supply except in special cases.

Fuel supply hazards and heat source hazards are more readily controlled. If heat sources are kept separated from fuel supplies, the condition

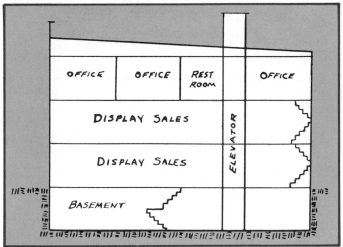

Figure 17.6 A sectional view sketch consisting of a cutaway view may be needed to illustrate elevation changes.

Photographs

Photographs show worthwhile detail for inspection reports, especially if they can be taken from more than one angle. One view which is especially good from a fire fighting standpoint is from an elevated position. An adjoining building or elevated tower can be used to take the photographs. Interior and close-up photographs are very effective aids in making a complete report. A good camera, equipped with flash attachments, should be considered as standard equipment for fire department inspectors.

Figure 17.8 During the inspection fire hazards should be noted and pointed out to the building owner or manager.

will be safe indefinitely. Some fuel supply hazards may be easily ignited and they may have some characteristics which make them dangerous from a fire safety standpoint. Any heat source may be dangerous. Some common fuel supply and heat source hazards are included in the following list.

Fuel Supply Hazards

- Ordinary Combustible Solids: Wood, cloth, or paper

- Flammable and Combustible Liquids: Gasoline, oils, or alcohol

- Combustible Gases: Natural, LPG, or manufactured

- Chemicals: Nitrates, oxides, or chlorates

- Dusts: Grain, wood, or coal

- Metals: Magnesium, sodium, or potassium

- Plastics: Resins, casein, cellulose

Heat Source Hazards

- Chemical Heat Energy: Heat of combustion, spontaneous heating, heat of decomposition, heat of solution

- Electrical Heat Energy: Resistance heating, dielectric heating

- Heat from arcing: Static electricity heating

- Heat generated by lightning

- Mechanical Heat Energy: Friction heat, heat of compression

- Nuclear Heat Energy

Common Hazards

A common fire hazard is a condition which is prevalent in almost all occupancies and will encourage a fire to start. Common fire hazards are listed in contrast to special hazards, which are usually peculiar to a given industry. Some common hazards are housekeeping, heating, lighting, power, floor cleaning compounds, packing materials, fumigation, and insecticides. The term "common" could be misleading to some individuals. It refers to the probable frequency of the hazard and not to the severity of the hazard.

Personal hazards are probably the most serious of all common hazards. The term "personal hazard" covers all of the individual traits, habits, and personalities of the people who work, live, or visit the property or building in question. Suppose an industrial property burns because someone carelessly threw a cigarette into a pile of trash. The fire extinguishers were found inoperative and the standpipe fire hose was rotten. Was the person smoking in a prohibited section of the building? Was the pile of trash a result of an incompetent janitor or did management fail to provide a place or a means to dispose of trash? What about the empty fire extinguishers and the rotten fire hose — whose fault were they? Personal hazards may be considered to be intangible but they are always present.

CHECK CLOSELY FOR PERSONAL SAFETY HAZARDS TO OCCUPANTS AND FIREFIGHTERS

Special Hazards

A special fire hazard may be defined as one which arises from the processes or operations that are peculiar to the individual occupancy. Some examples of special hazards are painting, welding, flammable liquids, processing chemicals, acids, and dusts. The widespread use of flammable liquids in dip tanks, ovens, driers, mixing, coating, spraying, and degreasing processes involves a variety of special hazards. Flammable gases are commonly used as a fuel and for special purposes, such as welding and cutting. The processing of certain materials, such as metal, rubber, cork, fertilizer, and drugs, produces flammable dusts which may present explosion problems and spontaneous ignition.

Target Hazards Properties

A target hazard may be defined as a facility or process which could produce or stimulate a fire that could cause a large loss of life or property. Some examples of target hazards are lumber yards, bulk

Figure 17.9 Target hazards are those with a potential large loss of life or property.

oil storage, area shopping centers, hospitals, theaters, fur storage vaults, rows of frame tenements, and schools (Figure 17.9).

DWELLING INSPECTIONS

Dwelling inspections consist of a house-to-house fire prevention activity performed by fire fighting forces not only to provide an inspection but to perform an educational and advisory service as well. A great deal of advanced planning and publicity is necessary to gain full acceptance of this program. It must be understood that the program is a fire prevention activity and not a police activity. In other words, the firefighter is to look for hazards, not violations. The objectives of home fire inspection programs are as follows:

- To obtain proper life safety conditions

- To keep fires from starting

- To help the owner or occupant understand and improve existing conditions

In addition to reducing loss of life and property damage, the fire department will realize other important benefits. Dwelling inspections give the firefighters a chance to impress people with the program and activities of the fire department. The citizens who support the fire department will feel that they are getting "more for their money." Inspections give them a complete service, not just emergency service, and they will become more familiar with the duties and responsibilities of firefighters. Personal visits by firefighters to households for dwelling inspections generally result in an improvement in community support of the department.

Dwelling inspections are also a good way to distribute fire prevention literature, home fire drill information, invalid markers, telephone stickers, and other fire safety information. The firefighter can explain each item of the literature and possibly tie in a "local angle" of fire experience. Many fire departments also print special cards or slips to compliment the homeowner when the dwelling is found to be in good fire safe condition. Other cards are used to notify absent households that the firefighters visited the property for inspection.

Firefighters who participate in dwelling inspections become acquainted with streets, hydrants and water supply location, area development, home construction, and pre-fire planning. Notes of these items and other useful information should be made and discussed during training sessions. Using fire apparatus will also improve driver proficiency. While these fringe benefits are helpful, the primary reason for making inspections is to reduce life and fire hazards.

Firefighter Responsibilities

Every firefighter must fully understand that a dwelling inspection campaign is a fire department effort to reduce the number of fire deaths and home fires. It is the firefighter's responsibility to represent their fire department and meet the citizens with dignity and pride. The public has every right to expect inspectors to be fully qualified to advise on matters pertaining to fire prevention.

The approach to each home should be made on sidewalks or a path (never cut across the lawn), and shoes should be cleaned before entering the house. If no one is at home and the inspector is to leave some material, the mail box should not be used. Firefighters should ask permission to make the inspection and state that they are from the local fire department. They should explain to the householder the purpose of dwelling inspection service and be courteous at all times. When hazards are found, corrections should not be ordered, the hazard should be explained, its potential danger, and the proper method for correction. Do not argue with the householder and be complimentary when favorable conditions are found. The inspection should only include the basement, attic, utility rooms, storage rooms, kitchens, and garages. However, other rooms of the house may be inspected at the request of the householder. It is important that the homeowner be sold on the idea of carrying out suggestions. The following are some very important pointers:

- Courteous approach on all inspections.
- Thank the owner or occupant for the invitation into the home.
- The primary interest is preventing a fire which could take the lives of the occupants and destroy the home.

- All comments are constructive and intended to eliminate hazardous conditions.
- As far as the inspector is concerned, the inspection is a confidential matter.
- Gossip should not be tolerated.
- Under no circumstances should any conditions noted on a tour be made available to an insurance carrier, repair service organization, sales promotion groups, or in any form of publicity which would identify a given home.

What To Look For

There are several items firefighters should look for when making inspections of homes. The use of an inspection form to be filled out for each dwelling may be helpful. It can serve as a guide for firefighters making the inspection and a copy can be given to the occupant or it can be used to make summaries of the inspections (Figure 17.10).

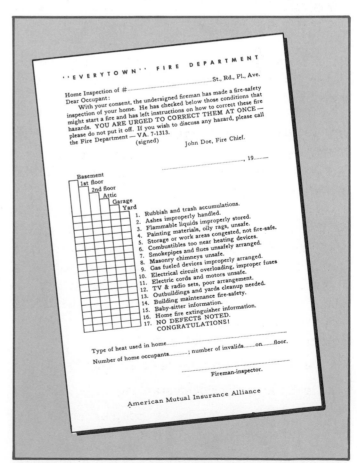

Figure 17.10 This sample home inspection form indicates the items that should be checked during dwelling inspections.

ITEMS TO BE CHECKED FROM OUTSIDE

Condition of roof. Roofing that is old and warped is easily ignited by sparks and flying brands.

Condition of chimneys. Chimneys supported on wooden posts or brackets are likely to crack from settling and allow sparks or hot flue gases to set fire to woodwork. Loose bricks, open joints, and cracks which can be seen indicate that similar defects may exist in other parts where they might start a fire. In such cases, a thorough investigation should be made.

Condition of yard. Dry grass, leaves, paper, boards, branches of trees, and other combustible waste materials in yards and under porches and houses are readily ignited and are a fire hazard to buildings.

Waste burners. The location and arrangements for outdoor burning of waste should be noted for conformity to local regulations.

Condition of garages and sheds. Cleanliness and good maintenance are important precautions against fire that apply to sheds and garages as well as other buildings.

Flammable liquids. Gasoline, and other similar flammable liquids should not be kept in dwellings. They should be stored in a safety type can in an outside storage area. Flammable liquids should never be used for home dry cleaning or for other purposes that would expose the dwelling to their explosive vapors (Figure 17.11).

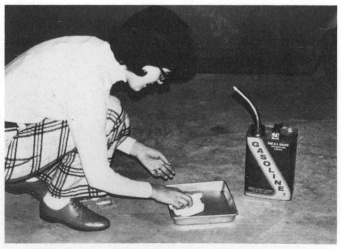

Figure 17.11 Watch for flammable liquids that are being used for household cleaning.

ITEMS TO BE CHECKED IN BASEMENT

Accumulations of waste and discarded material. Waste papers, discarded furniture, and partly-filled paint cans that are no longer worth keeping constitute a fire hazard. A suggestion from a firefighter may provide the necessary incentive to get rid of such accumulations (Figure 17.12). Rags soiled with oil or paint are especially hazardous because of the danger of spontaneous ignition. Occasionally other waste materials found in basements are subject to the same hazard.

Figure 17.12 Accumulations of combustible materials should be removed from around heating appliances.

Furnace and stove vent pipes. Firefighters should be familiar with local regulations governing the installation of vent pipes from heating and cooking appliances. Charring of wood and blistering of paint indicate exposure to excessive temperature.

Gas appliances. Corroded piping and rubber tubing may result in gas leaks. Automatic gas devices without provisions for automatically cutting off the supply of gas when the pilot flame is extinguished may produce an explosion. A separate shutoff should be provided on the supply line to every appliance.

Oil burning installations. Oil burners, supply tanks, and piping need to be properly installed to avoid danger of fire. Installations should be checked against local regulations and with the Standards for Oil Burning Equipment, NFPA No. 31.

Work rooms. Removal of shavings from work benches and the orderly storage of flammable liquids are features to be commented upon.

ATTICS

When permission is obtained to go into the attic, firefighters should check for faulty electrical equipment and other hazards. Occupants should be encouraged to clean out all combustible material that is no longer worth keeping.

Home Fire Safety

Home fire safety should include escape plans in case of fire. These plans should be carefully reviewed and practiced with the children at regular intervals. Some basic rules include:

- Have two escape exits from every room.

- Windows should be easily opened by children and doors should remain closed.

- Always stay low if awakened by smoke; do not raise up.

- Have a whistle by every bed to alert other family members if awakened by smell of smoke.

- Crawl out of bed and creep to the door. Feel the door and if it is warm, use the window for escape.

- Never return to house once outside.

- A fire escape ladder should be kept by the window in second-story houses and all family members should practice descending it.

- A meeting place outside the home should be agreed upon so all members will be together after escaping.

OPERATION EDITH

Operation Edith — Exit Drills in The Home — is a nationwide campaign designed to encourage families to prepare an escape plan and to practice evacuation of their home.

It is essential that each and every member of the family take an active part in planning and using of escape routes. The more familiar they become with the plan, the less likely they are to panic in an actual fire situation.

Devise two possible escape routes from each room in the home. The best and easiest method will be through rooms and hallways in the home. If these avenues are blocked then escape through windows will be necessary (Figure 17.13).

Figure 17.13 Encourage families during dwelling inspections to develop two escape routes from each room.

The following information should be considered when formulating your escape plan.

- Identify two escape routes.

- Shut all bedroom doors to delay flame spread.

- Inform the local fire department if there is an invalid in your home and what room they sleep in.

- Know the proper procedures for opening and breaking windows.

- Have a predetermined place for family members to meet outside the home.

The most important facts to remember in case of fire are to move quickly but intelligently over the planned escape routes and, once at the meeting place, do not leave until all family members are present. Alert the fire department, and above all DO NOT PANIC!

FIRE EXIT DRILLS FOR SCHOOLS

Fire exit drills are a matter of great importance when inspecting school buildings. Some states have laws which establish where the responsibility lies for holding school fire drills, their frequency, and other details. Every fire department inspector should be aware of the specific requirements of such regulations. Where no law, code, or ordinance exists that requires exit drills, the inspector should encourage voluntary adoption of a regular drill program.

The purpose of fire exit drills is to insure the efficient and safe use of exit facilities available. Proper drills insure orderly exit under controlled supervision and prevent panic which has been responsible for the majority of fatalities in major fire disasters. Order and control are the primary purposes of the drill. Speed in emptying buildings, while desirable, is not in itself an objective, and should be made secondary to keeping proper order and discipline (Figure 17.14).

Fire exit drills should be held with sufficient frequency to familiarize all occupants with the drill as a matter of established routine. Drills should be held at different times and under varying conditions to simulate the unusual conditions experienced during a fire. Drills should be executed at different hours of the day. Some of these periods are during the changing of classes, when the school is at assembly, during the recess or gymnastic periods, or during other special events. If a drill is called during the time classes are changing, the pupils should be instructed to form a line and immediately proceed to the nearest available exit in an orderly manner. Instruction cards should be conspicuously posted which describe the procedure of the drills.

Responsibility for the planning and conducting of drills should be assigned only to competent persons qualified to exercise leadership. Emphasis should be placed upon orderly evacuation under proper discipline rather than upon speed. Drills should include procedures which insure that all persons in the building, or all persons subject to the drill actually participate. If a fire exit drill is considered merely as a routine exercise from which some persons may be excused, there is grave danger that in an actual fire, the drill will fail in its intended purpose.

Fire exit drills should be held at least once a month. If climatic conditions may endanger the health of children during the winter months, weekly drills may be held at the beginning of the school term to complete the required number of drills before cold weather. Such drills would be held when the pupils are fully clothed, by using the exit drill alarm signal.

Figure 17.14 When inspecting schools, witness fire exit drills. Discuss the manner in which it was conducted and a regular exit drill program with the staff.

Each class or group should proceed to a predetermined point outside the building and remain there while a check is made to see that all are accounted for and that the building is safe to reenter. Such points should be sufficiently far from the building to avoid danger from fire in the building, interference with fire department operations, or confusion between classes or groups. No one should be permitted to reenter until after the drill is completed.

All fire exit drill alarms should be sounded on the fire alarm system and not on the signal system used to dismiss classes. Whenever any of the school authorities determine that an actual fire exists, they should immediately notify the local fire department. It should be a duty of principals and teachers to inspect all exit facilities daily. Particular attention should be given to keeping all doors unlocked, the paths of egress unblocked and all stairs and fire escapes free from all obstructions. Any condition likely to interfere with safe exit should be immediately corrected or reported to the appropriate authorities.

Fire inspectors should check on the frequency of exit drills and the time required to vacate the building. A stop watch should be considered as standard equipment for the evacuation practice of school buildings. The building fire alarm system should be examined to see if the alarm can be heard in all portions of the building and that it can be activated from each floor. Arrangements for promptly notifying the fire department should also be investigated.

APPENDIX A

TABLE 2-1
CHARACTERISTICS OF EXTINGUISHERS (Metrics)

Extinguishing Agent	Method of Operation	Capacity	UL or ULC Classification
Water	Stored Pressure	9.5 L	2-A
Water	Pump Tank	5.7 L	1-A
	Pump Tank	9.5 L	2-A
	Pump Tank	15.1 L	3-A
	Pump Tank	18.9 L	4-A
Water (Antifreeze Calcium Chloride)	Cartridge or Stored Pressure	4.73, 5.7 L	1-A
	Cartridge or Stored Pressure	9.5 L	2-A
	Cylinder	125 L	20 A
Water (Wetting Agent)	Stored Pressure	4.7 L	2-A
	Carbon Dioxide Cylinder	94.6 L (wheeled)	10-A
	Carbon Dioxide Cylinder	170.3 L (wheeled)	30-A
	Carbon Dioxide Cylinder	227 L (wheeled)	40-A
Water (Soda Acid)	Chemically Generated Expellant	4.73, 5.7 L	1-A
	Chemically Generated Expellant	9.5 L	2-A
	Chemically Generated Expellant	64.3 L (wheeled)	10-A
	Chemically Generated Expellant	125 L (wheeled)	20-A
Water (Loaded Stream)	Stored Pressure	9.5 L	3-A
	Cartridge or Stored Pressure	125 L (wheeled)	20-A
AFFF	Stored Pressure	9.5 L	3-A 20 B
	Nitrogen Cylinder	125 L (wheeled)	20-A 160 B
Carbon Dioxide	Self-Expellant	.9 to 2.26 kg	1 to 5-B:C
	Self-Expellant	4.5 to 6.8 kg	2 to 10-B:C
	Self-Expellant	9 kg	10-B:C
	Self-Expellant	22.6 to 45.3 kg (wheeled)	10 to 20-B:C
Dry Chemical (Sodium Bicarbonate)	Stored Pressure	.45 kg	1 to 2-B:C
	Stored Pressure	.68 to 1.13 kg	2 to 10-B:C
	Cartridge or Stored Pressure	1.24 to 2.26 kg	5 to 20-B:C
	Cartridge or Stored Pressure	2.7 to 13.6 kg	10 to 160-B:C
	Nitrogen Cylinder or Stored Pressure	34 to 158.7 kg (wheeled)	40 to 320-B:C
Dry Chemical (Potassium Bicarbonate)	Stored Pressure	.45 to .9 kg	1 to 5-B:C
	Cartridge or Stored Pressure	1.02 to 2.26 kg	5 to 20-B:C
	Cartridge or Stored Pressure	2.4 to 4.5 kg	10 to 80-B:C
	Cartridge or Stored Pressure	7.2 to 13.6 kg	40 to 120-B:C
	Cartridge	21.7 kg	120-B:C
	Nitrogen Cylinder or Stored Pressure	56.7 to 142.8 kg (wheeled)	80 to 640-B:C

Extinguishing Agent	Method of Operation	Capacity	UL or ULC Classification
Dry Chemical (Potassium Chloride)	Stored Pressure	1.13 to 3.8 kg	5 to 10-B:C
	Stored Pressure	2.26 to 4 kg	20 to 40-B:C
	Stored Pressure	4.5 to 9 kg	20 to 40-B:C
	Stored Pressure	61.2 kg	160-B:C
Dry Chemical (Ammonium Phosphate)	Stored Pressure	.45 to 2.26 kg	1 to 2-A and 2 to 10-B:C
	Stored Pressure or Cartridge	1.13 to 3.8 kg	1 to 4-A and 10 to 40-B:C
	Stored Pressure or Cartridge	4 to 7.7 kg	2 to 20-A and 10 to 80-B:C
	Stored Pressure or Cartridge	7.7 to 13.6 kg	3 to 20-A and 30 to 80-B:C
	Cartridge	20.4 kg	20-A and 80-B:C
	Nitrogen Cylinder or Stored Pressure	49.9 to 142.8 kg (wheeled)	20 to 40-A and 60 to 320-B:C
Dry Chemical (Foam Compatible)	Cartridge or Stored Pressure	2.1 to 4 kg	10 to 20-B:C
	Cartridge or Stored Pressure	4 to 12.2 kg	20 to 30-B:C
	Cartridge or Stored Pressure	8.2 to 13.6 kg	40 to 60-B:C
	Nitrogen Cylinder or Stored Pressure	68 to 158.7 kg	80 to 240-B:C
Dry Chemical (Potassium Chloride)	Cartridge or Stored Pressure	1.13 to 2.26 kg	10 to 20-B:C
	Cartridge or Stored Pressure	4.3 to 9 kg	40 to 60-B:C
	Cartridge or Stored Pressure	8.8 to 13.6 kg	60 to 80-B:C
	Stored Pressure	56.7 to 90.7 kg (wheeled)	160-B:C
Dry Chemical (Potassium Bicarbonate Urea Base)	Stored Pressure	2.26 to 4.98 kg	40 to 80-B:C
	Stored Pressure	4 to 10.4 kg	60 to 160-B:C
	Stored Pressure	79.3 kg	480-B:C
Bromotrifluoromethane	Stored Pressure	1.3 kg	2-B:C
Bromochlorodifluoromethane	Stored Pressure	.9 to 1.8 kg	2 to 5-B:C
	Stored Pressure	2.49 to 4 kg	1-A and 10-B:C
	Stored Pressure	7.25 to 9.97 kg	1 to 4-A and 20 to 80-B:C

APPENDIX B

TABLE 3.1
ROPE CHARACTERISTICS (Metrics)

	MANILA	SISAL	POLY-P	POLY-E	NYLON	DACRON (Polyester)
Moisture Regain	Up to 60%	Up to 60%	0%	0%	to 9%	Less than 1%
16 mm Dia. Strength	1995.8 kg	1596.6 kg	2812.3 kg	2540.1 kg	4717.4 kg	4535.9 kg
Elongation break in rope	13%	13%	24%	22%	35%	20%
Change of strength:	Up to +20%	Up to +20%	No change	No change	Less than -10%	No change
Floatability	No	No	Yes	Yes	No	No
Resistance to rot, mildew and attack by marine organisms	Poor	Very Poor	100% Resistant	100% Resistant	100% Resistant	100% Resistant
Resistance to Surface abrasion	Good	Fair	Good	Good	Very Good	Excellent
Acids	Very Poor	Very Poor	Excellent	Excellent	Fair	Very Good - Excellent
Alkalies	Very Poor	Very Poor	Good	Good	Excellent	Very Good
Solvents	Good	Good	Good	Good	Excellent	

TABLE 3.2
SAFE WORKING CAPACITIES FOR ROPES (Metrics)

Diameter (inches)	Manila (new)	(used)[1]	Sisal (new)	(used)[1]	Nylon (new)	(used)[1]	Dacron (new)	(used)[1]	Polypropylene (new)	(used)[1]	Braided Nylon Cover Nylon Core (new)	(used)[1]
.95 cm	122.5	61.2	97.9	48.9	164.6	82.1	130.2	64.8	185.1	92.5	381.1	190.5
1.27 cm	240.4	120.2	192.3	96.2	329.3	164.6	223.2	111.6	331.6	165.6	680.4	340.2
1.58 cm	399.2	199.6	319.3	159.6	508.9	254.5	364.2	181.9	508.9	254.5	1088.6	544.3
1.905 cm	489.9	244.9	391.9	195.9	673.6	336.6	498.9	249.5	745.7	372.8	1542.2	771.1
2.22 cm	698.5	349.3	558.8	279.4	972.9	486.3	678.6	339.3	871.4	435.4	2150.1	1075.1
2.54 cm	816.5	408.3	653.2	326.6	1197.5	598.7	898.1	449.1	1295.5	647.7	2585.5	1292.7
3.175 cm	1224.7	612.4	979.7	489.8	1796.2	898.1	1222.4	610.9	1750.4	874.9	3991.6	1995.8
3.81 cm	1678.3	839.2	1342.6	671.3	2544.6	1272.3	1721.4	860.5	2506.1	1252.8	5896.7	2948.3
4.44 cm	2404.1	1202.1	1923.2	961.6	3742.2	1871.1	2544.6	1272.3	3485.4	1742.7	8708.9	4354.5
5.08 cm	2812.3[2]	1406.1[2]	2249.8[2]	1124.9[2]	4465.6[2]	2234.8[2]	3043.6[2]	1521.8[2]	4171.7[2]	2085.6[2]	9525.4[2]	4762.7[2]

[1]According to manufacturers' information, manila and sisal rope are considered "used" rope after they have been in service for six months, providing the rope has had proper usage, care and storage. Persons using synthetic rope should consult the manufacturer for their interpretation of "used" rope.

[2]Based on manufacturer's recommendations for new rope.

Index

IFSTA MANUALS

RE SERVICE ORIENTATION & INDOCTRINATION

story, traditions, and organization of the fire service; operation of the fire depart-
ent and responsibilities and duties of firefighters; fire department companies and
eir functions; glossary of fire service terms.

RE SERVICE FIRST AID PRACTICES

ief explanations of the nervous, skeletal, muscular, abdominal, digestive, and
nitourinary systems; injuries and treatment relating to each system; bleeding
ntrol and bandaging; artificial respiration, cardiopulmonary resuscitation (CPR),
ock, poisoning, and emergencies caused by heat and cold; fractures, sprains,
d dislocations; emergency childbirth; short-distance transfer of patients; ambu-
nces; conducting a primary and secondary survey.

SENTIALS OF FIRE FIGHTING

is manual meets the objectives set forth in levels I and II of NFPA *Fire Fighter
ofessional Qualifications, 1981.* Included in the manual are the basics of: fire be-
vior, extinguishers, ropes and knots, self-contained breathing apparatus, lad-
rs, forcible entry, rescue, water supply, fire streams, hose, ventilation, salvage
d overhaul, fire cause determination, fire suppression techniques, communica-
ns, sprinkler systems, and fire inspection.

LF-INSTRUCTION FOR ESSENTIALS OF FIRE FIGHTING

vers the most important points of *Essentials,* including NFPA 1001 Fire Fighter I
d II requirements. Perfect supplement for formal training.

STA'S 500 COMPETENCIES FOR FIREFIGHTER CERTIFICATION

is manual identifies the competencies that must be achieved for certification as a
efighter for levels one and two. The text also identifies what the instructor needs
give the student, NFPA standards, and has space to record the student's score,
al standards, and the instructor's initials.

RE SERVICE GROUND LADDER PRACTICES

rious terms applied to ladders; types, construction, maintenance, and testing of
e service ground ladders; detailed information on handling ground ladders and
ecial tasks related to them.

RE HOSE PRACTICES

onstruction, care, and testing of hose and various fire hose accessories; prepara-
n and manipulation of hose for rolls, folds, connections, carries, drags, and spe-
al operations; loads and layouts for fire hose.

LVAGE AND OVERHAUL PRACTICES

anning and preparing for salvage operations, care and preparation of equipment,
ethods of spreading and folding salvage covers, most effective way to handle
ater runoff, value of proper overhaul and equipment needed, and recognizing and
eserving arson evidence.

RCIBLE ENTRY, ROPE AND PORTABLE TINGUISHER PRACTICES

pes of forcible entry tools and general building construction; use of tools in open-
g doors, windows, roofs, floors, walls, partitions and ceilings; types, uses, and
re of ropes, knots, and portable fire extinguishers.

LF-CONTAINED BREATHING APPARATUS

is manual is the most comprehensive self-contained breathing apparatus text
ailable. Beginning with the history of breathing apparatus and the reasons they
e needed, to how to use them, including maintenance and care, the firefighter is
ken step by step with the aid of programmed-learning questions and answers
roughout to complete knowledge of the subject. The donning, operation, and care
all types of breathing apparatus are covered in depth, as are training in SCBA
e, breathing-air purification, and recharging cylinders. There are also special
apters on emergency escape procedure and interior search and rescue.

RE VENTILATION PRACTICES

jectives and advantages of ventilation; requirements for burning, flammable liq-
d characteristics and products of combustion; phases of burning, backdrafts, and
e transmission of heat; construction features to be considered; the ventilation
ocess including evaluating and size up is discussed in length.

RE SERVICE RESCUE PRACTICES

STA's new *Rescue* has been enlarged and brought up to date. Sections include
ter and ice rescue, trenching, cave rescue, rigging, search-and-rescue techniques for inside structures and outside, and taking command at an incident. Also
included are vehicle extrication and a complete section on rescue tools. The book
covers all the information called for by the rescue sections of NFPA 1001 for Fire
Fighter I, II, and III, and is profusely illustrated.

THE FIRE DEPARTMENT COMPANY OFFICER

This manual focuses on the basic principles of fire department organization, working relationships, and personnel management. For the firefighter aspiring to become a company officer and the company officer who wishes to improve management skills this manual will be invaluable. This manual will help individuals develop
and improve the necessary traits to effectively manage the fire company.

FIRE CAUSE DETERMINATION

Covers need for determination, finding origin and cause, documenting evidence,
interviewing witnesses, courtroom demeanor, and more. Ideal text for company officers, firefighters, inspectors, investigators, insurance, and industrial personnel.

PRIVATE FIRE PROTECTION & DETECTION

Automatic sprinkler systems, special extinguishing systems, standpipes, detection
and alarm systems. Includes how to test sprinkler systems for the firefighter to meet
NFPA 1001.

INDUSTRIAL FIRE PROTECTION

Past loss experience shows, without a doubt, that devastating fires in industrial
plants do occur. They occur at a rate of 145 industrial fires every day. *Industrial Fire
Protection* is the single source document designed for training and managing industrial fire brigades.

This text is a must for all industrial sites, large and small, to meet the requirements
of the Occupational Safety and Health Administration's (OSHA) regulation 29 CFR
part 1910, Subpart L, concerning incipient industrial fire fighting.

FIRE SERVICE INSTRUCTOR

Characteristics of good instructor; determining training requirements and what to
teach; types, principles, and procedures of teaching and learning; training aids and
devices; conference leadership.

PUBLIC FIRE EDUCATION

A valuable contribution to your community's fire safety. Includes public fire education planning, target audiences, seasonal fire problems, smoke detectors, working
with the media, burn injuries, and resource exchange.

FIRE PREVENTION AND INSPECTION PRACTICES

Fire prevention bureau and inspecting agencies; fire hazards and causes; prevention and inspection techniques; building construction, occupancy, and fire load;
special-purpose inspections; inspection forms and checklists along with reference
sources; maps and symbols; records and reports.

WATER SUPPLIES FOR FIRE PROTECTION

Importance, basic components, adequacy, reliability, and carrying capacity of
water systems; specifications, installation, maintenance, and distribution of fire
hydrants; flow test and control valves; sprinkler and standpipe systems.

FIRE APPARATUS PRACTICES

Various types of fire apparatus classified by functions; driving and operating apparatus including pumpers, aerial ladders, and elevating platforms; maintenance and
testing of apparatus.

FIRE STREAM PRACTICES

Characteristics, requirements, and principles of fire streams; developing, comput-
ing, and applying various types of streams to operational situations; formulas for
application of hydraulics; actions and reactions created by applying streams under
different circumstances.

FIRE PROTECTION ADMINISTRATION

A reprint of the Illinois Department of Commerce and Community Affairs publication. A manual for trustees, municipal officials, and fire chiefs of fire districts and
small communities. Subjects covered include officials' duties and responsibilities,
organization and management, personnel management and training, budgeting
and finance, annexation and disconnection.

FIREFIGHTER SAFETY

Basic concepts and philosophy of accident prevention; essentials of a safety program and training for safety; station house facility safety; hazards enroute and at the emergency scene; personal protective equipment; special hazards, including chemicals, electricity, and radioactive materials; inspection safety; health considerations.

FIRE PROBLEMS IN HIGH-RISE BUILDINGS

Locating, confining, and extinguishing fires; heat, smoke, fire gases, and life hazards; exposures, water supplies and communications; pre-fire planning, ventilation, salvage and overhaul; smokeproof stairways and problems of building design and maintenance; tactical checklist.

AIRCRAFT FIRE PROTECTION AND RESCUE PROCEDURES

Aircraft types, engines, and systems, conventional and specialized fire fighting apparatus, tools, clothing, extinguishing agents, dangerous materials, communications, pre-fire planning, and airfield operations.

GROUND COVER FIRE FIGHTING PRACTICES

Ground cover fire apparatus, equipment, extinguishing agents, and fireground safety; organization and planning for ground cover fire; authority, jurisdiction, and mutual aid, techniques and procedures used for combating ground cover fire.

FIRE SERVICE PRACTICES FOR VOLUNTEER FIRE DEPARTMENTS

A general overview of material covered in detail in *Forcible Entry, Ladders, Hose, Salvage and Overhaul, Fire Streams, Apparatus, Ventilation, Rescue, Inspection,* and *Self-Contained Breathing Apparatus*.

INSTRUCTOR GUIDE SETS

Available for *Forcible Entry, Ladder, Hose, Salvage and Overhaul, Fire Streams, Apparatus, Ventilation, Rescue, First Aid, Inspection, Aircraft,* and for the slide program *Fire Department Support of Automatic Sprinkler Systems*. Basic lesson plan, tips for instructor, references. *Essentials* has NFPA Standard 1001 references and pertinent proficiency tests.

TRANSPARENCIES

Multicolored overhead transparencies to augment *Essentials of Fire Fighting* are now available. Since costs and availability vary with different chapters, contact IFSTA Headquarters for details. Units available:

Fire Behavior; Portable Extinguishers; Ropes and Knots; Hose Tools and Appliances; Handling Hose; Handling Ground Ladders; Ventilation; Fire Streams; Ladder Carries and Raises; Forcible Entry; Salvage and Overhaul; Inspection; Ground Cover Fires; Communications; Water Supply; Automatic Sprinkler Systems; Rescue; Protective Breathing Apparatus.

SLIDES

2-inch by 2-inch slides that can be used in any 35 mm slide projector; supplements to respective manuals and sprinkler guide sets.

Ladders
Sprinklers
Smoke Detectors Can Save Your Life
Matches Aren't For Children
Public Relations for the Fire Service
Public Fire Education Specialist (Slide/Tape)
Salvage*

*The complete package consists of the slides, instructor's manual, and instructor's guide sets.

MANUAL HOLDER

The fast, efficient way to organize your IFSTA manuals. These attractive heavy duty vinyl holders have specially designed side panels that allow easy access to all of your IFSTA manuals. Manual holders stand unsupported and will hold up to eight manuals.

GUIDE SHEET BINDERS

Free with purchase of complete guide set. Binders also available separately.

WATER FLOW TEST SUMMARY SHEETS

50 summary sheets and instructions on how to use; logarithmic scale to simplify the process of determining the available water in an area.

PERSONNEL RECORD FOLDERS

Personnel record folders should be used by the training officer for each member of the department. Such data as IFSTA training, technical training (seminars), and college work can be recorded in the file, along with other valuable information. Letter size or legal size.

Ship to:

Date _____

Name _____ Customer Number

Organization __**(FOR INVOICE TO ONLY)**__ Phone _____

Address _____

City _____ State _____ Zip _____

Send to
Fire Protection Publications
Oklahoma State University
Stillwater, Oklahoma 74078
(405) 624-5723
Or Contact Your Local Distributor

ORDER FORM

IFSTA MANUALS

WRITE THE NUMBER OF COPIES OF EACH MANUAL NEXT TO ITS TITLE.

	No of Each		No of Each		No of Each		No of Each
Indoctrination	_____	Rescue	_____	Fire Protection Administration	_____	Public Fire Education Specialists	_____
First Aid	_____	Company Officer	_____	Safety	_____	Smoke Detectors Can Save Your Life	_____
Essentials	_____	Fire Cause Determination	_____	Aircraft	_____	Matches Aren't For Children	_____
Essentials Self-Instruction	_____	Private Fire Protection	_____	High-Rise	_____		
500 Competencies for Essentials	_____	Industrial Fire Protection	_____	Volunteer	_____	Public Relations for the Fire Service	_____
Ladders	_____	Instructor	_____	Ground Cover	_____		
Hose	_____	Public Fire Education	_____	Manual Holder	_____		
Salvage and Overhaul	_____	Fire Prevention/Inspection	_____	**SLIDES**			
Forcible Entry	_____	Water Supplies	_____	Ladder	_____		
Self-Contained Breathing Apparatus	_____	Fire Streams	_____	Salvage	_____		
Ventilation	_____	Apparatus Practices	_____				

TRANSPARENCIES AND SLIDES
Multicolored overhead transparencies to augment each chapter of *Essentials of Fire Fighting* are now available. Also available are slide programs for each of the major fire pump manufacturers and slides for sprinkler systems. Since costs and availability vary with different sets, contact Fire Protection Publications for details.

OTHER MANUALS AND MATERIALS MAY BE ORDERED BELOW:

QUANTITY	TITLE	LIST PRICE	TOTAL

All Foreign Orders must be prepaid in U.S. currency and include 20% shipping and handling charges.

For Free Subscription to Speaking of Fire ☐

Obtain postage and prices from current IFSTA Catalog or they will be inserted by Customer Services.

Note: Payment with your order saves you postage and handling charges when ordering from Fire Protection Publications.

Payment Enclosed ☐   Bill Me Later ☐

Allow 4 to 6 weeks for delivery.

SUBTOTAL	$ _____
Discount, if applicable	$ _____
Postage and Handling, if applicable	$ _____
TOTAL	$ _____

FOR ORDERS
TOLL FREE NUMBER — 800-654-4055

Oklahoma, Hawaii, and Alaska call collect.

COMMENT SHEET   **ESSENTIALS OF FIRE FIGHTING**

DATE _____ NAME _____

ADDRESS _____

ORGANIZATION REPRESENTED _____

CHAPTER TITLE _____ NUMBER _____

SECTION/PARAGRAPH/FIGURE _____ PAGE _____

1. Proposal (include proposed wording, or identification of wording to be deleted), OR PROPOSED FIGURE:

2. Statement of Problem and Substantiation for Proposal:

RETURN TO: IFSTA Editor   SIGNATURE _____
                Fire Protection Publications
                Oklahoma State University
                Stillwater, OK 74078

Use this sheet to make any suggestions, recommendations, or comments. We need your input to make the manuals the most up to date as possible. Your help is appreciated. Use additional pages if necessary.